Lecture Notes in Mathematics

Edited by A. Dold and B. Eckmann

1367

Manfred Knebusch

Weakly Semialgebraic Spaces

Springer-Verlag

Berlin Heidelberg New York London Paris Tokyo

Author

Manfred Knebusch
Fakultät für Mathematik, Universität Regensburg
8400 Regensburg, Federal Republic of Germany

Mathematics Subject Classification (1980): 14 G 30, 54 E 99, 54 E 60, 55 Q 05, 55 N 10, 55 N 20, 55 P 05, 55 P 10

ISBN 3-540-50815-5 Springer-Verlag Berlin Heidelberg New York
ISBN 0-387-50815-5 Springer-Verlag New York Berlin Heidelberg

© Springer-Verlag Berlin Heidelberg 1989
Printed in Germany

Printing and binding: Druckhaus Beltz, Hemsbach/Bergstr.
2146/3140-543210

Introduction

This is the second in a chain of (hopefully) three volumes devoted to an explication of the fundamentals of semialgebraic topology over an arbitrary real closed field R. We refer the uninitiated reader to the preface of the first volume [LSA][1] and some other papers cited there to get an idea of the program we have in mind with the term "semialgebraic topology" as a basis of real algebraic geometry.

Let us roughly recall what has been achieved in the first volume and where we stand now.
As we explained in [LSA], the "good" locally semialgebraic spaces, which fortunately seem to suffice for most applications, are the regular paracompact ones. These are precisely those locally semialgebraic spaces which can be triangulated (I.4.8 and II.4.4)[2]. Moreover, any locally finite family of locally semialgebraic sets in such a space can be triangulated simultaneously (II.4.4). This fact seems to be the key result for many proofs in [LSA].

We accomplished less work on the triangulation of locally semialgebraic maps. Here our main result has been the triangulability of finite maps (II.6.13). Much more can probably be done, as is to be expected by the book [V] of Verona, but we do not pursue this line of investigation in the present volume. {Verona works over $\mathbb{R}$ and uses transcendental techniques.}

---

[1] cf. the references

[2] This refers to Example 4.8 in Chapter I and Theorem 4.4 in Chapter II of [LSA]. The main body of this volume starts with Chapter IV. The signs I, II, III refer to the chapters of [LSA].

On the other hand we obtained in Chapter II of [LSA] a fairly detailed picture of the various possibilities how to "complete" a regular paracompact space M, i.e. to embed M densely into a partially complete regular paracompact space. Partial completeness is a typical notion of semialgebraic topology which has no counterpart in classical topology, cf. I, §6.

In Chapters I and II of [LSA] we also obtained basic results on the structure of locally semialgebraic maps. But the theory of fibrations and covering maps (= Überlagerungen) had to be delayed since a certain amount of homotopy theory is needed here, not yet available in the first two chapters.

Some of that homotopy theory has been presented in the last Chapter III of [LSA]. Our central result there are the two "main theorems" in various versions (III.3.1, 4.2, 5.1, 6.3, 6.4). As a consequence of these theorems all the homotopy groups and various homotopy sets in the category of regular paracompact spaces over R are "equal" to homotopy groups (resp. sets) in the classical topological sense of such spaces over $\mathbb{R}$. This opens the possibility to transfer a considerable amount of classical homotopy theory to the locally semialgebraic setting, as has been illustrated in Chapter III by several examples.

The homotopy theory in [LSA] seems to be sufficient for studying (ramified) coverings of regular paracompact spaces. To some extent it also gives access to the theory of fibrations and fiber bundles for such spaces (although here something remains to be desired, see below). Nevertheless this homotopy theory has serious deficiencies compared with classical (= topological) homotopy theory, and this brings us to the contents of the present volume.

The main deficiencies are the following.

1) In the category LSA(R) of regular paracompact locally semialgebraic spaces over R we do not have infinite CW-complexes at our disposal.
2) In LSA(R) we do not have mapping spaces Map(X,Y) and prominent subspaces of them, as for example loop spaces ΩX , at our disposal.

One main goal in the present volume is to explain how the first deficiency can be overcome. We will construct "semialgebraic" CW-complexes over the field R. A CW-complex over R is a ringed space over R [LSA, p. 3] which is a suitable inductive limit of "polytopes" over R, together with a cell structure. Such inductive limits will generally be called "weak polytopes". (We briefly alluded to these spacec at the end of III, §6 and in [DK_6].) By a polytope over R we simply mean a complete affine semialgebraic space over R. This terminology is justified since these spaces are precisely all ringed spaces over R which are isomorphic to the underlying semialgebraic space of some closed finite simplicial complex over R, hence isomorphic to the union of finitely many closed simplices in some R^n.

We have to be careful which inductive systems of polytopes we admit in building weak polytopes. This is a delicate problem. If we are too restrictive then our weak polytopes will not be useful. On the other hand, if we are too permissive then we are in danger that our inductive limits become too wild spaces. (Recall that every real closed field different from $\mathbb{R}$ is totally disconnected in the topological sense.) Working in the category of ringed spaces over R gives us control which continuous functions we admit on a given space, and this gives us control on connectedness and other geometric properties implicitly.

Once we have defined weak polytopes in the right way and have established the basic properties of these spaces it will be an easy matter to define cell structures on some of them, which will be our CW-complexes. Then the door is open to transfer a really big amount of classical homotopy theory to the semialgebraic setting. In particular we can define spectra, in the sense of algebraic topology, and generalized homology and cohomology theories over R, and we can work with them nearly as easily as in classical homotopy theory (cf. Chapter VI).

Although the category of weak polytopes suffices to deal with infinite CW-complexes it is technically advisable to work in a slightly broader category, the category WSA(R) of "weakly semialgebraic spaces" over R. These spaces are inductive limits of affine semialgebraic spaces instead of just polytopes. For example, an open subspace (in the sense of locally ringed spaces) of a weak polytope is a weakly semialgebraic space, but usually is not a weak polytope. It would be cumbersome to exclude open subspaces of weak polytopes from our considerations. WSA(R) contains the category LSA(R) of regular paracompact locally semialgebraic spaces over R as a full subcategory.

The morphisms between weakly semialgebraic spaces will be called weakly semialgebraic maps. In Chapter IV we give the definition and basic properties of weakly semialgebraic spaces and maps. The key result for later use seems to be that in the category WSA(R) a space M can be glued to another space N along any closed subspace A of M by a "partially proper" weakly semialgebraic map $f : A \to N$ (Theorem IV.8.6). An analogous result had been proved in II, §10 within the category LSA(R) for proper maps. But the class of partially proper maps is much bigger than the class of proper maps and more useful (cf. I, §5-§6 and IV, §5 below). Most important, if the space M above is a weak polytope then also A is a weak polytope and every weakly semialgebraic map $f : A \to N$ is partially proper.

In general a weakly semialgebraic space M cannot be triangulated. But M still is isomorphic to a "patch complex". This is a very weak substitute of a simplicial complex which nevertheless is sufficient for some homotopy considerations.

Roughly one obtains a patch complex if one work with arbitrary affine semialgebraic spaces instead of simplices. The theory of patch complexes and their use in homotopy theory is displayed in Chapter V. Also some applications to open coverings (= Überdeckungen) of weakly semialgebraic spaces are given in V, §3.

Chapter V reveals that weakly semialgebraic spaces are beautiful from a homotopy viewpoint. For example, the two main theorems on homotopy sets from Chapter III in [LSA] extend to these spaces (V, §5) and there holds a strong "Whitehead theorem", stating that every weak homotopy equivalence is a genuine homotopy equivalence (Th. V.6.10). It is this chapter where the reader, having mastered the foundational labours of Chapter IV, will find out that weakly semialgebraic spaces are easy to handle and in some sense better natured, since "tamer", than topological spaces.

On the other hand, from a more geometric viewpoint, weakly semialgebraic spaces can be ugly. We shall demonstrate this in IV, §4 and Appendix C by rather simple examples. Various nice geometric properties we are accustomed to from locally semialgebraic spaces, as for instance the curve selection lemma, fail for these spaces. We do not know whether a weakly semialgebraic space M can be completed, i.e. densely embedded into a weak polytope. We do not know either whether M contains a weak polytope which is a strong deformation retract of M. In contrast to locally semialgebraic spaces there does not always exist a space N over the field R_o of real algebraic numbers such that

M is isomorphic to the base extension (cf. IV, §2) N(R) of N (cf. end of IV, §4). But still we can prove (V, §7) that M is homotopy equivalent to such a space N(R), even with N a CW-complex over R_o. Much later, in Chapter VII, §7, we shall see that M is homotopy equivalent to a closed simplicial complex.

Under the mild restriction that the base field R is sequential, i.e. R contains a sequence of positive elements converging to zero, things are even better. Then there exists, for every weakly semialgebraic space M over R, a <u>canonical</u> homotopy equivalence $p_M : P(M) \to M$ with P(M) a weak polytope. The space P(M) will be defined in Chapter IV, §9. It has the same underlying set as M but a "finer" space structure than M. On the set theoretic level, p_M is just the identity of M.

The space P(M) is the inductive limit of the system of all polytopes contained in M. It seems to be a very natural "simplification" of the space M (simplification for some purposes). If M is locally semialgebraic and locally complete then P(M) coincides with the space M_{loc} defined in I, §7. But already if M is a semialgebraic subset of some R^n which is not locally closed in R^n then P(M) is not locally semialgebraic.

More generally, given a weakly semialgebraic map $f : M \to N$, we shall define in IV, §10 a weakly semialgebraic space $P_f(M)$ together with a weakly semialgebraic map $p_f : P_f(M) \to M$ (if R is sequential) which has the following universal property. The map $f \circ p_f$ is partially proper, and every weakly semialgebraic map $q : L \to M$ with $f \circ q$ partially proper factors uniquely through p_f. If N is the one-point space then $P_f(M) = P(M)$.

These spaces $P_f(M)$, and in particular the spaces P(M), will do good service in homotopy theory at various places. They are typical for the somewhat different flavour of semialgebraic homotopy theory compared with classical homotopy theory.

A particularly good instance to see how the spaces $P_f(M)$ and similar ones can be used and how the various techniques we have developed in Chapters IV and V fit together is the proof of Theorem V.6.8 on d-equivalences (instead of just weak homotopy equivalences) which precedes and implies the Whitehead theorem mentioned above. The reader cannot do better than trying to obtain an impression of the main lines of this proof at an early stage in order to get a good feeling for the subject.

Of course, we try to proceed in semialgebraic homotopy theory as much as possible in a way parallel to the classical topological homotopy theory, as long as this is advisable. Here there comes up a dichotomy of goals and methods everyone working in this area will face.

On the other hand, one would like to obtain results in the semialgebraic theory by transfer from the topological theory, as already exercised in Chapter III. One wants to have available the enormous body of results of topological homotopy theory in the semialgebraic setting without much further labour.

On the other hand, there is a more radical viewpoint, to the best of my knowledge first expressed by Brumfiel in his book [B]: One should do algebraic topology from scratch over an arbitrary real closed field in such a way that the field $\mathbb{R}$ does not play any special role.

This is an ambitious program. While writing this volume I somewhat oscillated between the two viewpoints. Whenever the semialgebraic geometry was easy I avoided transfer principles. When not I gave preference to the first view point, but often I also tried to indicate how things can be done in the spirit of the second one.

Long passages in Chapter V may nourish the conviction that a homotopy theory in the sense of Brumfiel is already at hands. But there are still problems to be settled. As a testing ground I have chosen here - as already in [LSA], Chapter III - the homotopy excision theorem of Blakers and Massey. In topology there exists an elementary proof of this theorem going back to Boardman, cf. [DKP, p. 211ff]. This proof (as well as the proof of Blakers and Massey) strongly uses the axiom of Archimedes in the field of real numbers. We are able to prove the analogue of the theorem for weakly semialgebraic spaces (V, §7), but for that we need the Blakers-Massey theorem for topological CW-complexes and transfer techniques.

The homotopy theory developed in Chapter V suffices for studying generalized homology and cohomology groups of pairs of weakly semialgebraic spaces. {The word "generalized" means that we do not insist on the Eilenberg-Steenrod dimension axiom.} In Chapter VI we define generalized homology and cohomology theories on the category $\mathcal{P}(2,R)$ of weak polytopes over R in full analogy to the definition of such theories on the category $\mathcal{W}(2)$ of pairs of topological CW-complexes [W_2] (or [W], [Sw], etc.). We then explicate how every topological homology theory h_* or cohomology theory h^* on $\mathcal{W}(2)$ leads in a natural way to a homology theory respectively cohomology theory on $\mathcal{P}(2,R)$ which we denote again by h_* resp. h^*. We thus obtain a bijection, up to natural equivalence, between the homology and cohomology theories on $\mathcal{W}(2)$ and on $\mathcal{P}(2,R)$ for R fixed (VI, §2-4). We extend these theories in VI, §5 from $\mathcal{P}(2,R)$ to the category WSA(2,R) of pairs of weakly semialgebraic spaces over R, and we prove in VI, §6 a fairly general excision theorem for the groups $h_n(M,A)$

and $h^n(M,A)$. We also describe the theories h_* and h^* by spectra as one does in topology (VI, §8).

In this whole business it is important that we have weakly semialgebraic spaces at our disposal instead of just locally semialgebraic spaces. We mentioned already the need for infinite CW-complexes. But even suspensions pose a problem. They play an essential role in generalized homology theory, of course. Unfortunately we do not have suspensions for arbitrary weakly semialgebraic spaces but only for weak polytopes. This turns out to be sufficient. But if M is a locally semialgebraic (pointed) weak polytope then usually the suspension SM will not be locally semialgebraic.

If h_* is one of the prominent homology or cohomology theories in topology, as singular homology, singular cohomology, orthogonal, unitary, or symplectic K-theory, one of various cobordism theories, then there remains the important task to attach a geometric meaning to the elements of $h_n(M,N)$ or $h^n(M,A)$ for (M,A) a pair of weakly semialgebraic spaces. {In topology usually such a meaning is inherent in the definition of these groups.}

In the next volume [SFC] we shall solve this problem for the K-theories mentioned above. In the present one we solve it for ordinary homology $H_*(-,G)$ and ordinary cohomology $H^*(-,G)$ with coefficients in some abelian group G. These are those homology and cohomology theories which fulfill the Eilenberg-Steenrod dimension axiom. They arise from topological singular homology and cohomology theory with coefficients in G.

We prove in VI, §3 that if (M,A) is a pair of CW-complexes then the groups $H_n(M,A;G)$ and $H^n(M,A;G)$ have a description by cellular chains and cochains as in topology. It is then easy to conclude that for (M,A) a pair of locally semialgebraic spaces, these groups coincide with the groups $H_n(M,A;G)$ and $H^n(M,A;G)$ defined essentially by Delfs [D], [D_1], [DK_3]. {We described the groups $H_n(M,A;G)$ in III, §7.}

Here our theory reaches a remarkable point. To understand, why, let us recall the approach of Delfs to the homology groups, say, of a single polytope M. {We take $A = \emptyset$.} The polytope M can be triangulated. Choosing an isomorphism $\varphi : |K|_R \xrightarrow{\sim} M$ with K a finite abstract simplicial complex we "know" a priori what $H_n(M,G)$ should be: It should coincide with the abstract homology $H_n(K,G)$ of the simplicial complex K. The problem is, to prove that the groups $H_n(K,G)$ do not depend on the choice of the triangulation. Delfs solves this problem in an ingenious way. He looks at the simplicial cohomology groups $H^n(K,G)$ for the triangulations of M. He proves that they all are naturally isomorphic to the cohomology groups $H^n(M,G_M)$ of the constant sheaf G_M with stalk G. Knowing that the $H^n(K,G)$ are independent of the triangulation he concludes that the $H_n(K,G)$ also are independent of the triangulation.

In the course of this approach Delfs has to cope with some tedious geometric problems. {The main task is to prove the homotopy invariance of the groups $H^n(M,G_M)$. In [D_1] Delfs solves this problem brilliantly by using sheaf theory on abstract locally semialgebraic spaces.} The remarkable fact now is that we obtain the independence of the groups $H_n(K,G)$ from the choice of the triangulation in a much easier way. Once we have the homotopy theory of Chapter V at hands, which is a straightforward matter, we define the ordinary homology groups $H_n(M,G)$ almost by general categorial nonsense, and prove $H_n(M,G) \cong H_n(K,G)$ in the standard way (cf. VI, §3). Thus one may say that it is possible to circumvent the labours of Delfs by enlarging the category of affine semialgebraic spaces over R to a category of spaces which is more comfortable for homotopy considerations, namely WSA(R). {But notice that our approach does not give a connection of ordinary cohomology with sheaf cohomology.}

How about an interpretation of the elements of $H_n(M,A;G)$ by chains of singular simplices, as in topology? Of course, a singular simplex here means a semialgebraic map (= morphism) from the closed standard simplex $\nabla(n)$ in R^{n+1} to M. For any pair (M,A) of weakly semialgebraic spaces over R we can define the singular chain complex $C.(M,A;G)$ as in topology. The problem is to prove that the groups $H_n(C.(M,A;G))$ fit together to an ordinary homology theory and that $H_o(C.(*,\emptyset;G)) \cong G$, with $*$ denoting the one point space. This would imply a natural isomorphism from this homology theory to $H_*(-,G)$.

Delfs and I have tried for years in vain to find such a proof in a direct geometric way. The difficulty was always to prove an excision theorem for the groups $H_n(C.(M,A;G))$ in the case that the field R is not archimedean. We could not prove excision even for a triad of polytopes. As in classical theory one would like to make a given singular chain "small" with respect to a given finite open covering (with two open semialgebraic sets) by applying some iterated subdivision to the singular simplices in the chain. But the trouble is that, as long as one tries barycentric subdivision or some other sort of finite linear subdivision, the simplices have no reason to become small if R is not archimedean.

The last Chapter VII of the present book contains a solution of the problem - along very different lines. This solution is perhaps the most convincing single issue, up to now, to demonstrate that weakly semialgebraic spaces are really useful.

We proceed roughly as follows. Every simplicial set K (= semisimplicial set = semisimplicial complex, in other terminologies) can be "realized" as a weak polytope $|K|_R$ over R in much the same way as this is known in topology [Mi_1]. The space $|K|_R$ carries a natural structure of a CW-complex. If (K,L) is a pair of simplicial sets (of course, with L a simplicial subset of K), then it follows from the cellular description of ordinary homology mentioned above that the ordinary homology groups $H_n(|K|_R,|L|_R;G)$ can be identified with the well known (cf. [La] or [May]) "abstract" homology groups $H_n(K,L;G)$.

If M is a weakly semialgebraic space over R we can form the singular simplicial set Sin M consisting of the singular simplices of M. The realization $|\mathrm{Sin}\, M|_R$ comes with a canonical weakly semialgebraic map $j_M : |\mathrm{Sin}\, M| \to M$. We prove that j_M is a homotopy equivalence (VI, §7) following the book [LW] of Lundell and Weingram. {In topology j_M is only a weak homotopy equivalence. In most texts on simplicial methods - but non in [LW] - this is proved by already using the fact that the topological singular homology groups form an ordinary homology theory.}

More generally, if A is a subspace of M, then j_M gives a homotopy equivalence from the pair $(|\mathrm{Sin}\, M|_R, |\mathrm{Sin}\, A|_R)$ to (M,A). Thus

$$H_q(M,A;G) \cong H_q(|\mathrm{Sin}\, M|_R, |\mathrm{Sin}\, A|_R; G) \cong H_q(\mathrm{Sin}\, M, \mathrm{Sin}\, A; G),$$

and this group is $H_q(C.(M,A;G))$ by definition.

Since we know that the canonical maps j_M are homotopy equivalences the door is now wide open for the use of simplicial sets in semialgebraic geometry. Thus, finally, we can abolish our previous verdict "no simplicial sets, only simplicial complexes" [DK_3, p. 124].

Simplicial sets have proved to be enormously useful in many branches of topology, in particular in the theory of fibrations. Much of this material can now be used in semialgebraic geometry. Some applications to the theory of semialgebraic fibrations will be given in the next volume [SFC].

But one needs more. One needs simplicial spaces instead of just simplicial sets. By a simplicial space X over R we mean a simplicial object in WSA(R), i.e. a sequence $(X_n | n \in N_o)$ of weakly semialgebraic spaces over R with various weakly semialgebraic face and degeneracy maps between them (VII, §1). Simplicial sets may be regarded as discrete simplicial spaces over R.

Roughly half of our last Chapter VII is devoted to an explication of the fundamentals of simplicial spaces and their realizations. Difficulties for future application will arise from the fact that we are only able to construct the realization $|X|_R$ of a partially proper simplicial space X. By this we mean a simplicial space all whose face maps are partially proper. Fortunately discrete simplicial spaces are partially proper.

A reader having worked through the fundamentals of weakly semialgebraic spaces and maps in Chapter IV may feel bored to meet in Chapter VII similar stuff about simplicial spaces. To give such a reader some comfort we indicate now by an example that this stuff is really useful.

Let G be a complete semialgebraic group over R. {For instance think of some orthogonal group O(n,R).} If M is an affine semialgebraic space, then it is clear from the beginnings of semialgebraic geometry what is meant by a principal G-fibre bundle $\varphi : E \to M$ over M. The definition is exactly as in topology, of course with a finite trivializing covering of M by open semialgebraic subsets.

We now pose the following problem. Let S be a real closed overfield of

R and let $\psi : F \to M(S)$ be a principal $G(S)$-bundle over $M(S)$. Does there exist a principal G-bundle $\varphi : E \to M$ over M such that the base extension $\varphi_S : E(S) \to M(S)$ is isomorphic to ψ over $M(S)$?

It seems hard to solve this problem in a direct geometric way. We shall solve it in [SFC] in the affirmative as follows. Let $\mathcal{N}G$ denote the nerve of the group G. This is a simplicial space built as in topology, cf. Example VII.1.2.v below. $\mathcal{N}G$ is partially proper since G is complete (partially complete would suffice). Let BG denote the realization $|\mathcal{N}G|$. One finds as in topology that the isomorphism classes of G-principal bundles over M are in natural one-to-one correspondence with the elements of the homotopy set [M,BG]. By the first main theorem on homotopy sets the base extension map from [M,BG] to [M(S),(BG)(S)] is bijective (V.5.2.i; essentially this is already clear from III.3.1). By the canonical nature of the definition of $\mathcal{N}G$ it is evident that (BG)(S) = B(G(S)). Thus we have a natural bijection from [M,BG] to [M(S)B(G(S))]. We conclude that the isomorphism classes of principal G-bundles over M correspond uniquely with the isomorphism classes of principal G(S)-bundles over M(S) by base extension. The answer to the question above is "Yes".

At first glance the present book might convey the impression that in semialgebraic geometry one now has a homotopy theory at hands which is as good and easy as the topological one. But this impression is deceptive. In order to destroy it I come back to the two deficiencies of the homotopy theory in [LSA] listed above. While the first one disappears in the category WSA(R), the second one (existence of mapping spaces) remains serious.

One would like to have good substitutes (or "models") of the presumably not existing mapping spaces and their prominent subspaces. In VI, §7 we define "pseudo-mapping spaces" and "pseudo-loop spaces" which do <u>some</u>

of the service one expects from such substitutes. Our construction is based on an analogue of Brown's representation theorem [Bn]. This construction is canonical only up to homotopy. Using Chapter VII we are better off. If M and N are any weakly semialgebraic spaces then we can form a simplicial mapping set [May, p. 17] Map(Sin M, Sin N) and choose the realization $|\mathrm{Map}(\mathrm{Sin}\, M, \mathrm{Sin}\, N)|_R$ as a canonical substitute of the presumably not existing space Map(M,N). But this substitute and similar constructions are not sufficient for all purposes, as will become amply clear in the theory of fibrations. There one has to work with the notion of fibre homotopy equivalence instead of homotopy equivalence. The question for a substitute of the topological "path mapping space" ([W, 7.2], [DKP, 5.3]) which turns a given map $f : M \to N$ into a fibration, is a case in point.

Another strategy is to establish a space structure on a sufficiently big subset of the set Map(M,N) of weakly semialgebraic maps from M to N (sufficiently big for some purposes). In the last section of Chapter VII we do something like this for $M = N = [0,1]$, the unit interval of R, in the case that the field R is sequential.

I deviated from the original plan, announced in the preface of [LSA], to deal with fibrations in Chapter IV and with coverings (= Überlagerungen) in Chapter V. We shall do this only in the next volume [SFC]. Originally I intended to introduce weakly semialgebraic spaces at a much later stage. In the meantime I realized that this would cause a duplication of arguments, since many proofs in the theory of fibrations and coverings can be done in the same way for the categories LSA(R) and WSA(R). I also realized how well it pays in many other ways to introduce weakly semialgebraic spaces as early as possible.

I thank Professors Ronnie Brown, Rudolf Fritsch, J. Peter May, and Rainer Vogt for useful advice in tackling with simplicial homotopy. I further thank Hans Delfs, Roland Huber, Claus Scheiderer and Niels Schwartz for help with details of the proofs too numerous to be listed here. In particular, Appendix C is entirely due to Huber. I also thank these persons for proof reading and a successful search for mistakes in previous versions of this volume. Finally special thanks are due to my secretary Marina Franke for a very efficient typing of all these versions without losing patience in critical situations.

Regensburg, March 1988

Manfred Knebusch

TABLE OF CONTENTS

Chapter IV - Basic theory of weakly semialgebraic spaces

§1 - Definition and construction of weakly semialgebraic spaces

R is a fixed real closed field. As in I, §1 we consider a generalized topological space $M = (M, \mathring{\mathfrak{T}}(M), Cov_M)$. Here $\mathring{\mathfrak{T}}(M)$ means the set of (admissible) open subsets of M and Cov_M the set of (admissible) open coverings, cf. I, §1, Def. 1. Starting with such a space M we give a chain of definitions and examples leading to the definition of a weakly semialgebraic space and a weak polytope (Definitions 6,7 below).

Definition 1. We call a subset K of M small in M if, for every $U \in \mathring{\mathfrak{T}}(M)$ and every $(U_\lambda | \lambda \in \Lambda) \in Cov_M(U)$, the set $U \cap K$ is already the union of the sets $U_\lambda \cap K$ with λ running through a suitable finite subset of Λ.

Example 1.1. If M is a locally semialgebraic space over R then every semialgebraic subset of M is small in M.

Definition 2. a) A function ringed space M over R is a generalized topological space M equipped with a sheaf $\mathcal{O}_M$ of rings of R-valued functions.

b) A morphism between function ringed spaces M,N over R is a continuous map $f : M \to N$ (in the sense of generalized topological spaces), such that for every $V \in \mathring{\mathfrak{T}}(N)$ and $h \in \mathcal{O}_N(V)$ the composite function $h \circ f : f^{-1}(V) \to R$ is an element of $\mathcal{O}_M(f^{-1}(V))$.

c) We denote the category of function ringed spaces over R by Space(R).

Example 1.2. Every locally semialgebraic space over R is a function ringed space over R. The morphisms between such spaces are the locally semialgebraic maps.

Henceforth let M be a function ringed space (always over R). If K is a small subset of M then M induces on K the structure of a function ringed space over R as follows. $\mathring{\mathfrak{T}}(K)$ is the set of all intersections $U \cap K$ with $U \in \mathring{\mathfrak{T}}(M)$. If $(V_\lambda | \lambda\in\Lambda)$ is a family in $\mathring{\mathfrak{T}}(K)$ then $(V_\lambda | \lambda\in\Lambda) \in \mathrm{Cov}_K$ if and only if there exists a finite subset Λ' of Λ such that the set $V := \cup(V_\lambda | \lambda\in\Lambda)$ is already the union of all V_λ with $\lambda \in \Lambda'$. {As usual, we then write $(V_\lambda | \lambda\in\Lambda) \in \mathrm{Cov}_K(V)$. The axioms i-viii in I, §1 are clearly fulfilled.} $\mathcal{O}_K$ is the sheaf associated to the presheaf $\mathcal{O}_K^o$ defined as follows. A function $h : V \to R$ on some $V \in \mathring{\mathfrak{T}}(K)$ is an element of $\mathcal{O}_K^o(V)$ iff there exists some $U \in \mathring{\mathfrak{T}}(M)$ and some $g \in \mathcal{O}_M(U)$ with $U \cap K \supset V$ and $h = g|V$. {We then can make U a little smaller such that $U \cap K = V$.} Thus a function $h : V \to R$, with $V \in \mathring{\mathfrak{T}}(K)$, is an element of $\mathcal{O}_K(V)$ iff there exist finitely many sets $U_1,\dots,U_r \in \mathring{\mathfrak{T}}(M)$ and functions $g_i \in \mathcal{O}_M(U_i)$ such that $K \cap (U_1 \cup \dots \cup U_r) = V$ and $g_i|K \cap U_i = h|K \cap U_i$ for $1 \le i \le r$.

Definition 3. We call such a space $(K,\mathcal{O}_K)$ a small subspace of $(M,\mathcal{O}_M)$. Notice that we have K equipped with the "coarsest structure" of a function ringed space such that the inclusion map $K \hookrightarrow M$ is a morphism.

Definition 4. Of course, also every $U \in \mathring{\mathfrak{T}}(M)$ has a natural induced structure as a function ringed space over R. These are the open subspaces of M.

Example 1.3. If M is a locally semialgebraic space and K is a closed semialgebraic subset of M then K, with its usual structure as a semialgebraic subspace of M (I, §3), is a small subspace of M. This follows from Tietze's extension theorem for affine semialgebraic spaces [DK_5, Th. 4.5]. If in addition, M is regular (I, §3), then $\mathcal{O}_K^o = \mathcal{O}_K$ since now K has an open affine semialgebraic neighbourhood in M. Indeed, every semialgebraic subset of M is affine [R].

Important convention. From now on, in the whole book, a semialgebraic space always means an *affine* semialgebraic space, and a locally semialgebraic space means a *regular* locally semialgebraic space.

Definition 5. A subset K of M is called *closed semialgebraic in* M if K is closed in M, i.e. $M \setminus K \in \mathring{\mathcal{T}}(M)$, if K is small in M, and if the small subspace $(K,\mathcal{O}_K)$ of M is a semialgebraic space. K is called a *polytope in* M if, in addition, the semialgebraic space $(K,\mathcal{O}_K)$ is a polytope, i.e. complete [DK_2, §9]. The set of all closed semialgebraic subsets of M is denoted by $\overline{\gamma}(M)$ and the set of all polytopes in M is denoted by $\gamma_c(M)$.

Example 1.4. If M is locally semialgebraic then $\overline{\gamma}(M)$ and $\gamma_c(M)$ have the same meaning as in [LSA].

Now we are ready for the main definition of the whole book. By an *ordered family* of subsets $(X_\lambda \mid \lambda \in \Lambda)$ of a set X we mean a family of subsets with a partially ordered index set.

Definition 6. A *weakly semialgebraic* space over R is a function ringed space M over R which contains an ordered family $(M_\alpha \mid \alpha \in I)$ of closed semialgebraic subsets (Def. 5) such that the following properties hold.

E1) $M = \cup(M_\alpha \mid \alpha \in I)$.

E2) If $\alpha \le \beta$ then $M_\alpha \subset M_\beta$.

E3) For every $\alpha \in I$ there exist only finitely many $\beta \in I$ with $\beta < \alpha$.

E4) For any two indices $\alpha, \beta \in I$ there exists an index $\gamma \in I$ with $\gamma \le \alpha$, $\gamma \le \beta$, and $M_\alpha \cap M_\beta = M_\gamma$.

E5) I is directed, i.e. for any two indices $\alpha, \beta \in I$ there exists an index $\gamma \in I$ with $\alpha \le \gamma$, $\beta \le \gamma$.

E6) The function ringed space M is the inductive limit of the family of semialgebraic spaces $(M_\alpha \mid \alpha \in I)$ in the category Space(R). This means the following:

a) A subset U of M is an element of $\mathring{\mathfrak{T}}(M)$ iff $U \cap M_\alpha \in \mathring{\mathfrak{T}}(M_\alpha)$ for every $\alpha \in I$. {N.B. Using the notation of [LSA], $\mathring{\mathfrak{T}}(M_\alpha) = \mathring{\mathfrak{r}}(M_\alpha) =$ the set of all open semialgebraic subsets of M_α.}

b) A family $(U_\lambda | \lambda \in \Lambda)$ in $\mathring{\mathfrak{T}}(M)$ is an element of Cov_M iff $(U_\lambda \cap M_\alpha | \lambda \in \Lambda)$ is an element of Cov_{M_α} for every $\alpha \in I$. This means that, for every $\alpha \in I$, there exists a finite subset $\Lambda(\alpha)$ of Λ such that

$$\cup(U_\lambda \cap M_\alpha | \lambda \in \Lambda) = \cup(U_\lambda \cap M_\alpha | \lambda \in \Lambda(\alpha)).$$

c) If $h : U \to R$ is an R-valued function on some $U \in \mathring{\mathfrak{T}}(M)$, then $h \in \mathcal{O}_M(U)$ iff $h|U \cap M_\alpha \in \mathcal{O}_{M_\alpha}(U \cap M_\alpha)$ for every $\alpha \in I$.

Any such family $(M_\alpha | \alpha \in I)$ of subsets of M is called an <u>exhaustion</u> of M. {The letter "E" in the labels above refers to "exhaustion".}

<u>Definition 7.</u> A function ringed space M over R is called a <u>weak polytope</u> over R if M has an exhaustion $(M_\alpha | \alpha \in I)$ such that every M_α is a polytope in M (cf. Def. 5).

For the applications which we have in mind we are mainly interested in weak polytopes, but for technical reasons it is necessary to work in the more general class of weakly semialgebraic spaces.

<u>Example 1.5.</u> Every paracompact locally semialgebraic space M over R is a weakly semialgebraic space over R. Indeed, choose a weak triangulation $\varphi : X \xrightarrow{\sim} M$, i.e. a locally semialgebraic isomorphism of a locally finite simplicial complex (I, §2) to M. {There exists even a triangulation, i.e. an isomorphism with X strictly locally finite, cf. II, §4.} Let I denote the set of all finite subcomplexes Y of X which are closed in X, ordered by inclusion. Then $(\varphi(Y) | Y \in I)$ is an exhaustion of M. If M is partially complete then the $\varphi(Y)$ are poly-

topes, hence M is a weak polytope.

We now describe a method for constructing weakly semialgebraic spaces. This method is based on the results in II, §1 on gluing locally semialgebraic spaces, which will be exploited here only for semialgebraic spaces.

We start with the following situation. M is a set, and $(M_\alpha | \alpha \in I)$ is an ordered family of subsets of M such that conditions E1, E2, E3 (cf. Def. 6) hold and, instead of E4, the following weaker condition holds.

E4*) For any two indices $\alpha, \beta \in I$ there exist indices $\gamma_1, \dots, \gamma_r$ in I with $\gamma_i \leq \alpha$, $\gamma_i \leq \beta$ $(1 \leq i \leq r)$ and $M_\alpha \cap M_\beta = M_{\gamma_1} \cup \dots \cup M_{\gamma_r}$.

We do not assume anything like E5. We assume that every M_α is equipped with the structure of a semialgebraic space over R and that for $\alpha < \beta$ the space M_α is a closed semialgebraic subspace of M_β.

We equip M with that structure of a function ringed space over R which makes M the inductive limit of the family of spaces $(M_\alpha | \alpha \in I)$. This structure has the same description as given in Definition 6 (although perhaps I is not directed). For any subset J of I we denote by M_J the union of all M_α with $\alpha \in J$. By $\hat{I}$ we denote the set of all finite subsets J of I which enjoy the following property: If $\alpha \in J$ and $\beta \in I$, $\beta \leq \alpha$, then $\beta \in J$. The set $\hat{I}$ is partially ordered by the inclusion relation. It is a lattice, since for any two elements J,K of $\hat{I}$ also $J \cap K$ and $J \cup K$ are elements of $\hat{I}$. Moreover $\hat{I}$ has a smallest element, the set $\emptyset$.

Theorem 1.6. Every set M_α is closed semialgebraic in the function ringed space M, and M induces on M_α the given space structure. The family $(M_J | J \in \hat{I})$ is an exhaustion of M. Thus M is a weakly semialgebraic space. If the family $(M_\alpha | \alpha \in I)$ fulfills E4 and E5 then already this family is an exhaustion of M.

Before proving the theorem we give simple examples which already show that weakly semialgebraic spaces form a more general class of spaces than paracompact locally semialgebraic spaces. For these examples the full strength of the theorem is not needed.

Example 1.7 (Simplicial complexes). Let X be a simplicial complex*) over R. As in [LSA] we denote by $\Sigma(X)$ the set of open simplices of X, ordered by the face relation. Let M be the underlying set of X and let $(M_\alpha|\alpha\in I)$ be the family $(\bar{\sigma}\cap M|\sigma\in\Sigma(X))$ of closures in M of all open simplices. Every set $\bar{\sigma}\cap M$ carries a natural structure of a semialgebraic space. By our theorem we obtain a natural structure of a weakly semialgebraic space on X, an exhaustion being given by the finite subcomplexes of X which are closed in X. We denote this weakly semialgebraic space again by the letter X. If the complex X is closed, i.e. $\Sigma(X)$ contains every open face of every $\sigma\in\Sigma(X)$, then the space X is a weak polytope.

Example 1.8 (Wedge of semialgebraic spaces). Let $(M_\lambda|\lambda\in\Lambda)$ be a family of disjoint semialgebraic spaces. We choose in every M_λ a base point x_λ and we identify all these base points into one point x_o. Let M be the set $\cup(M_\lambda|\lambda\in\Lambda)$ after this identification. We add to the set Λ a symbol "O" and we order the set $I := \Lambda\cup\{O\}$ as follows. $O<\lambda$ for every $\lambda\in\Lambda$. No two elements of Λ are comparable. We put $M_O := \{x_o\}$. Applying the theorem to the family $(M_\alpha|\alpha\in I)$ of semialgebraic spaces we obtain on the set M a natural structure of a weakly semialgebraic space, such that every M_λ - in its given structure - is a small sub-

---

*) In this chapter - and the next one - a simplicial complex always means a geometric simplicial complex, as defined in I, §2.

space of M. We call this space M, with base point x_o, the <u>wedge</u> of the pointed spaces (M_λ, x_λ) and write $M = V(M_\lambda | \lambda \in \Lambda)$, or more precisely

$$(M, x_o) = V((M_\lambda, x_\lambda) | \lambda \in \Lambda) .$$

The sets $\cup(M_\lambda | \lambda \in K)$, with K running through the finite subsets of Λ, form an exhaustion of M. If the M_λ are polytopes then M is a weak polytope.

<u>Example 1.9.</u> We consider the family $(\mathbb{P}_n(R) | n \in \mathbb{N}_o)$ of sets of real points of the standard projective spaces $\mathbb{P}_{n,R}$ over R. Each $\mathbb{P}_n(R)$ is a polytope. We embed $\mathbb{P}_n(R)$ into $\mathbb{P}_{n+1}(R)$ in the usual way as the hyperplane $x_{n+1} = 0$. Let $\mathbb{P}_\infty(R)$ be the union of all $\mathbb{P}_n(R)$. By the theorem, $\mathbb{P}_\infty(R)$ is a weak polytope in a natural way, with exhaustion $(\mathbb{P}_n(R) | n \in \mathbb{N}_o)$.

<u>Example 1.10</u> (Direct sums). For every family $(M_\lambda | \lambda \in \Lambda)$ of function ringed spaces over R there exists the direct sum $\sqcup(M_\lambda | \lambda \in \Lambda) = M$ in the category of function ringed spaces over R. The set M is the disjoint union of the sets M_λ. The elements of $\mathring{\mathfrak{T}}(M)$ are the unions $\cup(U_\lambda | \lambda \in \Lambda)$ of families $(U_\lambda | \lambda \in \Lambda)$ with $U_\lambda \in \mathring{\mathfrak{T}}(M_\lambda)$, and also Cov_M, $\mathcal{O}_M$ are defined in the obvious way. Assume now that every M_λ is weakly semialgebraic. We choose an exhaustion $(M_{\lambda,\alpha} | \alpha \in I_\lambda)$ of M_λ. Let $I := \sqcup(I_\lambda | \lambda \in \Lambda)$ be the direct sum of the ordered sets I_λ. For every $\alpha \in I$ we define $M_\alpha :=$ $M_{\lambda,\alpha}$ if $\alpha \in I_\lambda$. The ordered family $(M_\alpha | \alpha \in I)$ of semialgebraic spaces fulfills the conditions of Theorem 1.6. Clearly the function ringed space M is the inductive limit of this family. Thus, by Theorem 1.6, M is a weakly semialgebraic space.

We start out to prove Theorem 1.6. The ordered family $(M_J | J \in \hat{I})$ of subsets of M filfills E1 - E5, as is easily checked. We equip every M_J with the function ringed space structure as the inductive limit of the family of semialgebraic spaces $(M_\alpha | \alpha \in J)$. For every $J \in \hat{I}$ we claim that the space M_J is semialgebraic and that, whenever $K \subset J$ and $K \in \hat{I}$, then M_K is, in its given space structure, a closed semialgebraic subspace of M_J.

For any $\alpha \in I$ we denote by $m(\alpha)$ the number of elements $\beta \in I$ with $\beta < \alpha$. For any $J \in \hat{I}$ we denote by $m(J)$ the maximum of all $m(\alpha)$ with $\alpha \in J$.

We prove the claim by induction on $m(J)$. If $m(J) = 0$, then it follows from E4* that the sets M_α with $\alpha \in J$ are pairwise disjoint. M_J is the direct sum of the finitely many semialgebraic spaces M_α (I, 2.4), and the claim is obvious.

Let now $m(J) = n \geq 1$. If α and β are two different indices in J then, by E4*, $M_\alpha \cap M_\beta = M_L$ with some $L \in \hat{I}$ such that $\gamma \leq \alpha$, $\gamma \leq \beta$ for every $\gamma \in L$. Of course, $m(L) \leq n$. Suppose there exists some $\gamma \in L$ with $m(\gamma) = n$. This would force $\gamma = \alpha$ and $\gamma = \beta$. But $\alpha \neq \beta$. Thus $m(L) < n$. By induction hypothesis, the space M_L is semialgebraic, and every M_γ, $\gamma \in L$, is, in its given structure, a closed semialgebraic subspace of M_L. On the other hand, since these M_γ are closed semialgebraic subsets of M_α, the set M_L is closed semialgebraic in M_α. It now follows, say from the uniqueness statement in Th. II.1.3, that the subspace structure of M_L in M_α coincides with the given structure. The same holds for M_L and M_β. Thus we learn, that $M_\alpha \cap M_\beta$ is closed semialgebraic in M_α and in M_β and that the subspace structures on $M_\alpha \cap M_\beta$ with respect to M_α and M_β are the same. Theorem II.1.3 tells us, that our space M_J is semialgebraic, and that the given space structure on M_J is in fact the unique one such that every M_α, $\alpha \in J$, is a closed semialgebraic subspace of M_J in its given structure. By the same uniqueness argument as above we see that, for any $K \in \hat{I}$ with $K \subset J$, the space M_K in its given structure, is a closed semialgebraic subspace in M_J, and our claim is proved.

Of course, the inductive limit space structure on M with respect to the new family $(M_J|J\in\hat{I})$ is the same as with respect to the old family $(M_\alpha|\alpha\in I)$.

For any two subsets J and K of I, with $J\in\hat{I}$ but K perhaps infinite, we have $M_J\setminus M_K\in\mathring{\mathfrak{T}}(M_J)$, since $M_K\cap M_J = M_L$ with some $L\in\hat{I}$, $L\subset J$. Thus $M\setminus M_K\in\mathring{\mathfrak{T}}(M)$ for every $K\subset I$. It is also evident from the definitions that, for every finite $J\subset I$, the set M_J is small in M.

Certainly the proof of Theorem 1.6 will be finished if we verify that, for every $J\in\hat{I}$, the presheaf of functions $\mathcal{O}^o_{M_J}$ induced by $\mathcal{O}_M$ on M_J coincides with the sheaf of semialgebraic functions on M_J. We prove a proposition which contains this fact and something more.

Proposition 1.11. Let K be any subset of I, and let $\tilde{K}$ denote the set of all $\alpha\in I$ with $\alpha\le\beta$ for some $\beta\in K$. We equip the set $A := M_K = M_{\tilde{K}}$ with the inductive limit space structure of the ordered family $(M_J|J\in\hat{I},J\subset\tilde{K})$ of semialgebraic spaces. Then the following holds.

a) A is closed in M, i.e. $M\setminus A\in\mathring{\mathfrak{T}}(M)$.

b) For every $V\in\mathring{\mathfrak{T}}(A)$ there exists some $U\in\mathring{\mathfrak{T}}(M)$ with $U\cap A = V$.

c) Given such sets V,U and a function $h\in\mathcal{O}_A(V)$ there exists some $g\in\mathcal{O}_M(U)$ with $g|V = h$.

Proof. a) has been verified above.

b) Let $U := (M\setminus A)\cup V = M\setminus(A\setminus V)$. For every $\alpha\in I$ the set $(A\setminus V)\cap M_\alpha$ is closed semialgebraic in M , since the properties E3 and E4* hold. Thus $U\in\mathring{\mathfrak{T}}(M)$. Clearly $U\cap A = V$.

c) We consider the set of all pairs (J,f) with $K\subset J\subset I$, $f\in\mathcal{O}_{M_J}(U\cap M_J)$ and $f|A = g$. We order this set in the obvious way: $(J,f)\le(J',f')$

iff $J \subset J'$ and $f'|U \cap M_J = f$. If $((J_\lambda, f_\lambda)|\lambda \in \Lambda)$ is a chain in this ordered set, then the f_λ automatically glue to a function $f \in \mathcal{O}_{M_J}(U \cap M_J)$ with $J = \cup(J_\lambda|\lambda \in \Lambda)$, and we have $(J_\lambda, f_\lambda) \leq (J,f)$ for every $\lambda \in \Lambda$. Thus, by Zorn's lemma, our set contains a maximal element (L,g). We prove that $L = I$ and then will be done. Suppose $L \neq I$. We choose some $\alpha \in I \setminus L$. The set $M_L \cap M_\alpha \cap U$ is closed and semialgebraic in $M_\alpha \cap U$. By Tietze's extension theorem [DK_5, Th. 4.5] there exists a semialgebraic function u on $M_\alpha \cap U$ which extends $g|M_L \cap M_\alpha \cap U$. The functions u and g glue to a function $v \in \mathcal{O}_{M_S}(U \cap M_S)$ with $S = L \cup \{\alpha\}$. Then $(L,g) < (S,v)$ in contradiction to the maximality of (L,g). Thus $L = I$. q.e.d.

<u>Remark 1.12.</u> Assertion c) in Proposition 1.11 can be improved. If h takes values in a generalized interval $L \subset R$ (= convex semialgebraic subset of R) then g can be chosen such that g takes values in L.

This is proved in the same way as above using Tietze's extension theorem for semialgebraic spaces and functions with values in L, cf. [DK_5, Thm. 4.5].

<u>Remark 1.13.</u> In the situation of Theorem 1.6 assume that in addition the following holds.

(F) For every $\alpha \in I$ there exist only finitely many $\beta \in I$ such that $M_\alpha \cap M_\beta \neq \emptyset$.

Then the weakly semialgebraic space M is a paracompact locally semialgebraic space.

This follows from Lemma 1.2 and Theorem 1.3 in Chapter II about gluing of locally semialgebraic spaces. There one has to assume that, for every $\alpha \in I$, the family $(M_\alpha \cap M_\beta|\beta \in I)$ is locally finite in the semialgebraic space M_α. This means just (F).

We close this section with some easy observations on exhaustions of a given weakly semialgebraic space M. An exhaustion $(M_\alpha|\alpha\in I)$ of M may have some redundancy which can be eliminated.

Definition 8. An exhaustion $(M_\alpha|\alpha\in I)$ of M is called **faithful** if, instead of E2, the following stronger property E2' holds.[*]
E2') For any two indices $\alpha,\beta \in I$, $\alpha \leq \beta \Leftrightarrow M_\alpha \subset M_\beta$.

Proposition 1.14. If $(M_\alpha|\alpha\in I)$ is an exhaustion of a weakly semialgebraic space M then there exists a subset I' of I such that the family $(M_\alpha|\alpha\in I')$ is a faithful exhaustion.

Proof. Throw out every index $\alpha \in I$ such that there exists some $\gamma \in I$ with $\gamma < \alpha$ and $M_\gamma = M_\alpha$, and check properties E1, E2', E3-E6 for the new family! In this check the properties E3 and E4 for the old family play a crucial role.

We can find exhaustions with even better formal properties.

Definition 9. A **lattice exhaustion** of M is an exhaustion $(M_\alpha|\alpha\in I)$ of M with the following additional properties.
E7) For any two indices $\alpha,\beta \in I$ there exists some $\gamma \in I$ with $M_\alpha \cup M_\beta = M_\gamma$.
E8) For the smallest index $\sigma \in I$ we have $M_\sigma = \emptyset$.

Notice that then the set $\{M_\alpha|\alpha\in I\}$ is a sublattice of the lattice $\overline{\gamma}(M)$ of closed semialgebraic subsets of M, hence the name.

Proposition 1.15. Every faithful exhaustion $(M_\alpha|\alpha\in I)$ of a weakly semialgebraic space M can be enlarged to a faithful lattice exhaustion

_____

[*] Such a family may be regarded as a **set** of subsets of M, each set being indexed by itself.

$(M_\alpha | \alpha \in \tilde{I})$, with $\tilde{I}$ an ordered set containing I and inducing the given ordering on I.

<u>Proof.</u> We regard the given exhaustion as a set H of subsets of M, ordered by inclusion. The set $\tilde{H}$ consisting of all unions of finitely many elements of H, ordered by inclusion, fulfills E1, E2', E3-E6, and also E7, E8.

It is often comfortable to work with faithful lattice exhaustions. (We shall use them in a crucial way in §7 and §8.) On the other hand, we should not always insist to use faithful or lattice exhaustions, since this would lead to unnecessary notational complications. For example, if $(M_\alpha | \alpha \in I)$ and $(N_\beta | \beta \in J)$ are two exhaustions of M then $(M_\alpha \cap N_\beta | (\alpha,\beta) \in I \times J)$ is again an exhaustion of M, as will become clear in §2, the ordering on $I \times J$ being the product of the orderings on I and J $\{(\alpha,\beta) \leq (\gamma,\delta)$ iff $\alpha \leq \gamma$ and $\beta \leq \delta\}$. But, even if $(M_\alpha | \alpha \in I)$ and $(N_\beta | \beta \in J)$ are faithful, perhaps $(M_\alpha \cap N_\beta | (\alpha,\beta) \in I \times J)$ is not faithful.

In later sections a faithful exhaustion of a space M will most often be regarded as a <u>set</u> of subsets of M, as has already been done above. We then denote such an exhaustion by something like H, $\tilde{H}$, H', or H with a subscript, if several faithful exhaustions are considered at the same time.

We now describe a partition of M into "patches" starting from a given exhaustion $(M_\alpha | \alpha \in I)$.

<u>Lemma 1.16.</u> For any subset F of M, which is contained in some set M_α, there exists, in the ordered set I, the infimum γ of all indices $\alpha \in I$ with $F \subset M_\alpha$, and $F \subset M_\gamma$.

This follows easily from E3 and E4.

Definition 10. The index $\eta(x)$ of a point $x \in M$, with respect to the given exhaustion $(M_\alpha | \alpha \in I)$, is the infimum of all indices $\alpha \in I$ with $x \in M_\alpha$.

Notice that the index function $\eta : M \to I$ fully describes the exhaustion, namely for every $\alpha \in I$,

$$M_\alpha = \{x \in M | \eta(x) \leq \alpha\}.$$

Definition 11. For every $\alpha \in I$ we define

$$M_\alpha^o := \{x \in M | \eta(x) = \alpha\}.$$

An index $\alpha \in I$ is called primitive, if M_α^o is not empty, which means that $\alpha \in \eta(M)$. The set $\eta(M)$ of primitive indices is denoted by I^o. The sets M_α^o, with α running through I^o, are called the patches of M with respect to the given exhaustion.

$(M_\alpha^o | \alpha \in I^o)$ is a partition of the set M into non empty disjoint subsets. Clearly, for every $\alpha \in I$,

(1.17) $$M_\alpha^o = M_\alpha \smallsetminus \cup (M_\beta | \beta \in I, \beta < \alpha)$$

and

(1.18) $$M_\alpha = \cup (M_\gamma^o | \gamma \leq \alpha, \gamma \in I^o) .$$

This partition of M will be very useful in the following.

Example 1.19. Let X be a simplicial complex, regarded as a weakly semi-algebraic space (Ex. 1.7). We consider the exhaustion of X by the finite subcomplexes Y which are closed in X, every Y being indexed by itself. Here the patches are just the open simplices of X. The index $\eta(x)$ of a point x in a given open simplex σ of X is the complex $\bar{\sigma} \cap X$.

§2 - Morphisms

In the following M is a weakly semialgebraic space and $(M_\alpha | \alpha \in I)$ is a fixed exhaustion of M. We seek an understanding of the morphisms from another weakly semialgebraic space L to M in the category of function ringed spaces over R (cf. §1, Def. 2). The crucial fact is

Theorem 2.1. Assume that the space L is semialgebraic. Let $f : L \to M$ be a map from the set L to the set M. The following are equivalent.

a) f is a morphism.

b) There exists some $\alpha \in I$ such that $f(L) \subset M_\alpha$ and the map $f_\alpha : L \to M_\alpha$, obtained from f by restriction of the range, is semialgebraic (i.e. a morphism between the semialgebraic spaces L and M_α).

If f is a morphism then the map f_α is semialgebraic for every $\alpha \in I$ with $f(L) \subset M_\alpha$.

Proof. Every map f with property b) is a morphism since the inclusion map $M_\alpha \hookrightarrow M$ is a morphism from the small subspace M_α of M to M. Assume now that $f : L \to M$ is a morphism. If we have found some $\alpha \in I$ with $f(L) \subset M_\alpha$ then we are done, since it is evident from the space structure of M_α as a small subspace of M that f_α is again a morphism.

Let $\eta : M \to I$ be the index function of our exhaustion (cf. §1, Def. 10). Suppose that $\eta(f(L))$ is an infinite subset J of the set of primitive indices I^o. For every $\alpha \in J$ we choose a point $x_\alpha \in L$ such that the point $y_\alpha := f(x_\alpha)$ has index α, i.e. $y_\alpha \in M_\alpha^o$. Let $S := \{y_\alpha | \alpha \in J\}$. For every $\gamma \in I$ the set $S \cap M_\gamma$ is finite, since M_γ is the union of finitely many patches M_α^o. Thus $(M \setminus S) \cap M_\gamma \in \mathring{\gamma}(M_\gamma)$, and we conclude that $M \setminus S \in \mathring{\mathfrak{T}}(M)$. Since f is a morphism this implies that $f^{-1}(M \setminus S) = L \setminus f^{-1}(S) \in \mathring{\gamma}(L)$. Thus $A := f^{-1}(S)$ is a closed semialgebraic subset of L. For the same reason the preimage $f^{-1}(S')$ of any subset S' of S is a closed semialgebraic subset of L, hence of A. If $S' \neq \emptyset$ then also $f^{-1}(S') \neq \emptyset$ since $f^{-1}(S')$ contains the points x_α with $y_\alpha \in S'$.

The semialgebraic space A has a finite number r of connected components. We choose a partition

$$S = S_1 \sqcup S_2 \sqcup \ldots \sqcup S_{r+1}$$

of S into r+1 disjoint non empty sets. This is possible since S is infinite. Then

$$A = f^{-1}(S_1) \sqcup f^{-1}(S_2) \sqcup \ldots \sqcup f^{-1}(S_{r+1})$$

is a partition of A into r+1 disjoint non empty closed semialgebraic subsets. But such a partition is impossible since A has only r connected components. This contradiction proves that J is finite. It follows, by properties E2 and E5 of the exhaustion of M, that $f(L) \subset M_\alpha$ for some $\alpha \in I$. q.e.d.

Definition 1. In view of condition b) in the theorem we call a morphism from a semialgebraic space L to a weakly semialgebraic space M a semialgebraic map from L to M (over R).

Here is an application of Theorem 2.1.

Proposition 2.2. Let $A \in \overline{\gamma}(M)$ (cf. §1, Def. 5). Then there exists some $\alpha \in I$ with $A \subset M_\alpha$. For every such index α we have $A \in \overline{\gamma}(M_\alpha)$, and the structure on A as a closed semialgebraic subspace of M_α coincides with the structure as a small subspace of M.

Proof. The inclusion map $A \hookrightarrow M$ is a morphism from A, with its structure as a small subspace of M, to M. Applying Theorem 2.1 to this morphism we obtain the result. (Recall Ex. 1.3 for the last statement).

Theorem 2.3. Let N be a second weakly semialgebraic space and let $(N_\beta | \beta \in J)$ be an exhaustion of N. Let $f : M \to N$ be a map from the set M to the set N. The following are equivalent.

a) f is a morphism

b) For every $\alpha \in I$ there exists some $\beta \in J$ such that $f(M_\alpha) \subset N_\beta$, and that the map $f_{\alpha\beta} : M_\alpha \to N_\beta$ obtained from f by restriction is semialgebraic.

If f is a morphism, and if $\alpha \in I$, $\beta \in J$ are indices with $f(M_\alpha) \subset N_\beta$, then the restriction $f_{\alpha\beta} : M_\alpha \to N_\beta$ is always semialgebraic.

All this is evident from Theorem 2.1. Indeed, since M carries the inductive limit space structure of the function ringed spaces M_α, a map $f : M \to N$ is a morphism iff all restrictions $f|M_\alpha : M_\alpha \to N$ are semialgebraic.

Definition 2. In view of condition b) in the theorem we call a morphism between weakly semialgebraic spaces a weakly semialgebraic map (over R).

Example 2.4. Let $f : X \to Y$ be a weakly simplicial map (cf. [LSA, p. 23]) between simplicial complexes X and Y. We regard X and Y as weakly semialgebraic spaces, cf. Ex. 1.7. Then f is a weakly semialgebraic map. If f is even simplicial (cf. [LSA, p. 24]), then the closure $\overline{f} : \overline{X} \to \overline{Y}$ of f (loc.cit.) is again a weakly semialgebraic map.

Example 2.5. A map $f : M \to N$ from M to a semialgebraic space N is weakly semialgebraic iff, for every $\alpha \in I$, the restriction $f|M_\alpha : M_\alpha \to N$ is semialgebraic.

In particular, the weakly semialgebraic maps from M to the real line R are precisely the elements of $\mathcal{O}_M(M)$. This special case of Theorem 2.3 is already evident from the definition of an exhaustion.

Definition 3. From now on we call the elements of $\mathcal{O}_M(M)$ the weakly semialgebraic functions on M. More generally we call, for every $U \in \mathring{\mathcal{T}}(M)$, the elements of $\mathcal{O}_M(U)$ the weakly semialgebraic functions on U. This is justified since U, as an open subspace of M, is again weakly semialgebraic with exhaustion $(U \cap M_\alpha | \alpha \in I)$.

Remark 2.6. In the situation of Theorem 2.3, let $f : M \to N$ be a weakly semialgebraic map. f induces a map $\kappa : I \to J$ as follows (cf. Lemma 1.16).

$$\kappa(\alpha) := \inf(\beta \in J | f(M_\alpha) \subset N_\beta) .$$

This map κ is monotonic, i.e. $\alpha \leq \beta$ implies $\kappa(\alpha) \leq \kappa(\beta)$. Moreover, for every $\alpha \in I$, $f(M_\alpha) \subset N_{\kappa(\alpha)}$.

Applying this remark to the case $N = M$, $f = id_M$, we see that any two exhaustions of a given space are "comparable". More precisely,

Proposition 2.7. If $(N_\beta | \beta \in J)$ is a second exhaustion of M, then there exist monotonic maps $\kappa : I \to J$ and $\mu : J \to I$ with $M_\alpha \subset N_{\kappa(\alpha)}$ for every $\alpha \in I$, and $N_\beta \subset M_{\mu(\beta)}$ for every $\beta \in J$.

Of course, M_α is closed semialgebraic in $N_{\kappa(\alpha)}$, and N_β is closed semialgebraic in $M_{\mu(\beta)}$.

Corollary 2.8. If M is a weak polytope and $(N_\beta | \beta \in J)$ is any exhaustion of M, then every N_β is a polytope.

Let now N be a locally semialgebraic space over R. We want to analyse the morphisms from M to N and from N to M in the category of function ringed spaces over R. We choose an admissible covering $(U_\lambda | \lambda \in \Lambda)$ of N by open semialgebraic subsets. For every finite subset J of λ we denote by U_J the union of all U_λ with $\lambda \in J$. Notice that N is the inductive limit of the family of semialgebraic spaces $(U_J \mid J \subset \Lambda, J$ finite).

Proposition 2.9. A map $f : M \to N$ is a morphism iff, for every $\alpha \in I$, the restriction $f|M_\alpha : M_\alpha \to N$ is a morphism, i.e. a semialgebraic map (cf. [LSA, p. 9]).

This is evident since M is the inductive limit of the spaces M_α. (We observed it in a special case already in Ex. 2.5.)

Definition 4. We call the morphisms from M to a locally (instead of weakly) semialgebraic space N again the weakly semialgebraic maps from M to N. This is justified by Proposition 2.9. (N.B. There is no conflict with Definition 2 if N is at the same time weakly and locally semialgebraic.)

Example 2.10. The weakly semialgebraic maps from M to R_{loc} (cf. [LSA, p. 15, p. 81]) are precisely those weakly semialgebraic functions on M which are bounded on every M_α.

Proposition 2.11. A map $f : N \to M$ is a morphism iff, for every $\lambda \in \Lambda$, the restriction $f|U_\lambda : U_\lambda \to M$ is a semialgebraic map.

Proof. The condition is necessary since the inclusions $U_\lambda \hookrightarrow N$ are morphisms. Assume now that $f|U_\lambda$ is semialgebraic, i.e. a morphism, for every $\lambda \in \Lambda$. Then also for every finite subset J of Λ the restriction $f|U_J$ is a morphism from U_J to M. Since N is the inductive limit of the spaces U_J we conclude that f is a morphism. q.e.d.

Definition 5. We call the morphisms from a locally semialgebraic space N to M the locally semialgebraic maps from N to M.

If M is weakly semialgebraic *and* locally semialgebraic then the weakly semialgebraic maps from M to N are the same as the locally semialgebraic maps from M to N, as considered in [LSA], since both kinds of maps are just the morphisms from M to N in the category of function ringed spaces over R. Applying this remark to the case M = N and the identity map id_M we obtain from Propositions 2.9 and 2.11 the following fact.

Proposition 2.12. Assume that M is also locally semialgebraic and let $(U_\lambda | \lambda \in \Lambda)$ be an admissible covering of M by open semialgebraic subsets (in the terminology of [LSA]). For every $\alpha \in I$ there exists some finite subset J of Λ with $M_\alpha \subset U_J$. Then $M_\alpha \in \overline{\gamma}(U_J)$. For every finite subset J of Λ there exists some $\beta \in I$ with $U_J \subset M_\beta$. Then $U_J \in \mathring{\gamma}(M_\beta)$.

Definition 6. We denote by WSA(R) the category whose objects are the weakly semialgebraic spaces over R and whose morphisms are the weakly semialgebraic maps over R. It is a full subcategory of the category of function ringed spaces over R, and it contains, as a full subcategory, the category LSA(R) of paracompact locally semialgebraic spaces over R.

Let S be a real closed overfield of R. There exists a natural and very useful functor "base extension" from LSA(R) to LSA(S), cf. [LSA, p. 19f and p. 309] and, for semialgebraic spaces, [DK_3, §4]. We want to extend this functor to a functor from WSA(R) to WSA(S).

Given, as before, a weakly semialgebraic space M over R and an exhaustion $(M_\alpha | \alpha \in I)$ of M we obtain by base extension an ordered family $(M_\alpha(S) | \alpha \in I)$ of semialgebraic spaces over S. For any two indices $\beta < \alpha$ the inclusion map $M_\beta \to M_\alpha$ gives, by base extension, an embedding $M_\beta(S) \to M_\alpha(S)$ of $M_\beta(S)$ into $M_\alpha(S)$ as a closed semialgebraic subspace [DK_3, p. 142]. Let N denote the inductive limit of the family of sets

$(M_\alpha(S)|\alpha\in I)$, with these transition maps. Every $M_\alpha(S)$ injects into N. We regard the $M_\alpha(S)$ as subsets of N. The family of subsets $(M_\alpha(S)|\alpha\in I)$ of N fulfills E1-E5. We equip N with the inductive limit space structure of the semialgebraic spaces $M_\alpha(S)$. By Theorem 1.6 this function ringed space N over S is weakly semialgebraic, and every $M_\alpha(S)$ is, in its given space structure, a closed semialgebraic subspace of N, and $(M_\alpha(S)|\alpha\in I)$ is an exhaustion of N.

Definition 7. We denote the weakly semialgebraic space N by M(S). We call M(S) the space obtained from M by base field extension from R to S. We always regard M as a subset of M(S).

It is evident from Proposition 2.7 that the space M(S) does not depend on the choice of the exhaustion $(M_\alpha|\alpha\in I)$. If M is also a locally semialgebraic space then M(S) is the same space as defined in [LSA]. This follows easily from Proposition 2.12. In particular, then M(S) is again locally semialgebraic.

Let N be a second weakly semialgebraic space over R and $(N_\beta|\beta\in J)$ an exhaustion of N. Let $f : M \to N$ be a weakly semialgebraic map over R. We obtain a weakly semialgebraic map $f_S : M(S) \to N(S)$ over S as follows. We choose a monotonic map $\kappa : I \to J$ such that, for every $\alpha \in I$, $f(M_\alpha) \subset N_{\kappa(\alpha)}$ (cf. Remark 2.6). The restrictions $f_\alpha : M_\alpha \to N_{\kappa(\alpha)}$ yield, by base field extension from R to S, semialgebraic maps $(f_\alpha)_S : M_\alpha(S) \to N_{\kappa(\alpha)}(S)$. All these maps fit together to a weakly semialgebraic map $f_S : M(S) \to N(S)$. This map f_S neither depends on the choice of the exhaustions of M and N nor on the choice of κ.

Definition 8. We call f_S the map obtained from f by base field extension from R to S.

If N is a locally semialgebraic space over R we obtain, in a similar way, from any morphism $f : M \to N$ (resp. $g : N \to M$) over R a morphism $f_S : M(S) \to N(S)$ (resp. $N(S) \to M(S)$) over S. This is the same map as in [LSA] if both M and N are locally semialgebraic.

§3 - Subspaces and products

As before, M is a weakly semialgebraic space over R and $(M_\alpha | \alpha \in I)$ is a fixed exhaustion of M.

Definition 1. A subset X of M is called weakly semialgebraic in M if, for every $\alpha \in I$, the set $X \cap M_\alpha$ is semialgebraic in the semialgebraic space M_α, i.e., in the notation of [LSA], $X \cap M_\alpha \in \mathscr{Y}(M_\alpha)$. The set of weakly semialgebraic subsets of M is denoted by $\mathcal{T}(M)$.

This set $\mathcal{T}(M)$ does not depend on the choice of the exhaustion $(M_\alpha | \alpha \in I)$, as is evident from Proposition 2.7. Clearly $\mathring{\mathcal{T}}(M) \subset \mathcal{T}(M)$. Also $\overline{\mathcal{T}}(M) \subset \mathcal{T}(M)$. In particular, $M_\alpha \in \mathcal{T}(M)$ for every $\alpha \in I$.

If X and Y are weakly semialgebraic subsets of M then $X \cup Y$, $X \cap Y$, and $X \setminus Y$ are again weakly semialgebraic in M. It is also easily verified that the preimage $f^{-1}(X)$ of any $X \in \mathcal{T}(M)$ under a weakly semialgebraic map $f : N \to M$ is weakly semialgebraic in N.

Weakly semialgebraic sets can be obtained by "collecting" semialgebraic sets as follows. Recall that $(M_\alpha^o | \alpha \in I^o)$ denotes the family of patches of the exhaustion $(M_\alpha | \alpha \in I)$. Every M_α^o is an (open) semialgebraic subset of M_α (cf. 1.17), hence a semialgebraic space.

Remarks 3.1. i) Assume that, for every $\alpha \in I$, there is given a semialgebraic subset X_α of M_α such that, whenever $\beta < \alpha$, the set $X_\alpha \cap M_\beta$ is contained in X_β. Let X denote the union of all X_α. Using E4 we see that, for every $\alpha \in I$,

$$X \cap M_\alpha = \cup (X_\beta | \beta \leq \alpha).$$

By E3 this is a semialgebraic subset of M_α. Thus $X \in \mathcal{T}(M)$.

ii) Assume that, for every primitive index $\alpha \in I^o$, there is given a semialgebraic subset Z_α of M_α^o. Let X denote the union of all Z_α. Clearly $X \cap M_\alpha^o = Z_\alpha$ for every $\alpha \in I^o$. Thus we have, for every $\alpha \in I$,

$$X \cap M_\alpha = \cup(Z_\beta \mid \beta \leq \alpha, \beta \in I^o) .$$

This is a semialgebraic subset of M_α. We conclude that $X \in \mathcal{T}(M)$.

If X is a weakly semialgebraic subset of M then we regard, for every $\alpha \in I$, the set $X \cap M_\alpha$ as a semialgebraic space, namely a semialgebraic subspace of M_α. Applying Theorem 1.6 to the set X and the family of spaces $(X \cap M_\alpha \mid \alpha \in I)$ we obtain a structure of a weakly semialgebraic space on X which is the inductive limit of the spaces $X \cap M_\alpha$. Every space $X \cap M_\alpha$ is a closed semialgebraic subspace of X and $(X \cap M_\alpha \mid \alpha \in I)$ is an exhaustion of X. It is evident from Proposition 2.7 that the space structure on X does not depend on the choice of the exhaustion $(M_\alpha \mid \alpha \in I)$.

Definition 2. We call such a space X a (weakly semialgebraic) subspace of M.

This terminology is justified by the following fact, which follows immediately from Theorem 2.3 and the subspace theory for semialgebraic spaces [DK_2, p. 186].

Proposition 3.2. Let $X \in \mathcal{T}(M)$. The inclusion map $j : X \hookrightarrow M$ is weakly semialgebraic. If $g : N \to X$ is a map from a weakly semialgebraic space N to X then g is weakly semialgebraic iff $j \circ g$ is weakly semialgebraic.

Also the following fact is rather obvious from the theory of semialgebraic spaces.

<u>Proposition 3.3.</u> Let $X \in \mathcal{T}(M)$ and let Y be a subset of X. Then $Y \in \mathcal{T}(X)$ iff $Y \in \mathcal{T}(M)$. In this case the subspace structures on Y with respect to X and to M are equal.

<u>Definition 3.</u> A subset X of M is called <u>semialgebraic</u> in M if $X \in \mathcal{T}(M)$ and the subspace X of M is a semialgebraic space. We denote the set of all semialgebraic subsets of M by $\gamma(M)$.

Notice that $\bar{\gamma}(M) \subset \gamma(M)$. In particular, $M_\alpha \in \gamma(M)$ for every $\alpha \in I$.

<u>Proposition 3.4.</u> A weakly semialgebraic subset X of M is semialgebraic iff there exists some $\alpha \in I$ with $X \subset M_\alpha$.

<u>Proof.</u> If $X \in \mathcal{T}(M)$ and $X \subset M_\alpha$ for some $\alpha \in I$ then $X = X \cap M_\alpha$ is a semialgebraic subset of M_α, and the subspace structures on X with respect to M and M_α are equal. Thus X is semialgebraic in M. Conversely, assume that $X \in \gamma(M)$. Then $\{X\}$ is an exhaustion of X. But also $(X \cap M_\alpha | \alpha \in I)$ is an exhaustion of X. By Proposition 2.7 there exists some $\alpha \in I$ with $X \cap M_\alpha = X$, i.e. $X \subset M_\alpha$. q.e.d.

By this proposition it is obvious that for any two semialgebraic sets X,Y in M the sets $X \cup Y$, $X \cap Y$, $X \setminus Y$ are again semialgebraic in M. Using Theorem 2.3 and the semialgebraic theory we also see that the image f(X) of any $X \in \gamma(M)$ under a weakly semialgebraic map $f : M \to N$ is semialgebraic in N.

We introduce on M the <u>strong topology</u>, defined as follows.

<u>Definition 4.</u> a) A subset U of M is open in the strong topology if, for every $\alpha \in I$, the set $U \cap M_\alpha$ is open in M_α in the strong topology of M_α, as defined in [DK_2,§7]. In other words, $U \cap M_\alpha$ is a - perhaps in-

finite - union of open semialgebraic subsets of M_α.

b) We denote the set M together with the strong topology by M_{top}. This is a topological space in the classical sense. M_{top} is the inductive limit of the family $((M_\alpha)_{top} | \alpha \in I)$ in the category of topological spaces, in fact even in the category of generalized topological spaces. M_{top} does not depend on our choice of the exhaustion $(M_\alpha | \alpha \in I)$. We shall see soon (Cor. 3.12) that M_{top} is Hausdorff.

Caution. Every $U \in \mathring{\mathfrak{T}}(M)$ is open in M_{top}, but in general $\mathring{\mathfrak{T}}(M)$ is not a basis of open sets in M_{top}, cf. Appendix C.

Proposition 3.5. i) $\mathring{\mathfrak{T}}(M)$ is the set of all $X \in \mathfrak{T}(M)$ which are open in the strong topology. For every $X \in \mathring{\mathfrak{T}}(M)$ the subspace structure in M, as defined now, coincides with the structure as an open subspace of the function ringed space M (cf. §1, Def. 4).

ii) $\overline{\mathfrak{F}}(M)$ is the set of all $X \in \mathfrak{F}(M)$ which are closed in M in the strong topology. For every $X \in \overline{\mathfrak{F}}(M)$ the subspace structure in M, as defined now, coincides with the structure as a small subspace of the function ringed space M (cf. §1, Def. 3).

Proof. Both assertions hold if M is semialgebraic. (Recall 1.3 and 1.4 for ii.) Starting from that it is easy to extend them to the general case. (Recall 2.2 for ii.)

A new terminology. From now on, topological notions like "open", "closed", "dense", "continuous", "closure" and "interior" of a set, usually refer to the strong topology. The sets $U \in \mathring{\mathfrak{T}}(M)$ are now called "open weakly semialgebraic sets" instead of "open sets". Our previous terminology "closed semialgebraic" for the sets in $\overline{\mathfrak{F}}(M)$ can be maintained.

<u>Proposition 3.6.</u> The closure $\overline{X}$ of a semialgebraic subset of M is again semialgebraic in M, hence $\overline{X} \in \overline{\gamma}(M)$.

<u>Proof.</u> This is evident since $X \subset M_\alpha$ for some $\alpha \in I$ and since $\overline{X}$ is also the closure of X in M_α, which is well known to be semialgebraic [DK_2, Th. 7.7].

<u>Definition 5.</u> A subset X of M is called <u>closed weakly semialgebraic</u> in M, if X is closed in M (in the strong topology) and $X \in \mathcal{T}(M)$. The set of all these sets is denoted by $\overline{\mathcal{T}}(M)$.

By Prop. 3.5.i, $\overline{\mathcal{T}}(M)$ is the set of complements $M \setminus U$ of all $U \in \mathring{\mathcal{T}}(M)$. Notice also that $\overline{\mathcal{T}}(M) \cap \gamma(M) = \overline{\gamma}(M)$ and

$$\overline{\mathcal{T}}(M) = \{X \in \mathcal{T}(M) \mid X \cap M_\alpha \in \overline{\gamma}(M_\alpha) \text{ for every } \alpha \in I\}.$$

<u>Definition 6.</u> $\mathcal{T}_c(M)$ denotes the set of all $X \in \mathcal{T}(M)$ such that the subspace X of M is a weak polytope.

<u>Proposition 3.7.</u> A subset X of M is an element of $\mathcal{T}_c(M)$ iff $X \cap M_\alpha \in \gamma_c(M_\alpha)$ for every $\alpha \in I$. Then $X \in \overline{\mathcal{T}}(M)$.

<u>Proof.</u> Let $X \in \mathcal{T}_c(M)$ be given. Since $(X \cap M_\alpha \mid \alpha \in I)$ is an exhaustion of the space X, every space $X \cap M_\alpha$ is a polytope in M_α. In particular, $X \cap M_\alpha \in \overline{\gamma}(M_\alpha)$. Thus $X \in \overline{\mathcal{T}}(M)$. Conversely, if X is a subset of M with $X \cap M_\alpha \in \gamma_c(M_\alpha)$ for every $\alpha \in I$, then $X \in \mathcal{T}(M)$ and the space X is a weak polytope.

<u>Remark 3.8.</u> Assume that our weakly semialgebraic space M is also locally semialgebraic. Then the strong topology in M defined here coincides with the strong topology defined in I, §3. The set $\mathcal{T}(M)$

of weakly semialgebraic subsets coincides with the set of locally semialgebraic subsets defined in I, §3. For every $X \in \mathcal{T}(M)$ the subspace structures on X defined here and there are equal. All this follows immediately from Proposition 2.12. Our symbols $\mathcal{T}(M)$, $\mathring{\mathcal{T}}(M)$, $\overline{\mathcal{T}}(M)$, $\mathring{\gamma}(M)$, $\overline{\gamma}(M)$, $\gamma_c(M)$, $\mathcal{T}_c(M)$ have the same meaning as in [LSA]. Now $\mathring{\mathcal{T}}(M)$ and $\mathring{\gamma}(M)$ are both bases of open sets for the strong topology.

Remark 3.9. Let $X \in \mathcal{T}(M)$. In general, X_{top} is not a topological subspace of M_{top}, cf. 4.8.e below. But this holds obviously if $X \in \mathring{\mathcal{T}}(M)$. It also holds if $X \in \overline{\mathcal{T}}(M)$. Indeed, let V be a subset of X which is open in X_{top}. Then $U := (M \setminus X) \cup V$ is open in M_{top}, since $(M \setminus U) \cap M_\alpha = (X \setminus V) \cap M_\alpha$ is closed in $(M_\alpha)_{top}$ for every $\alpha \in I$. Clearly $U \cap X = V$. If $V \in \mathcal{T}(X)$ then $U \in \mathring{\mathcal{T}}(M)$.

Proposition 3.10 ("Tietze's extension theorem"). Let $A \in \overline{\mathcal{T}}(M)$. Every weakly semialgebraic function $f : A \to L$ with values in a generalized interval $L \subset R$ can be extended to a weakly semialgebraic function $g : M \to L$.

This follows from Proposition 1.11.c and Remark 1.12 if we have proved

Lemma 3.11. Given a closed weakly semialgebraic subset A of M there exists an exhaustion $(A_\lambda | \lambda \in \Lambda)$ of M and a subset J of Λ such that $A = \bigcup (A_\lambda | \lambda \in J)$.

Proof. We start with the given exhaustion $(M_\alpha | \alpha \in I)$. We consider the direct product $\Lambda := I \times \{0,1\}$ of the ordered sets I and $\{0,1\}$, the second set being ordered by $0 < 1$. For any $(\alpha,t) \in \Lambda$ we define $A_{(\alpha,t)} = M_\alpha$ if $t = 1$ and $A_{(\alpha,t)} = A \cap M_\alpha$ if $t = 0$. Then $(A_{(\alpha,t)} | (\alpha,t) \in \Lambda)$ is an exhaustion with the desired property.

Corollary 3.12 ("Urysohn's Lemma"). Given two disjoint closed weakly semialgebraic subsets A and B of M there exists a weakly semialgebraic function $g : M \to [0,1]$ with $g^{-1}(0) \supset A$ and $g^{-1}(1) \supset B$.*)

In particular, A and B can be separated by the open weakly semialgebraic sets $g^{-1}([0,\frac{1}{4}[)$ and $g^{-1}(]\frac{3}{4},1])$. Choosing A and B as one point sets we see that the strong topology on M is Hausdorff.

Remark 3.13. It is even possible to find a weakly semialgebraic function $g : M \to [0,1]$ with $g^{-1}(0) = A$ and $g^{-1}(1) = B$.

Starting from the semialgebraic case [DK_1, Th. 1.6] this can be proved by an inductive procedure similar to the proof of Prop. 1.11.c, using the following observation. If X_1, X_2 are closed semialgebraic subsets of M and $g_1, g_2 : X_1 \cup X_2 \rightrightarrows [0,1]$ are semialgebraic functions with $g_i^{-1}(0) \cap X_i = A \cap X_i$, $g_i^{-1}(0) \cap X_j \supset A \cap X_j$, $g_i^{-1}(1) \cap X_i = B \cap X_i$, $g_i^{-1}(1) \cap X_j \supset B \cap X_j$, for $(i,j) = (1,2)$ and $= (2,1)$, then $g := \frac{1}{2}(g_1+g_2)$ is a semialgebraic function on $X_1 \cup X_2$ with values in $[0,1]$ and $g^{-1}(0) = A \cap (X_1 \cup X_2)$, $g^{-1}(1) = B \cap (X_1 \cup X_2)$.

Definition 7. A family $(X_\lambda | \lambda \in \Lambda)$ in $\mathcal{T}(M)$ is called an admissible covering of M if every $B \in \gamma(M)$ is contained in the union of finitely many X_λ. Of course, it suffices to check this property with B running through the sets M_α.

Notice that, for any $U \in \mathring{\mathcal{T}}(M)$, the elements of $\mathrm{Cov}_M(U)$ are just the admissible coverings $(X_\lambda | \lambda \in \Lambda)$ of U, in the sense of this definition, with $X_\lambda \in \mathring{\mathcal{T}}(U)$ for every $\lambda \in \Lambda$.

---

*) As always, $[0,1]$ denotes the closed unit interval in R.

Admissible coverings behave well under taking preimages.

Proposition 3.14. If $f : N \to M$ is a weakly semialgebraic map and if $(X_\lambda | \lambda \in \Lambda)$ is an admissible covering of M then $(f^{-1}(X_\lambda) | \lambda \in \Lambda)$ is an admissible covering of N.

Proof. Let $B \in \gamma(N)$. Then $f(B) \in \gamma(M)$. There exists a finite subset J of Λ with $f(B) \subset \cup(X_\lambda | \lambda \in J)$. This implies that $B \subset \cup(f^{-1}(X_\lambda) | \lambda \in J)$. q.e.d.

Admissible coverings of M by closed semialgebraic sets, or even by closed weakly semialgebraic sets, can be useful. They are less special than exhaustions, and nevertheless they sometimes do the sames service. This is indicated by the following theorem, which can be verified in a straightforward manner.

Theorem 3.15. Let $(A_\lambda | \lambda \in \Lambda)$ be an admissible covering of M by closed weakly semialgebraic subsets.

a) A subset X of M is an element of $\mathring{\mathcal{T}}(M)$ (resp. $\overline{\mathcal{T}}(M)$, resp. $\mathcal{T}(M)$) iff, for every $\lambda \in \Lambda$, the intersection $X \cap A_\lambda$ is an element of $\mathring{\mathcal{T}}(A_\lambda)$ (resp. $\overline{\mathcal{T}}(A_\lambda)$, resp. $\mathcal{T}(A_\lambda)$).
b) If $U \in \mathring{\mathcal{T}}(M)$ and $(X_\beta | \beta \in J)$ is a family of subsets of U, then $(X_\beta | \beta \in J) \in \mathrm{Cov}_M(U)$ iff, for every $\lambda \in \Lambda$, $(X_\beta \cap A_\lambda | \beta \in J) \in \mathrm{Cov}_{A_\lambda}(U \cap A_\lambda)$.
c) If $U \in \mathring{\mathcal{T}}(M)$, then a function $f : U \to R$ is weakly semialgebraic iff for every $\lambda \in \Lambda$, the function $f | U \cap A_\lambda$ on A_λ is weakly semialgebraic.
d) (Gluing principle for weakly semialgebraic maps). A map $f : M \to N$ into a function ringed space N over R is a morphism iff, for every $\lambda \in \Lambda$, the restriction $f | A_\lambda : A_\lambda \to N$ is a morphism.

Corollary 3.16. Let $(A_\lambda | \lambda \in \Lambda)$ be an ordered family in $\overline{\gamma}(M)$ with the properties E2-E5. Assume that every $B \in \gamma(M)$ is contained in some A_λ. Then $(A_\lambda | \lambda \in \Lambda)$ is an exhaustion of M. (Again it suffices to let B run through the sets M_α.)

Indeed, Theorem 3.15.d) {as well as Th. 3.15.a)-c)} tells us that the space M is the inductive limit of the family of spaces $(A_\lambda \mid \lambda \in \Lambda)$.

Definition 8. A *path* in M is a semialgebraic map from the unit interval [0,1] in R to M. For any point x of M the *path component* C(x,M) of x in M is the set of all $y \in M$ such that there exists a path $\gamma : [0,1] \rightarrow M$ with $\gamma(0) = x$ and $\gamma(1) = y$.

Proposition 3.17. Every path component C(x,M) is closed and also open weakly semialgebraic in M. The space M is the direct sum (cf. 1.10) of the different path components of M, considered, of course, as subspaces of M.

Proof. Clearly, for every $\alpha \in I$, the intersection $C(x,M) \cap M_\alpha$ is a union of path components of the semialgebraic space M_α. Thus $C(x,M) \cap M_\alpha \in \mathring{\mathcal{T}}(M_\alpha) \cap \overline{\mathcal{T}}(M_\alpha)$. The claim follows.

It is easily seen that a path component X is not the union of two disjoint non empty open weakly semialgebraic subsets.

Definition 9. We call such a space X *connected*. Justified by Proposition 3.17 we call the path components of M also the *connected components* of M.

Let N be a second weakly semialgebraic space over R and $(N_\beta \mid \beta \in J)$ an exhaustion of N. We want to construct the *direct product* of the spaces M and N. We equip the cartesian product $M \times N$ of the sets M,N with the inductive limit space structure of the ordered family of semialgebraic spaces $(M_\alpha \times N_\beta \mid (\alpha,\beta) \in I \times J)$, where $I \times J$ is the direct product of the ordered sets I and J. This family fulfills the assumptions in Theorem 1.6 and in addition E4 and E5. Thus, by Theorem 1.6, the space $M\times N$ is weakly semialgebraic with exhaustion $(M_\alpha \times N_\beta \mid (\alpha,\beta) \in I\times J)$ and the subspace structure of $M_\alpha \times N_\beta$ in $M\times N$ coincides with the given structure as the direct product of the semialgebraic spaces M_α and N_β. Using Theorem 2.3 it is easily checked that the natural projections $pr_1 : M\times N \to M$, $pr_2 : M\times N \to N$ are weakly semialgebraic maps, and that $M\times N$, with these projections, is the direct product of M and N in the category WSA(R) of weakly semialgebraic spaces over R.

Caution. The strong topology on $M\times N$ may have more open sets than the direct product of the strong topologies on M and N.

Proposition 3.18. Let $X \in \mathcal{T}(M)$ and $Y \in \mathcal{T}(N)$. Then the set $X\times Y$ is weakly semialgebraic in the space $M\times N$, and the subspace structure on $X\times Y$ in $M\times N$ coincides with the structure as the direct product of the subspaces X and Y of M and N. If X and Y are semialgebraic then $X\times Y$ is semialgebraic. If X and Y are closed (resp. open) in M and N then $X\times Y$ is closed (resp. open) in $M\times N$.

All this is obvious from the definitions. Using Theorem 2.3 also the following proposition is easily verified.

Proposition 3.19. Let $f : M \to N$ be a weakly semialgebraic map. The graph $\Gamma(f)$ of f is a closed weakly semialgebraic subset of $M\times N$. The natural projection $pr_1 \mid \Gamma(f)$ from $\Gamma(f)$ to M is an isomorphism of the subspace $\Gamma(f)$ of $M\times N$ to M.

We finally construct fibre products in the category WSA(R). Let $f : M \to L$ and $g : N \to L$ be weakly semialgebraic maps over R. Then

$f \times g : M \times N \to L \times L$ is again a weakly semialgebraic map and

$$M \times_L N := \{(x,y) \in M \times N \mid f(x) = g(y)\}$$

is a closed weakly semialgebraic subset of $M \times N$, since $M \times_L N$ is the preimage of the diagonal $\Delta_L = \Gamma(id_L)$ of $L \times L$ under $f \times g$. We equip $M \times_L N$ with the subspace structure in $M \times N$. The following can now be verified in a straightforward manner.

Theorem 3.20. The commutative diagram

$$\begin{array}{ccc} M \times_L N & \xrightarrow{q} & N \\ {\scriptstyle p}\downarrow & & \downarrow{\scriptstyle g} \\ M & \xrightarrow{f} & L \end{array},$$

with p and q the natural projections, i.e. the restrictions of pr_1 and pr_2 to $M \times_L N$, is a cartesian square in WSA(R).

Caution. We do not claim that this diagram is cartesian in the category of function ringed spaces over R.

Remark 3.21. If M and N are also locally semialgebraic spaces, then $M \times_L N$ is again locally semialgebraic. If all three spaces M,N,L are locally semialgebraic then our space $M \times_L N$ is the same as the fibre product $M \times_L N$ in [LSA].

This is rather evident from the definitions and Proposition 2.12.

All objects defined in this section behave well under base field extension (cf. Def. 7 and Def. 8 in §2) to some real closed field $S \supset R$.

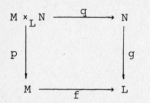

Remarks 3.22. a) Let $X \in \mathfrak{T}(M)$. For every $\alpha \in I$, the semialgebraic subset $X \cap M_\alpha$ of M_α yields by base field extension a semialgebraic subset $(X \cap M_\alpha)(S)$ of $M_\alpha(S)$ [DK$_3$, p. 142]. Let $X(S)$ denote the union of these sets in $M(S)$. We have $X(S) \cap M_\alpha(S) = (X \cap M_\alpha)(S)$, hence $X(S) \in \mathfrak{T}(M(S))$. The subspace structure on $X(S)$ in the space $M(S)$ is the same as the base field extension of the subspace structure on X in M. Thus the notion $X(S)$ has no ambiguity. If X is closed (resp. open) in M, then $X(S)$ is closed (resp. open) in $M(S)$. If $X \in \gamma(M)$ then $X(S) \in \gamma(M(S))$. We have $X(S) \cap M = X$.

b) If $(X_\lambda | \lambda \in \Lambda)$ is an admissible covering of M by weakly semialgebraic subsets, then $(X_\lambda(S) | \lambda \in \Lambda)$ is an admissible covering of $M(S)$.

c) If $(C_\lambda | \lambda \in \Lambda)$ is the family of connected components of M, then $(C_\lambda(S) | \lambda \in \Lambda)$ is the family of connected components of $M(S)$.

d) The space $(M \times N)(S)$ is the same as $M(S) \times N(S)$. If $f : M \to N$ is a weakly semialgebraic map, then the subsets $\Gamma(f_S)$ and $\Gamma(f)(S)$ of $M(S) \times N(S)$ are equal.

e) If two weakly semialgebraic maps $f : M \to L$, $g : N \to L$ are given then the two subsets $M(S) \times_{L(S)} N(S)$ (coming from the maps f_S, g_S) and $(M \times_L N)(S)$ of $(M \times N)(S)$ are equal. We conclude that the base field extension functor $WSA(R) \to WSA(S)$ preserves cartesian squares.

The easy proofs of a), b), d), e) and of similar statements may safely be left to the reader. c) is also easy once we know that the spaces $C_\lambda(S)$ are connected. This means the following claim:

If M is connected then $M(S)$ is connected.

The claim can be proved as follows: We fix a point $p \in M$. Let $x \in M(S)$ be given. Then $x \in M_\alpha(S)$ for some $\alpha \in I$. There exists a path γ in $M_\alpha(S)$ from x to some point $q \in M_\alpha$. (This is evident since, say, M_α is isomorphic to a simplicial complex over R.) There exists a path $\delta : [0,1] \to M$ with $\delta(0) = q$ and $\delta(1) = p$. Then δ_S is a path in $M(S)$

from q to p. The composite path $\gamma * \delta_S$ connects x to p.

§4 - Spaces of countable type

In this small and, up to some examples, almost trivial section we introduce a class of weakly semialgebraic spaces, the "spaces of countable type", which admit particularly simple exhaustions. This class of spaces behaves well under most constructions in this paper. Unfortunately it does not suffice for all purposes in semialgebraic topology.

In the following a "space" always means a weakly semialgebraic space over R. Also, if we write down an ordered family of sets $(X_n | n \in \mathbb{N})$ then we always mean that the set $\mathbb{N}$ of natural numbers is equipped with its natural total ordering.

Definition 1. A space M is of countable type if M has an admissible covering $(X_\lambda | \lambda \in \Lambda)$ by semialgebraic sets (cf. §3, Def. 7) with countable index set Λ.

Remarks 4.1. i) Of course, every semialgebraic space is of countable type.

ii) If M is of countable type, then also every subspace of M is of countable type.

iii) The direct product $M \times N$ of two spaces M,N of countable type is again of countable type. Indeed, if $(X_\lambda | \lambda \in \Lambda)$ and $(Y_\kappa | \kappa \in K)$ are admissible coverings of M and N by semialgebraic sets, with Λ and K countable, then $(X_\lambda \times Y_\kappa | (\lambda,\kappa) \in \Lambda \times K)$ is an admissible covering of $M \times N$ by semialgebraic sets with countable index set $\Lambda \times K$.

iv) The direct sum $\sqcup(M_\lambda | \lambda \in \Lambda)$ of a family $(M_\lambda | \lambda \in \Lambda)$ of non empty spaces of countable type is of countable type iff Λ is countable.

v) If S is a real closed overfield of R and M is a space over R of countable type then M(S) is again of countable type (cf. 3.22.b).

<u>Proposition 4.2.</u> Assume that M is of countable type. Let $(M_\alpha | \alpha \in I)$ be an exhaustion of M. Then there exists a sequence $(\alpha(n) | n \in \mathbb{N})$ in I, with $\alpha(n) \leq \alpha(n+1)$ for every $n \in \mathbb{N}$, such that $(M_{\alpha(n)} | n \in \mathbb{N})$ is an exhaustion of M.

<u>Proof.</u> M has an admissible covering $(X_n | n \in \mathbb{N})$ by semialgebraic sets. We successively find indices $\alpha(n)$ in I for $n = 1,2,\ldots$ such that $M_{\alpha(n)}$ contains X_n and $\alpha(n+1) \geq \alpha(n)$. The ordered family $(M_{\alpha(n)} | n \in \mathbb{N})$ of subsets of M fulfills trivially E2-E5, and it is an admissible covering of M. Thus, by Cor. 3.16, it is an exhaustion of M. q.e.d.

<u>Corollary 4.3.</u> A space M is of countable type iff M has an exhaustion with index set $\mathbb{N}$. This means just a family $(M_n | n \in \mathbb{N})$ of closed semi-algebraic subsets of M, with $M_n \subset M_{n+1}$ for every $n \in \mathbb{N}$, such that every $A \in \overline{\gamma}(M)$ is contained in some M_n.

<u>Proposition 4.4.</u> Assume that M is of countable type and $(M_\alpha | \alpha \in I)$ is a faithful exhaustion of M. Then I is countable.

<u>Proof.</u> By Prop. 4.2 there exists an isotonic sequence $(\alpha(n) | n \in \mathbb{N})$ in I such that $(M_{\alpha(n)} | n \in \mathbb{N})$ is an exhaustion of M. For every $\beta \in I$ there exists some $n \in \mathbb{N}$ with $M_\beta \subset M_{\alpha(n)}$. Since the exhaustion $(M_\alpha | \alpha \in I)$ is faithful this implies $\beta \leq \alpha(n)$. The sets $J(n) := \{\alpha \in I | \alpha \leq \alpha(n)\}$ are finite and their union is I. Thus I is countable. q.e.d.

<u>Corollary 4.5.</u> Let $(M_\alpha | \alpha \in I)$ be an exhaustion of M. Then M is of countable type iff the set I^o of primitive indices (cf. end of §1) is countable.

<u>Proof.</u> The family of patches $(M_\alpha^o | \alpha \in I^o)$ is an admissible covering of M by semialgebraic sets. Thus, if I^o is countable then M is certainly

of countable type. Assume now that M is of countable type. Let J be the set of all indices $\alpha \in I$ such that there does not exist an index $\beta < \alpha$ with $M_\beta = M_\alpha$. Then $(M_\alpha | \alpha \in J)$ is a faithful exhaustion of M (cf. proof of 1.14) and $J^o = I^o$. By Proposition 4.4, the set J is countable, and this implies that I^o is countable. q.e.d.

Proposition 4.6. Assume that the space M is locally semialgebraic. Then M is paracompact (as defined in I, §4), and every connected component of M is of countable type.

Proof. We assume without loss of generality that M is connected. We choose an exhaustion $(M_\alpha | \alpha \in I)$ of M and an admissible covering $(U_\lambda | \lambda \in \Lambda)$ of M by open semialgebraic subsets. For any finite subset K of Λ let U_K denote the union of all U_λ with $\lambda \in K$. We inductively choose a sequence $(K(n) | n \in \mathbb{N})$ of finite subsets of Λ and a sequence $(\alpha(n) | n \in \mathbb{N})$ in I such that

$$U_{K(n)} \subset M_{\alpha(n)} \subset U_{K(n+1)}$$

for every $n \in \mathbb{N}$. {We may start with $K(1) = \emptyset$.} This is possible since every set M_α is contained in some U_K and every set U_K is contained in some M_β (as already observed in 2.12). Let

$$X := \cup(U_{K(n)} | n \in \mathbb{N}) = \cup(M_{\alpha(n)} | n \in \mathbb{N}).$$

If $\alpha \in I$ is given then we conclude from the exhaustion axioms E3 and E4 (cf. §1, Def. 6) that $X \cap M_\alpha = M_\beta$ for some $\beta \leq \alpha$. Thus $X \in \overline{\mathfrak{T}}(M)$. But X is also open. Since M is connected we have X = M. This proves that M is of countable type and Lindelöf (cf. I, §4, Def. 3). We conclude by Proposition 3.6 (which states that, in the terminology of [LSA], M is taut) and I.7.15 that M is paracompact. q.e.d.

Example 4.7 (The countable comb). a) Let M be the subset of R^2 which is the union of $M_o := [0,1]\times\{0\}$ and the countably many sets $U_k := \{\frac{1}{k}\}\times]0,1]$ ("the teeth of the comb", $k \in \mathbb{N}$). We regard, for every $n \in \mathbb{N}$, the set $M_n := M_o \cup U_1 \cup \ldots \cup U_n$ as a polytope in R^2. Using Theorem 1.6 we equip the set M with the unique structure of a space such that every M_n, in its given structure, is a closed subspace of M and $(M_n | n\in\mathbb{N})$ is an exhaustion of M. Then M is a weak polytope of countable type.

b) This space M is useful to exhibit some pathologies which commonly are met with weak polytopes. The union U of all teeth U_n is an element of $\mathring{\mathcal{J}}(M)$, since $U = M \setminus M_o$. But the closure $\bar{U}$ of U is not weakly semialgebraic. Indeed, $\bar{U} \cap M_o$ consists of the infinitely many points $x_n := (\frac{1}{n},0)$, with $n \in \mathbb{N}$, and perhaps the point $x_o := (0,0)$, and this is not a semialgebraic subset of M_o. The interior of the semialgebraic set M_o in M is the complement of $\bar{U} \cap M_o$ in M_o, and this set is again not semialgebraic.

c) The space M cannot be triangulated, i.e. is not isomorphic to a simplicial complex (regarded as a space). Indeed, assume a triangulation of M is given. Then all simplices have dimension ≤ 1. Every point x_n has a fundamental system $\mathfrak{N}$ of connected semialgebraic neighbourhoods such that, for every $N \in \mathfrak{N}$, the set $N \setminus \{x_n\}$ has three connected components. Thus x_n must be a vertex of the triangulation. But the semialgebraic subset M_o of M cannot contain infinitely many vertices.

d) The space U is the direct sum (cf. 1.10) of the semialgebraic spaces $U_n (n \in \mathbb{N})$. Assume that R is archimedean. Then the point x_o lies in the closure $\bar{U}$ of the set U in M_{top}, but there does not exist a path $\alpha : [0,1] \to M$ with $\alpha([0,1[) \subset U$ and $\alpha(1) = x_o$. Thus the curve selection lemma fails in M for weakly semialgebraic spaces.

e) The subspace $X := U \cup \{x_o\}$ of M is the direct sum of the spaces U_n and $\{x_o\}$. The set $\{x_o\}$ is open in X_{top}. If R is archimedean then $\{x_o\}$ is not open in the subspace topology of X with respect to M_{top}. Thus, in this case, X_{top} is not a topological subspace of M_{top}.

Example 4.8 (Uncountable combs). Let, more generally, Λ be any infinite subset of $]0,1]$, and let M denote the subset $([0,1] \times \{0\}) \cup (\Lambda \times]0,1])$ of R^2. Further let, for any finite subset J of Λ,

$$M_J := ([0,1] \times \{0\}) \cup J \times]0,1],$$

equipped with the subspace structure in the semialgebraic standard space R^2. Using Theorem 1.6 we equip the set M with the unique structure of a weak polytope such that every M_J, in its given structure, is a closed subspace of M and $(M_J \mid J \subset \Lambda,\ J \text{ finite})$ is an exhaustion of M. If Λ is uncountable, then M is not of countable type by Proposition 4.4.

Choosing Λ as a set which contains 0 in its closure, e.g. $\Lambda =]0,1]$, we can observe the phenomena described in 4.7.d and 4.7.e above also for R not archimedean.

Remark 4.9. If the set Λ in Example 4.8 is uncountable then there does not exist a space N over R_o such that M is isomorphic to N(R).

Proof. Assume there is given an isomorphism $\varphi : M \xrightarrow{\sim} N(R)$ with N a space over R_o. We choose some $A \in \gamma(N)$, with $\varphi([0,1] \times \{0\}) \subset A(R)$. Let $\Lambda' := \Lambda \setminus \{1\}$. We shall prove that $\varphi(\Lambda' \times \{0\}) \subset A$. This will be the desired contradiction since the set A is countable.

Let $\lambda \in \Lambda'$ be given, and let $\xi := \varphi(\lambda,0)$. We have to verify that $\xi \in A$. We choose a closed semialgebraic subset X of N such that the image $\varphi(M_{\{\lambda\}})$ of the set

$$M_{\{\lambda\}} = ([0,1]_R \times \{0\}) \cup (\{\lambda\} \times [0,1]_R)$$

is contained in X(R). Now X is a polytope of dimension 1. We call a point y of a one-dimensional polytope Y a branching point of Y, if y has some semialgebraic neighbourhood V_o in Y such that for every semialgebraic neighbourhood $V \subset V_o$ the space $V \setminus \{y\}$ has at least three components. Clearly ξ is a branching point of $\varphi(M_{\{\lambda\}})$. Choosing a simultaneous triangulation of $X(R)$ and $\varphi(M_{\{\lambda\}})$ we see that ξ is a branching point of $X(R)$. Then choosing a triangulation of X we see that $\xi \in X$. We conclude that $\xi \in X \cap A(R) \subset A$. q.e.d.

Example 4.10. Assume that the field R is not archimedean but still contains a strictly decreasing sequence $(\varepsilon_n | n \in \mathbb{N})$ of positive elements converging to zero (cf. §9, Def. 3 below). We consider the space M in Example 4.8 with $\Lambda := \{\varepsilon_n | n \in \mathbb{N}\}$. Then M is of countable type. We claim that there does not exist a space N over R_o such that M is isomorphic to $N(R)$.

Proof. Suppose there exists an isomorphism $\varphi : M \xrightarrow{\sim} N(R)$ with N a space over R_o. As in the preceding proof we choose some $A \in \gamma(N)$ with $\varphi(M_o) \subset A(R)$, and we see that $\varphi(\Lambda \times \{0\}) \subset A$. We identify $M_o = [0,1]_R \times \{0\}$ with the unit interval $[0,1]_R$. We think of A as a semialgebraic subset in some R_o^d. Let $\| \ \|$ denote the euclidean norm in R^d. By the Łojasiewicz inequality [BCR, p. 39], applied to the functions $(x,y) \mapsto |x-y|$ and $(x,y) \mapsto \|\varphi(x) - \varphi(y)\|$ on the unit square $[0,1]_R \times [0,1]_R$, there exist some constants $C > 0$ in R and N in $\mathbb{N}$ such that

$$\|\varphi(x) - \varphi(y)\|^N \leq C|x-y|$$

for all x and y in $[0,1]_R$. Let $\xi_n := \varphi(\varepsilon_n)$. The ξ_n are pairwise different points in R_o^d with $\|\xi_n - \xi_m\|^N < C\varepsilon_n$ if $m > n$. But this is impossible, since for n large $C\varepsilon_n$ is smaller than every positive element of R_o. Thus an isomorphism φ as above does not exist. q.e.d.

§5 - Proper maps and partially proper maps

In this section a "space" means a weakly semialgebraic space over R, and a "map" means a weakly semialgebraic map between spaces, if nothing else is said.

In the following M and N are spaces and $(M_\alpha | \alpha \in I)$, $(N_\beta | \beta \in J)$ are fixed exhaustions of M and N respectively. $(M_\alpha^o | \alpha \in I^o)$ is the family of patches of M with respect to the given exhaustion (cf. end of §1).

Definition 1. A map $f : M \to N$ is called semialgebraic if the preimage $f^{-1}(Y)$ of every semialgebraic subset Y of N is semialgebraic in M.

Remarks 5.1. i) It suffices to check this property with Y running through the sets N_β. If f is semialgebraic then, by Prop. 3.14 and Cor. 3.16, the ordered family $(f^{-1}(N_\beta) | \beta \in J)$ is an exhaustion of M.

ii) If $f : M \to N$ is semialgebraic then the image f(X) of every $X \in \mathcal{T}(M)$ is a weakly semialgebraic subset of N.

iii) For every subspace X of M the inclusion map $X \hookrightarrow M$ is semialgebraic.

iv) The pull back $f' : M \times_N N' \to N'$ of a semialgebraic map $f : M \to N$ by an arbitrary (weakly semialgebraic) map $g : N' \to N$ is semialgebraic.

We call a set $A \in \mathcal{T}(M)$ discrete (in M) if the subspace A of M is discrete, i.e. is the direct sum (cf. 1.10) of one-point spaces. This means that the connected components (cf. §3, Def. 9) of A are one point sets. Analogously to Prop. I.5.4 we have the following criterion for a map to be semialgebraic.

Proposition 5.2. Let $f : M \to N$ be a map. Assume that

a) all fibres of f are semialgebraic,

b) every closed discrete weakly semialgebraic subset A of M has a weakly semialgebraic image f(A).

Then f is semialgebraic.

N.B. Conversely, if f is semialgebraic then a) and b) hold (cf. Remark 5.1.ii for b)).

The proof of Proposition 5.2 is fully analogous to the proof of the corresponding fact for locally semialgebraic spaces [LSA, p. 56f.] but easier. We may assume that N is semialgebraic. Then we have to prove that M is semialgebraic. For every $\alpha \in I^{o}$ we choose a point $x_\alpha \in M_\alpha^{o}$. The set $A := \{x_\gamma | \gamma \in I^{o}\}$ has a finite intersection with every M_α. Thus $A \in \overline{\mathfrak{T}}(M)$. The same goes for any subset A' of A. Thus A is discrete in M. By assumption b), $f(A) \in \gamma(N)$. Moreover, $f(A') \in \gamma(N)$ for every subset A' of A. Thus every subset of the semialgebraic space $B := f(A)$ is semialgebraic, and we conclude that B is a finite set (cf. [LSA, p. 57]). For every $y \in B$, the set $f^{-1}(y) \cap A$ is a semialgebraic subset of the discrete space A, hence a finite set. We conclude that A is a finite set, i.e. I^{o} is finite. This means that M is semialgebraic.

Definition 2. A map $f : M \to N$ is called proper, if, for every map $g : N' \to N$, the pull back $f' : M \times_N N' \to N'$ by g enjoys the following property: The image $f'(X)$ of every $X \in \overline{\mathfrak{T}}(M \times_N N')$ is closed weakly semialgebraic in N'. The space M is called complete, if the map from M to the one point space is proper.

The following facts are easily verified.

Remarks 5.3. i) Proper maps between weakly semialgebraic spaces enjoy the usual formal properties, cf. I.5.5.

ii) If M is semialgebraic, then a map $f : M \to N$ is proper iff $f(M)$ is closed in N and the map "f" : $M \to f(M)$ between semialgebraic spaces is proper in the category of semialgebraic spaces, cf. [DK_2, §9].

iii) Let $(B_\lambda | \lambda \in \Lambda)$ be an admissible covering of N by closed weakly semialgebraic subsets (e.g. an exhaustion of N). Then a map $f : M \to N$ is proper iff, for every $\lambda \in \Lambda$, the restriction $f^{-1}(B_\lambda) \to B_\lambda$ of f is proper.

iv) If M and N are weakly semialgebraic and also locally semialgebraic then a map $f : M \to N$ is proper in the present sense iff f is proper in the category of locally semialgebraic spaces (I, §5).

Theorem 5.4. Every proper map $f : M \to N$ is semialgebraic.

Proof. a) We first consider the special case that N is the one point space. Now M is a complete space, and we want to prove that M is semialgebraic. For every $\alpha \in I^o$ we choose a point $x_\alpha \in M_\alpha^o$. The set $A := \{x_\alpha | \alpha \in I^o\}$ is closed and weakly semialgebraic in M, and the same holds for every subset of A. Thus A is a complete discrete space. We conclude by an easy argument, cf. [LSA, p. 59], that A is a finite set. This means that I^o is finite, and M is indeed semialgebraic.

b) In the general case we now know that all fibres of f are semialgebraic. We conclude by Prop. 5.2 that M is semialgebraic. q.e.d.

Definition 3. A map $f : M \to N$ is called partially proper if, for every $A \in \overline{\gamma}(M)$, the restriction $f|A : A \to N$ is proper. The space M is called partially complete if the map from M to the one point space is partially proper. This means that every $A \in \overline{\gamma}(M)$ is complete.

Of course, it suffices to check these properties for A running through the sets M_α of our exhaustion of M. Notice that the partially complete spaces (resp. complete spaces) are the same objects as the weak polytopes (resp. polytopes) defined in §1.

Partially proper maps will be in the center of our interest. Thus we describe their formal properties rather explicitly. {The theory here is easier than in I, §5 since we can use Prop. 3.6.}

Remarks 5.5. Let $f : M \to N$ and $g : N \to L$ be maps.

i) If f and g are partially proper then $g \circ f$ is partially proper.

ii) If $g \circ f$ is partially proper then f is partially proper.

iii) If $g \circ f$ is partially proper and f is surjective and semialgebraic then g is partially proper. {It suffices to assume that f is "strongly surjective" (cf. Def. 3 in §8) instead of surjective and semialgebraic.}

All this follows immediately from the definitions and formal properties of proper maps.

Remark 5.6. A map $f : M \to N$ is proper iff f is partially proper and semialgebraic.

This is evident from the definitions, Remark 5.3.iii and Theorem 5.4.

Definition 4. An incomplete path in M is a semialgebraic map $\gamma : [0,1[\to M$ from the half open unit interval in R to M.

We are interested whether a given incomplete path γ in M can be completed, i.e. extended to a path $\bar{\gamma} : [0,1] \to M$. Notice that there can be at most one completion $\bar{\gamma}$.

Theorem 5.7 (Relative path completion criterion). For a map $f : M \to N$ the following are equivalent.

a) f is partially proper.

b) If γ is an incomplete path in M, such that $\delta := f \circ \gamma$ can be completed in N, then γ can be completed in M.

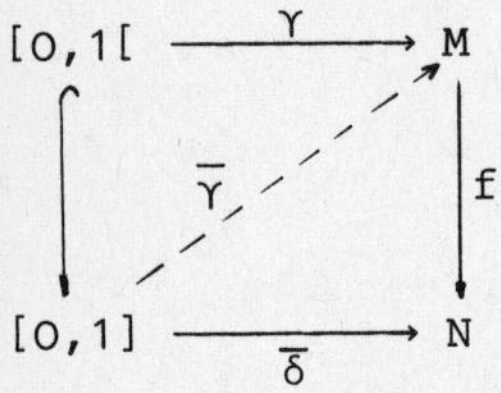

Proof. a) ⇒ b): The closure A of $\gamma([0,1[)$ in M is semialgebraic in M (Prop. 3.6). Thus $f(A) \in \overline{\gamma}(N)$ and the map $h : A \to f(A)$ obtained from A by restriction is proper. The path $\bar{\delta}$ runs in $f(A)$. The claim b) now follows from the semialgebraic relative path completion criterion (I.6.8, [DK_4, 2.3]).

b) ⇒ a): The semialgebraic relative path completion criterion implies that $f|A$ is proper for every $A \in \overline{\gamma}(M)$. This means that f is partially proper.

Corollary 5.8 (Absolute path completion criterion). The space M is a weak polytope iff every incomplete path in M can be completed.

Proposition 5.9. The pull back $f' : M \times_N N' \to N'$ of a partially proper map $f : M \to N$ by an arbitrary map $g : N' \to N$ is partially proper.

This can be proved as in I, §5 by using the relative path completion criterion. In contrast to [LSA] also a proof directly from the definitions is possible.

We indicate this second proof. Let $M' := M \times_N N'$ and let $g' : M' \to M$

denote the pull back of g by f. Let some $A \in \overline{\gamma}(M')$ be given. We denote the closure of $g'(A)$ in M by B and the closure of $f'(A)$ in N' by C. By Proposition 3.6 we have $B \in \overline{\gamma}(M)$ and $C \in \overline{\gamma}(N')$. Then $D := f(B) \in \overline{\gamma}(N)$ and $B \times_D C \in \overline{\gamma}(M')$. We have a cartesian square of semialgebraic maps

$$\begin{array}{ccc} B \times_D C & \xrightarrow{f_1'} & C \\ {\scriptstyle g_1'}\downarrow & & \downarrow{\scriptstyle g_1} \\ B & \xrightarrow[f_1]{} & D \end{array}$$

with f_1, g_1, f_1', g_1' obtained from f, g, f', g' by restriction. f_1 is proper, hence f_1' is proper. Since A is closed semialgebraic in $B \times_D C$, also $f_1' | A$ is proper. Since $C \in \overline{\gamma}(N)$ this implies that $f' | A$ is proper. q.e.d.

Proposition 5.10. For a map $f : M \to N$ the following are equivalent.

a) f is partially proper.

b) If P is a weak polytope in N then $f^{-1}(P)$ is a weak polytope in M.

c) If Q is a polytope in N with $\dim Q \leq 1$ then $f^{-1}(Q)$ is a weak polytope in M.

Proof. The implications a) ⇒ b) ⇒ c) are trivial. c) ⇒ a) follows from Theorem 5.7 and its Corollary 5.8 (relative and absolute path completion criterion), since, for every path δ in N, the set $\delta([0,1])$ is a polytope in N of dimension ≤ 1.

Remark 5.11. Assume that the spaces M,N are paracompact locally semialgebraic. A map $f : M \to N$ is partially proper (resp. proper) in the sense defined here iff f is partially proper (resp. proper) in the sense of [LSA].

Indeed, partially proper maps from M to N are characterized here and in [LSA] by the same relative path completion criterion (Th. 5.7,

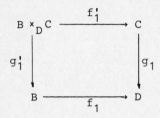

Th. I.6.8). Evidently also the property "semialgebraic" for f means the same here and there. Thus this also holds for "proper", since "proper" means "partially proper and semialgebraic" in both theories (Remark 5.6, [LSA, p. 70]).

We finally spell out in combinatorial terms what the main definitions of this section mean in the case of simplicial maps between simplicial complexes. The proofs are similar to those in [LSA, p. 71f].

Examples 5.12. Let $f : X \to Y$ be a simplicial map between simplicial complexes. We regard f as a map between spaces (cf. 2.4).

a) f is semialgebraic iff, for every open simplex $\rho \in \Sigma(Y)$ the complex $f^{-1}(\rho)$ is finite.

b) f is partially proper iff for every $\sigma \in \Sigma(X)$ the following holds: If τ is an open face of σ with $\overline{f}(\tau) \in \Sigma(Y)$, then $\tau \in \Sigma(X)$.

c) Thus f is proper iff f fulfills the combinatorial properties stated in a) and b).

§6 - Polytopic spaces; the one-point completion

In this section, as in the preceding one, a "space" means a weakly semialgebraic space over R, and a "map" between spaces means a weakly semialgebraic map.

Let M be a space and $(M_\alpha|\alpha\in I)$ an exhaustion of M.

Definition 1. M is called polytopic if, for every $\alpha \in I$, the semialgebraic space M_α is locally complete. This means (I, §7) that every point $x \in M_\alpha$ has a neighbourhood in M_α which is a polytope. Clearly this property does not depend on the choice of the exhaustion of M.

Remarks 6.1. i) Every weak polytope is a polytopic space.

ii) Every closed and every open subspace of a polytopic space is again polytopic.

iii) The direct product $M_1 \times M_2$ of two polytopic spaces M_1, M_2 is again polytopic.

Examples 6.2. a) Let R^∞ denote the set of all infinite sequences $(x_1,x_2,x_3,\ldots) = x$ of elements of R with only finitely many coordinates $x_i \neq 0$. We identify, for any $n \in \mathbb{N}$, the standard space R^n with the subset $\{x \in R^\infty | x_i = 0 \text{ for } i > n\}$ of R^∞. Theorem 1.6 gives us a structure of a space on R^∞ such that $(R^n|n\in\mathbb{N})$ is an exhaustion of R^∞ and each R^n, with its standard space structure, is a closed subspace of R^∞. This space R^∞ is polytopic.

b) For every $i \in \mathbb{N}$, the coordinate function T_i on R^∞, which on a point $(x_1,x_2,\ldots)$ has the value x_i, is weakly semialgebraic. Thus also all polynomials $F \in R[T_1,T_2,\ldots]$ are weakly semialgebraic functions on R^∞. If $F_1,\ldots,F_r$, $G_1,\ldots,G_s$ are finitely many such polynomials then we conclude from Remark 6.1.ii that the subspace

$$\{x \in R^{\infty} \mid F_1(x) \geq 0, \ldots, F_r(x) \geq 0,\ G_1(x) > 0, \ldots, G_s(x) > 0\}$$

of R^{∞} is polytopic. More generally, if $(F_{\lambda} \mid \lambda \in \Lambda)$ and $(G_{\kappa} \mid \kappa \in K)$ are two families in $R[T_1, T_2, \ldots]$ such that, for every $n \in \mathbb{N}$, the variable T_n occurs in F_{λ} for only finitely many $\lambda \in \Lambda$ and in G_{κ} for only finitely many $\kappa \in K$ then the set

$$X := \{x \in R^{\infty} \mid F_{\lambda}(x) \geq 0 \text{ for every } \lambda \in \Lambda,$$
$$G_{\kappa}(x) > 0 \text{ for every } \kappa \in K\}$$

is a polytopic subspace of R^{∞}. (More precisely, X is weakly semialgebraic in R^{∞} and, regarded as a subspace of R^{∞}, is polytopic.)

Example 6.3. Assume that the space M is locally semialgebraic (and, as before, weakly semialgebraic). Then M is polytopic iff M is locally complete (as defined in I, §7).

Indeed, the interior U_{α} of every set M_{α} is an element of $\mathring{\gamma}(M)$, and $(U_{\alpha} \mid \alpha \in I)$ is an admissible covering of M (cf. 2.12). If M is polytopic, then every M_{α} is locally complete. Thus also every U_{α} is locally complete, and we conclude that M is locally complete. Conversely if M is locally complete then every M_{α} is locally complete which means that M is polytopic.

Definition 2. A completion of the space M is a dense embedding $\varphi : M \to P$ into a weak polytope P, i.e. an isomorphism of M onto a dense subspace M' of P, followed by the inclusion map $M' \hookrightarrow P$.

Open Question A. Does every space have a completion?

In II, §1 we have constructed a completion for any paracompact locally semialgebraic space (even in the category LSA(R)). Also, in I, §7, we have constructed the one-point completion N^+ of any locally complete semialgebraic space N. Starting from this special case we now construct a completion $M \hookrightarrow M^+$ of our space M under the assumption that M is polytopic.

Let M^+ denote the <u>set</u> which is the disjoint union of the set M and one further point ∞. For every $\alpha \in I$ we equip the subset $M_\alpha^+ := M_\alpha \cup \{\infty\}$ of M^+ with the structure of the one-point completion of the space M_α, as described in I,§7. If $\beta < \alpha$ then M_β^+ is a closed subspace of the polytope M_α^+. By Theorem 1.6 we have a unique space structure on the set M^+ such that every M_α^+ is a closed subspace of M^+ and $(M_\alpha^+ | \alpha \in I)$ is an exhaustion of M. The space M^+ is a weak polytope, which contains M as an open subspace. If M is not yet a weak polytope, then M is dense in M^+.

<u>Definition 3.</u> We call the space M^+ or, more precisely, the inclusion $j : M \hookrightarrow M^+$ the <u>one-point completion of</u> M.

<u>Remark 6.4.</u> If M is already a weak polytope then the space M^+ is the direct sum (cf. 1.10) of M and the one-point space $\{\infty\}$. Thus, in this case, the definition is an abuse of language.

In I, §7 we gave an explicit description of the space structure of M^+ in the case that M is semialgebraic and locally complete. This implies a similar description of M^+ if M is polytopic.

<u>Proposition 6.5.</u> a) A subset U of $M \cup \{\infty\}$ is an element of $\mathring{\mathfrak{T}}(M^+)$ iff either $U \in \mathring{\mathfrak{T}}(M)$ or $U = (M \setminus K) \cup \{\infty\}$ with some $K \in \mathfrak{T}_c(M)$.

b) A family $(U_\lambda | \lambda \in \Lambda)$ in $\mathring{\mathfrak{T}}(M^+)$ is an element of Cov_{M^+} iff $(U_\lambda \setminus \{\infty\} | \lambda \in \Lambda)$ is an element of Cov_M.

c) If $U \in \mathring{\mathfrak{T}}(M)$, then $\mathcal{O}_M(U) = \mathcal{O}_{M^+}(U)$. If $U \in \mathring{\mathfrak{T}}(M^+)$ and $\infty \in U$ then a function $f : U \to R$ is an element of $\mathcal{O}_{M^+}(U)$ iff $f|U \setminus \{\infty\} \in \mathcal{O}_M(U \setminus \{\infty\})$ and $f(\infty)$ is the limit of the values $f(x)$ for $x \to \infty$, $x \in U \setminus \{\infty\}$ {i.e., for every $\varepsilon > 0$ in R there exists some $K \in \mathcal{T}_c(M)$ such that $M \setminus K \subset U$ and $|f(x) - f(\infty)| < \varepsilon$ for every $x \in M \setminus K$}.

Proposition 6.6. Assume that M is polytopic. Let Q be a weak polytope and let $f : V \to M$ be a partially proper map from an open subspace V of Q to M. Then f extends to a (weakly semialgebraic) map $g : Q \to M^+$ with $g(x) = \infty$ for every $x \in Q \setminus V$.

Proof. We choose an exhaustion $(Q_\beta | \beta \in J)$ of Q. There exists a monotonic map $\kappa : J \to I$ such that $f(V \cap Q_\beta) \subset M_{\kappa(\beta)}$ for every $\beta \in J$ (cf. 2.6). The restrictions $f_\beta : V \cap Q_\beta \to M_{\kappa(\beta)}$ are proper maps between locally complete semialgebraic spaces. Thus they extend to maps $g_\beta : Q_\beta \to M^+_{\kappa(\beta)}$ with $g_\beta(x) = \infty$ for $x \in Q_\beta \setminus (V \cap Q_\beta)$, cf. I.7.6. These maps fit together to the desired map $g : Q \to M^+$.

Corollary 6.7. Every partially proper map $f : N \to M$ between polytopic spaces extends to a map $f^+ : N^+ \to M^+$ with $f^+(\infty) = \infty$.

Conversely, if $g : N^+ \to M^+$ is a map with $g^{-1}(\infty) = \{\infty\}$ then the restriction $f : N \to M$ of g is partially proper since $g^{-1}(M) = N$ and g is partially proper. We have $f^+ = g$.

From the existence of a completion we can draw an important consequence on the structure of a polytopic space.

Theorem 6.8. Assume that M is polytopic. Let $U \in \mathring{\mathcal{T}}(M)$ be a neighbourhood of a given point $x \in M$. Then there exists a neighbourhood K of x in U which is a weak polytope.

<u>Proof.</u> Since $U \in \mathring{\mathfrak{I}}(M^+)$ there exists a weakly semialgebraic function $f : M^+ \to [0,1]$ with $f(x) = 0$ and $f(M^+ \setminus U) = \{1\}$. The set $K := f^{-1}([0,\frac{1}{2}])$ has the required properties.

§7 - A theorem on inductive limits of spaces

In this section (and a part of the next one) we abandon our convention that spaces and maps are always assumed to be weakly semialgebraic if nothing else is said.

The following technical result on inductive limits of weakly semialgebraic spaces will be very useful later on. It is a generalization of Theorem 1.6. Its proof seems to be the "hard point" in the elementary theory of weakly semialgebraic spaces, and will occupy nearly the whole section. (It is not so very hard, as we shall see.)

Theorem 7.1. Let M be a set, and let $(M_\alpha | \alpha \in I)$ be an ordered family of subsets of M fulfilling the conditions E1-E3 and E4* (cf. §1). We assume that every M_α carries the structure of a weakly semialgebraic space over R and that, whenever $\beta < \alpha$, the set M_β is closed weakly semialgebraic in M_α and carries the subspace structure (cf. §3) with respect to the space M_α. We equip M with the inductive limit structure of the spaces M_α in the category of function ringed spaces over R, as described in §1, Def. 6. Then the space M is weakly semialgebraic. Every M_α is, with its given space structure, a closed subspace of M, and $(M_\alpha | \alpha \in I)$ is an admissible covering of M (cf. §3, Def. 7). Moreover there exists a faithful lattice exhaustion H(M) of M (cf. §1, Def. 8, Def. 9) such that, for every M_α, the following holds.

P1. $\{X \in H(M) | X \subset M_\alpha\}$ is a lattice exhaustion of M_α.

P2. If $X \in H(M)$ then $X \cap M_\alpha \in H(M)$.

The proof will consist of two steps. In the first step we construct faithful lattice exhaustions H_α of the spaces M_α such that the following properties hold for every $\alpha \in I$.

Q1. If $\beta < \alpha$, then $H_\beta \subset H_\alpha$.

Q2. If $\beta < \alpha$ and $X \in H_\alpha$, then $X \cap M_\beta \in H_\beta$.

For every $\alpha \in I$ we denote by $m(\alpha)$ the number of elements $\beta \in I$ with $\beta < \alpha$. In order to construct the exhaustions H_α we proceed by induction on $m(\alpha)$. For the unique index σ of with $m(\sigma)=0$ we choose an arbitrary lattice exhaustion H_σ of M_σ. This is all right, since the conditions Q1 and Q2 are empty.

Let now $m(\alpha) = n > 0$. We choose a lattice exhaustion L of the subspace $M_\alpha^o := M_\alpha \setminus \cup(M_\beta | \beta<\alpha)$ of M_α. (Of course, if $M_\alpha^o = \emptyset$ then $L = \{\emptyset\}$.) Then we define H_α as the set of all sets $X = S \cup \bigcup(S_\beta | \beta<\alpha)$ with $S \in L$, $S_\beta \in H_\beta$, and

$$\overline{S} \cap [\cup(M_\beta | \beta<\alpha)] \subset \cup(S_\beta | \beta<\alpha).$$

This last condition means that the semialgebraic subset X of M_α is closed in M_α. {$\overline{S}$ denotes the closure of S in M_α.}

If $X, Y \in H_\alpha$ then clearly $X \cup Y \in H_\alpha$. Using the induction hypothesis and property E4* for the sets M_β it is easily checked that also $X \cap Y \in H_\alpha$. Thus H_α is a sublattice of $\overline{\gamma}(M_\alpha)$. It contains the empty set. The property Q1 is evident, and Q2 is easily checked, again by use of the induction hypothesis and E4*. It remains to verify the properties E1, E3, E6 for the family H_α (each set in H_α being indexed by itself).

We start with E3. Let $X = S \cup \bigcup(S_\beta | \beta<\alpha)$ be an element of H_α, as described above, and let $Y = T \cup \bigcup(T_\beta | \beta<\alpha)$ be a second element of H_α with T and the T_β also fulfilling the above conditions. Assume that $Y \subset X$. Intersecting with M_α^o we obtain $T \subset S$. Thus we have only finitely many possibilities for $T = Y \cap M_\alpha^o$. Intersecting with M_β for some $\beta < \alpha$ we obtain $Y \cap M_\beta \subset X \cap M_\beta$. By Q2 both $X \cap M_\beta$ and $Y \cap M_\beta$ are elements of H_β.

Thus we have only finitely many possibilities for $Y \cap M_\beta$. We conclude that we have only finitely many possibilities for Y, which proves E3.

We finally verify that H_α is an admissible covering of M_α. The family $(X \cap M_\alpha^o \mid X \in H_\alpha)$ is the same as L up to iteration of elements. Thus this family is an admissible covering of M_α^o. For every $\beta < \alpha$ the family $(X \cap M_\beta \mid X \in H_\alpha)$ is the same as H_β up to iterations. Thus this family is an admissible covering of M_β. This proves that indeed H_α is an admissible covering of M_α. Now E1 and, by Th. 3.15, also E6 are evident for the family H_α. This finishes the first step of our proof.

We define H as the set of all subsets $X = \cup(S_\alpha \mid \alpha \in I)$ of M with $S_\alpha \in H_\alpha$ for every $\alpha \in I$ and $S_\alpha \neq \emptyset$ for only finitely many $\alpha \in I$. We have $\emptyset \in H$. If $X \in H$ and $Y \in H$ then obviously $X \cup Y \in H$. Using property E4* for the family $(M_\alpha \mid \alpha \in I)$ and property Q2 for the families H_α one checks easily that also $X \cap Y \in H$. Thus H is a lattice of subsets of M. It contains all the lattices H_α (in fact is generated by them).

Using Q1 and Q2 for the families H_α and E4* for the family $(M_\alpha \mid \alpha \in I)$ one verifies that, for every $\alpha \in I$,

$$\{X \cap M_\alpha \mid X \in H\} \subset H_\alpha .$$

Since $H_\alpha \subset H$ it follows that

$$(*) \qquad H_\alpha = \{X \cap M_\alpha \mid X \in H\} = \{X \in H \mid X \subset M_\alpha\} .$$

This implies the statements P1 and P2 in the theorem.

Our lattice H trivially fulfills E1, E2, E4, E5 (E2 is tautological). It also fulfills E3. Indeed, if $X \in H$ then $X \subset \cup(M_\alpha \mid \alpha \in J)$ for some finite subset J of I. Assume that $Y \in H$ and $Y \subset X$. We have, for every $\alpha \in J$, $Y \cap M_\alpha \subset X \cap M_\alpha$, and both $Y \cap M_\alpha$, $X \cap M_\alpha$ are elements of H_α.

Thus there are only finitely many possibilities for $Y \cap M_\alpha$. Since this holds for every $\alpha \in J$ we conclude that there are only finitely many possibilities for Y, which proves E3.

We look for a space structure on a given set $X \in H$. We choose a finite set $J \subset I$ with the following properties.

i) $\alpha \in J,\ \beta < \alpha \Rightarrow \beta \in J$

ii) $X \subset \bigcup(M_\alpha \mid \alpha \in J)$.

For every $\alpha \in J$ the set $X \cap M_\alpha$ is closed semialgebraic in M_α, since $X \cap M_\alpha \in H_\alpha$. We equip $X \cap M_\alpha$ with the structure as a closed subspace of M_α. Then $X \cap M_\alpha$ is a semialgebraic space. If $\beta < \alpha$ then $X \cap M_\beta$ is a closed subspace of $X \cap M_\alpha$. We now equip X with the inductive limit function ringed space structure of the family $(X \cap M_\alpha \mid \alpha \in J)$ of semialgebraic spaces. This family fulfills E1-E3 and E4*. Thus, by Theorem 1.6, X is a weakly semialgebraic space, and every $X \cap M_\alpha$, in its given structure, is a closed subspace of X. Since J is finite, the space X is semialgebraic.

If K is another finite subset of I with the properties i) and ii) above, then we find a third finite subset L of I with i), ii) such that $L \supset J$ and $L \supset K$. It is clear that the three families $(X \cap M_\alpha \mid \alpha \in J)$, $(X \cap M_\alpha \mid \alpha \in L)$, $(X \cap M_\alpha \mid \alpha \in K)$ all yield the same inductive limit space structure on the space X. Thus the space structure on X is independent of the choice of J.

It is now clear, say from II, Lemma 1.2, that if $Y \in H$, $Y \subset X$, then the space Y is a closed subspace of X. We equip M with the inductive limit function ringed space structure of the family H of semialgebraic spaces. By Theorem 1.6 this makes M a weakly semialgebraic space with exhaustion H, and every $X \in H$ is a closed subspace of H in its given space structure.

It follows from the relation (*) above that $M_\alpha \in \bar{\mathfrak{T}}(M)$ for every $\alpha \in I$ and that M_α, in its given space structure, is a closed subspace of M. The family $(M_\alpha | \alpha \in I)$ is an admissible covering of M since H is so. Thus, by Theorem 3.15, our space structure on M coincides with the inductive limit of the family of spaces $(M_\alpha | \alpha \in I)$. This finishes the proof of the theorem.

We state two rather weak, but nevertheless interesting consequences of Theorem 7.1 and its proof.

Corollary 7.2. Let M be a weakly semialgebraic space and A a closed subspace of M. Let H_o be a faithful lattice exhaustion of A. Then there exista a faithful lattice exhaustion H of M with

$$\{X \in H | X \subset A\} = \{Y \cap A | Y \in H\} = H_o \ .$$

This is evident by our procedure to construct H in the proof of Theorem 7.1. There the first lattice exhaustion H_σ could be chosen arbitrarily. One takes for I the set $\{0,1\}$ with the ordering $0 < 1$ and puts $M_o := A$, $M_1 := M$.

Corollary 7.3. Let M be a set and let $M_1,\dots,M_r$ be finitely many subsets of M with $M_1 \cup \dots \cup M_r = M$. Assume that every M_i carries the structure of a (weakly semialgebraic) space (over R). Assume also that $M_i \cap M_j$ is a closed weakly semialgebraic subset of M_i and M_j and that the subspace structures on $M_i \cap M_j$ in M_i and M_j are equal $(1 \le i < j \le r)$. Then there exists on M a unique structure of a space such that every M_i, with its given space structure, is a closed subspace of M.

<u>Proof.</u> Let Λ denote the set of all non empty subsets of $\{1,\ldots,r\}$, partially ordered by the opposite of the inclusion relation ($J \leq K$ iff $J \supset K$). For any $J \in \Lambda$ we put $M_J := \cap(M_i \mid i \in J)$. We equip M_J with its closed subspace structure in M_i for some $i \in J$. This structure does not depend on the choice of i by the assumption on the intersections $M_i \cap M_j$ above. Theorem 7.1 applies to the ordered family $(M_J \mid J \in \Lambda)$ of weakly semialgebraic spaces and yields the result.

In the next section we will use Theorem 7.1 only in the easier case that the system $(M_\alpha \mid \alpha \in I)$ fulfills E1-E5 instead of E1-E3 and E4*.

§8 - Strong quotients; gluing of spaces

We start with some generalities on quotients in the categories Space(R) and WSA(R) (cf. §1, Def. 2 and §2, Def. 6).

Definition 1. A morphism $f : M \to N$ between function ringed spaces over R is called identifying if the following four properties hold.

ID1. f is surjective

ID2. If U is a subset of N with $f^{-1}(U) \in \mathring{\mathfrak{T}}(M)$ then $U \in \mathring{\mathfrak{T}}(N)$.

ID3. If $(U_\lambda | \lambda \in \Lambda)$ is a family in $\mathring{\mathfrak{T}}(N)$ with $(f^{-1}(U_\lambda) | \lambda \in \Lambda) \in \mathrm{Cov}_M$ then $(U_\lambda | \lambda \in \Lambda) \in \mathrm{Cov}_N$.

ID4. If $h : U \to R$ is a function on a set $U \in \mathring{\mathfrak{T}}(N)$ with $h \circ (f | f^{-1}(U)) \in \mathcal{O}_M(f^{-1}(U))$ then $h \in \mathcal{O}_N(U)$.

If these properties hold then we say also that N is a strong quotient of M (via f).

This terminology is justified by the following proposition which implies that a strong quotient is a categorial quotient (= strict epimorphism, cf. [Gr, Def 2.2]) in the category Space(R).

Proposition 8.1. Let $f : M \to N$ and $g : M \to L$ be morphisms in Space(R). Assume that f is identifying and that g, as a map between sets, is constant on every fibre of f. Then there exists a unique morphism $h : N \to L$ with $h \circ f = g$.

Proof. Since f is surjective the map h exists and is unique on the set theoretic level. One now checks directly that h is a morphism (cf. §1, Def. 2b) using properties ID2-ID4.

Definition 2. If M is a function ringed space over R and $f : M \to N$

is a surjection from the set M to a set N, then there exists on N a unique structure of a function ringed space over R, such that f is identifying. {The properties ID2-ID4 tell us how to define the structure.} We call N, with this structure, the <u>strong quotient of</u> M <u>by</u> f.

In the following (as already in §5 and §6) a "space" means a weakly semialgebraic space over R and a "map" means a weakly semialgebraic map between spaces over R, unless something else is said.

Every proper surjective map $f : M \to N$ is identifying. This can be proved similarly as in the theory of locally semialgebraic spaces (cf. [LSA, p. 179] for property ID4). We want to extend this observation to a suitable class of partially proper maps.

<u>Definition 3.</u> A map $f : M \to N$ is called <u>strongly surjective</u> if, for every $Y \in \gamma(N)$, there exists some $X \in \gamma(M)$ with $f(X) \supset Y$.

<u>Remarks 8.2.</u> i) Let $(M_\alpha | \alpha \in I)$ and $(N_\beta | \beta \in J)$ be exhaustions of M and N. The map f is strongly surjective iff, for every $\beta \in J$, there exists some $\alpha \in I$ with $f(M_\alpha) \supset N_\beta$. This means that $(f(M_\alpha) | \alpha \in I)$ is an admissible covering (cf. §3, Def. 7) of N.

ii) If f is strongly surjective and $(X_\lambda | \lambda \in \Lambda)$ is an admissible covering of M by semialgebraic sets, then $(f(X_\lambda) | \lambda \in \Lambda)$ is an admissible covering of N by semialgebraic sets. In particular, if M is of countable type then also N is of countable type.

iii) If f is semialgebraic and surjective then f is strongly surjective.

<u>Theorem 8.3.</u> Every strongly surjective partially proper map $f : M \to N$ is identifying.

<u>Proof.</u> Let $(M_\alpha | \alpha \in I)$ be an exhaustion of M. Then $(f(M_\alpha) | \alpha \in I)$ is an

admissible covering of N by closed semialgebraic sets. The restrictions $f_\alpha : M_\alpha \twoheadrightarrow f(M_\alpha)$ of f are proper maps between semialgebraic spaces, hence they are identifying. One now concludes in a straightforward way, by use of Theorem 3.15, that f is identifying.

Definition 4. If a map $f : M \to N$ is surjective and proper then we say that N is a *proper quotient* of M (via f). If f is strongly surjective and partially proper, then we say that N is a *partially proper quotient* of M. Notice that in both cases N is a strong quotient of M.

Given spaces M,N, a closed weakly semialgebraic subset A of M and a map $f : A \to N$ we want to "glue M to N along A by the map f". This means to construct a cocartesian square (= "push out")

$$(8.4) \qquad \begin{array}{ccc} A & \xrightarrow{\;f\;} & N \\ {\scriptstyle i}\big\downarrow & & \big\downarrow{\scriptstyle j} \\ M & \xrightarrow[\;g\;]{} & M \cup_f N \end{array}$$

in the category WSA(R) of weakly semialgebraic spaces over R, with i the inclusion map from A to M.

It has been proved in II, §10 that the analogous push out exists in the category LSA(R) of paracompact locally semialgebraic spaces over R if f is *proper*. Our main goal in this section is to construct a push out (8.4) in our broader category for f *partially proper.* This greater flexibility of weakly semialgebraic spaces compared with paracompact locally semialgebraic spaces is perhaps the most important reason why weakly semialgebraic spaces seem to be useful in semialgebraic topology.

Our gluing problem can be regarded as a question about the existence of suitable quotients. Let $L := M \cup_f N$ denote the <u>set</u> obtained from the disjoint union $M \sqcup N$ of M and N by identifying every $a \in A$ with $f(a) \in N$. Let $p : M \sqcup N \to L$ denote the natural projection and $g : M \to L$, $j : N \to L$ its restrictions to M and N. Notice that $g|M \setminus A$ and j are injective and that L is the union of the images $g(M \setminus A)$ and $j(N)$. We equip L with the strong quotient structure of the direct sum $M \sqcup N$ of the spaces M and N by the map p (cf. Def. 2 above). Then (8.4) is a push out in the category Space(R).

The preimage of $p(M \setminus A) = g(M \setminus A)$ under p is the set $M \setminus A \in \mathring{\mathcal{T}}(M \sqcup N)$. Thus $p(M \setminus A) \in \mathring{\mathcal{T}}(L)$ and $p|M \setminus A$ is an isomorphism from the open subspace $M \setminus A$ of M to the open subspace $p(M \setminus A)$ of L.

We will prove that L is weakly semialgebraic under a condition on f which is even slightly weaker than partial properness. We start with an easy consequence of Theorem II.10.7 (needed here only for semialgebraic spaces).

<u>Proposition 8.5.</u> Assume that M is semialgebraic and $f : A \to N$ is proper. Then the function ringed space L is weakly semialgebraic and is, via p, a proper quotient of $M \sqcup N$. The map $j = p|N$ is an isomorphism of N onto a closed subspace of L.

<u>Proof.</u> We choose an exhaustion $(N_\alpha | \alpha \in I)$ of N. Let $J := \{\alpha \in I | N_\alpha \supset f(A)\}$. Then $(N_\alpha | \alpha \in J)$ is again an exhaustion of N. We regard $L := M \cup_f N$ as a <u>set</u>. By Theorem II.10.7 the subset $L_\alpha := M \cup_f N_\alpha$ of L is, for every $\alpha \in J$, equipped with the structure of a semialgebraic space such that $p|M \sqcup N_\alpha$ is a proper map from $M \sqcup N_\alpha$ onto L_α. If $\beta \in J$ is an index with $\beta < \alpha$, then L_β is a closed subspace of L_α. The system $(L_\alpha | \alpha \in J)$ of subsets of L fulfills the conditions E1-E5. Thus, by Theorem 1.6, the inductive limit structure on L of the family of spaces $(L_\alpha | \alpha \in J)$ makes

L a weakly semialgebraic space such that every space L_α is a closed subspace of L and $(L_\alpha | \alpha \in J)$ is an exhaustion of L. The map $p : M \sqcup N \to L$ is proper for this space structure on L since all the restrictions $p^{-1}(L_\alpha) \to L_\alpha$ are proper. Thus L is the strong quotient of $M \sqcup N$ by p. This means that our space structure on L is the same as the one chosen above.

We have $p^{-1}p(N) = A \sqcup N \in \overline{\mathcal{T}}(M \sqcup N)$. Thus $p(N) \in \overline{\mathcal{T}}(L)$. The restriction $A \sqcup N \to p(N)$ of p is a proper map from $A \sqcup N$ to the closed subspace $p(N)$ of L. Restricting p further to N we obtain a bijective proper map from N to $p(N)$. This is an isomorphism. q.e.d.

Definition 5. Let X be a weakly semialgebraic subset of M. We call the map $f : A \to N$ partially proper near X if, for every $Y \in \gamma(X)$, the restriction of f to the intersection $A \cap \overline{Y}$ of A with the closure of Y in M is a proper map from $A \cap \overline{Y}$ to N. {N.B. This certainly holds if f is partially proper.}

Theorem 8.6. Assume that f is partially proper near $M \setminus A$. Then the strong quotient L of $M \sqcup N$ by p is weakly semialgebraic. Thus the diagram (8.4) is cocartesian in the category WSA(R). The map p is strongly surjective. The restriction $j = p|N$ of p is an isomorphism of N onto a closed subspace of L. If f is partially proper then also p is partially proper, hence L is a partially proper quotient of $M \sqcup N$ in this case.

Proof. As in the proof of the preceding proposition we do not equip the set L with the strong quotient structure by p. We will construct a space structure on L in a different way and will verify a posteriori that this space structure is the strong quotient of $M \sqcup N$ by p.

If E is a weakly semialgebraic subset of M then $p(E \sqcup N)$ is the set $E \cup_g N$ with $g : A \cap E \to N$ obtained from f by restriction. For simplicity we denote this subset of L by $E \cup_f N$.

We choose an exhaustion $(X_\alpha \mid \alpha \in I)$ of $M \setminus A$. For every $\alpha \in I$ let L_α denote the subset $\overline{X}_\alpha \cup_f N = p(X_\alpha) \cup p(N)$ of L. By the preceding proposition and our assumption on f we have on L_α the structure of a weakly semialgebraic space such that L_α is a proper quotient of $\overline{X}_\alpha \sqcup N$ under $p_\alpha := p \mid \overline{X}_\alpha \sqcup N$. If $\beta \in I$ and $\beta < \alpha$ then

$$p_\alpha^{-1}(L_\beta) = [\overline{X}_\beta \cup (A \cap \overline{X}_\alpha)] \sqcup N .$$

This is a closed weakly semialgebraic subset of $\overline{X}_\alpha \sqcup N$. Thus $L_\beta \in \overline{\mathfrak{T}}(L_\alpha)$. The map p_α gives by restriction a proper map from $p_\alpha^{-1}(L_\beta)$ to the space L_β. Thus L_β is, in its given structure, a closed subspace of L_α. The family $(L_\alpha \mid \alpha \in I)$ fulfills E1-E5. We equip the set L with the inductive limit space structure of this family of spaces. Then, by Theorem 7.1, L is weakly semialgebraic, every L_α is a closed subspace of L, and $(L_\alpha \mid \alpha \in I)$ is an admissible covering of L. The family $(\overline{X}_\alpha \mid \alpha \in I)$ together with A and N is an admissible covering of $M \sqcup N$. The restrictions $p_\alpha = p \mid \overline{X}_\alpha \sqcup N$ of p are weakly semialgebraic maps. Thus also $p \mid N$ and $p \mid A = (p \mid N) \circ f$ are weakly semialgebraic. We conclude that p is weakly semialgebraic (cf. 3.15).

We have to verify the axioms ID2-ID4 for p. Then we know that p is identifying, which means that our space structure on L really is the strong quotient structure by p. Let U be a subset of L with $p^{-1}(U) \in \mathring{\mathfrak{T}}(M \sqcup N)$. We have $p_\alpha^{-1}(U \cap L_\alpha) = p^{-1}(U) \cap (\overline{X}_\alpha \sqcup N) \in \mathring{\mathfrak{T}}(\overline{X}_\alpha \sqcup N)$. Thus $U \cap L_\alpha \in \mathring{\mathfrak{T}}(L_\alpha)$ for every $\alpha \in I$, and we conclude that $U \in \mathring{\mathfrak{T}}(L)$. This proves ID2. The proofs of ID3 and ID4 are similar.

We choose an index $\alpha \in I$. Then we know from Proposition 8.5 that

$p(N) = p_\alpha(N)$ is a closed weakly semialgebraic subset of L_α, hence of L. We also know from 8.5 that $p|N = p_\alpha|N = j$ is an isomorphism from N onto p(N). Moreover, it is easily seen, that $p|M \setminus A$ is an isomorphism from $M \setminus A$ onto the open subspace $p(M \setminus A)$ of L. Since L is the (disjoint) union of $p(M \setminus A)$ and p(N) it is now evident that p is strongly surjective. For every $\alpha \in I$ the restriction $p|\overline{X}_\alpha$ is proper. Also $p|N$ is proper. If f is partially proper then clearly $p|A$ is partially proper. Since the $\overline{X}_\alpha$ together with A and N form an admissible covering of $M \sqcup N$ by closed weakly semialgebraic sets we conclude that p is partially proper in this case. q.e.d.

In the following we always assume that $f : A \to N$ is partially proper near $M \setminus A$. We call $M \cup_f N$ the <u>space obtained by gluing</u> M <u>to</u> N <u>along</u> A <u>by</u> f. We usually regard $M \setminus A$ as an open subspace and N is a closed subspace of $M \cup_f N$ identifying $M \setminus A$ with $p(M \setminus A)$ and N with p(N).

<u>Remarks 8.7.</u> i) If Y is a subspace of N with $f(A) \subset Y$ then the map $g : A \to Y$ obtained from f by restriction of the range space is partially proper near $M \setminus A$ and $M \cup_g Y$ is a subspace of $M \cup_f N$. If Z is a closed subspace of M then the restriction $h : Z \cap A \to N$ of f is partially proper near $Z \setminus (Z \cap A)$ and $Z \cup_h N$ is a closed subspace of $M \cup_f N$. Thus if $Z \in \mathcal{T}(M)$, $Y \in \mathcal{T}(N)$ are given with $f(Z \cap A) \subset Y$ then the map $h : Z \cap A \to Y$ obtained from f by restriction is partially proper near $Z \setminus (Z \cap A)$ and $Z \cup_h Y$ is a subspace of $M \cup_f N$.

ii) If T is any space then $f \times id_T : A \times T \to N \times T$ is partially proper near $(M \setminus A) \times T$, and the natural map from $(M \times T) \cup_{f \times id} (N \times T)$ to $(M \cup_f N) \times T$ is an isomorphism. In short

$$(M \times T) \cup_{f \times id} (N \times T) = (M \cup_f N) \times T.$$

iii) If M and N are of countable type then $M \cup_f N$ is again of countable type.

iv) If M and N are weak polytopes then $M \cup_f N$ is again a weak polytope. Notice that, in this case, f is <u>automatically</u> partially proper.

v) Assume that M and N are polytopic and that f is partially proper. Then $M \cup_f N$ is again polytopic and

$$(M \cup_f N)^+ = M^+ \cup_{f^+} N^+ \quad .$$

<u>Definition 6.</u> Let M be a space and let X and A be two weakly semialgebraic subsets of M with A closed in M. We say that A is <u>partially complete near</u> X if the map from A to the one point space $\{*\}$ is partially proper near X. This means that $A \cap \overline{Y}$ is a polytope for every $Y \in \gamma(X)$.

<u>Example 8.8.</u> Let M be a space and A a closed weakly semialgebraic subset of M which is partially complete near $M \smallsetminus A$. According to Theorem 8.6 we can glue M along A to the one point space $\{*\}$. We denote the resulting weakly semialgebraic space by M/A. If A is empty then M/A is the direct sum of M and $\{*\}$. Otherwise M/A is the strong quotient of the space M by the natural projection $\pi : M \to M/A$. {π is the map g in the diagram (8.4) in the present special case.} The map π identifies the set A into one point ξ and maps $M \smallsetminus A$ isomorphically onto $(M/A) \smallsetminus \{\xi\}$. If $(M_\alpha | \alpha \in I)$ is an exhaustion of M then $(\pi(M_\alpha) | \alpha \in I)$ is an exhaustion of M/A, and the restriction $\pi_\alpha : M_\alpha \to \pi(M_\alpha)$ of π exhibits $\pi(M_\alpha)$ as the strong quotient $M_\alpha / A \cap M_\alpha$ of M_α, whenever $A \cap M_\alpha \neq \emptyset$.

Notice that, in order to construct M/A as a weakly semialgebraic space, it suffices to apply Theorem 1.6. The more difficult Theorem 7.1 is not needed here.

<u>Subexample 8.9.</u> Let M be polytopic and let $M \hookrightarrow P$ be a completion of M (§6, Def. 2), with M regarded as a subspace of P. Assume that M is open in P. {Such completions exist in abundance if M is paracompact

locally semialgebraic, cf. Chap. II.} The natural map $f : P \to M^+$ with $f(x) = x$ for $x \in M$ and $f(x) = \infty$ for $x \notin M$ (cf. Prop. 6.6) yields an isomorphism $P/P \setminus M \xrightarrow{\sim} M^+$.

We return to the general situation (8.4) assuming that f is partially proper near $M \setminus A$. We want to say a little more about the relation between the spaces $M \cup_f N$, M and N.

Definition 7. Let $\varphi : X \to Y$ be a map between spaces and let C be a weakly semialgebraic subset of X. We say that φ is partially proper relative C if, for every $Z \in \gamma(X \setminus C)$, the map $\varphi|\overline{Z}$ from $\overline{Z}$ to Y is proper.

Proposition 8.10. If $f : A \to N$ is partially proper near $M \setminus A$ then the map $g : M \to M \cup_f N$ (cf. 8.4) is partially proper relative A.

Proof. Let again $(X_\alpha | \alpha \in I)$ be an exhaustion of $M \setminus A$. By Prop. 8.5 the map $g|\overline{X}_\alpha : \overline{X}_\alpha \to \overline{X}_\alpha \cup_f N$ is proper for every $\alpha \in I$. Thus $g|\overline{Y}$ is proper for every $Y \in \gamma(M \setminus A)$. q.e.d.

Definition 8. a) A closed pair of spaces is a pair (X,C) of spaces with C a closed subspace of X.
b) A relative weak polytope is a closed pair of spaces (X,C) such that the map from X to the one point space {*} is partially proper relative C. We then also say that X is partially complete relative C.

This terminology may look over-sophisticated at present. It will turn out to be very convenient in the next chapter.

Proposition 8.11. If (M,A) is a relative weak polytope then any map $f : A \to N$ into a space N is partially proper near $M \setminus A$, and $(M \cup_f N, N)$ is again a relative weak polytope. In particular $(N = \{*\})$, A is partially complete near $M \setminus A$ and M/A is a weak polytope. The map $\bar{g} : M/A \to M \cup_f N/N$ induced by $g : M \to M \cup_f N$ is an isomorphism.

The proof may be safely left to the reader.

We mention that - similar to partial properness - also partial properness relative to a weakly semialgebraic set can be characterized by a path completion criterion.

Proposition 8.12. Let $f : M \to N$ be a map between spaces and let $C \in \Upsilon(M)$. Then f is partially proper relative C iff the following holds. If $\alpha : [0,1[\to M$ is an incomplete path such that $\alpha([0,1[) \subset M \setminus C$ and $f \circ \alpha$ can be completed then α can be completed.

Proof of the nontrivial direction. We assume that f fulfills the path lifting condition just stated, and we have to prove, for a given $Z \in \gamma(M \setminus C)$ that the restriction $f|B$ with $B := \bar{Z}$ is proper. We choose a closed semialgebraic subset D of N which contains $f(B)$. Let $h : B \to D$ be the map obtained from f by restriction. The set Z is semialgebraic and dense in B. We choose a proper extension of h, i.e. a commutative triangle of semialgebraic maps

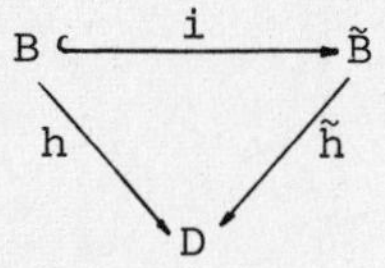

with $\tilde{h}$ proper and i a dense embedding of B, which we regard as an inclusion (cf. II.12.10). We want to prove that $B = \tilde{B}$. Suppose not. We choose a point $x \in \tilde{B} \setminus B$. Since Z is dense in $\tilde{B}$ there exists a path $\bar{\alpha} : [0,1] \to \tilde{B}$ with $\bar{\alpha}(1) = x$ and $\bar{\alpha}([0,1[) \subset Z$. Now consider the restriction $\alpha : [0,1[\to Z$ of $\bar{\alpha}$. The incomplete path $h \circ \alpha$ in D can be completed by $h \circ \bar{\alpha}$. By our assumption α can be completed to a path

$\beta : [0,1] \to B$. We must have $\overline{\alpha}(1) = \beta(1) = x$. But $x \notin B$. This contradiction proves that $f|B$ is indeed proper.

<u>Example 8.13.</u> A closed pair of spaces (M,A) is a relative weak polytope iff every incomplete path in $M \setminus A$ can be completed in M.

§9 - The weak polytope P(M)

We continue working in the category Space(R) of function ringed spaces over R instead of just WSA(R).

In the following M is a weakly semialgebraic space over R and $(M_\alpha|\alpha\in I)$ is an exhaustion of M. We want to construct a weak polytope P over R together with a weakly semialgebraic map $p : P \to M$ which has the following universal property. For every morphism (= weakly semialgebraic map) $f : Q \to M$ from a weak polytope Q (over R, as always) to N there exists a unique morphism $g : Q \to P$ with $p \circ g = f$.

(9.0)

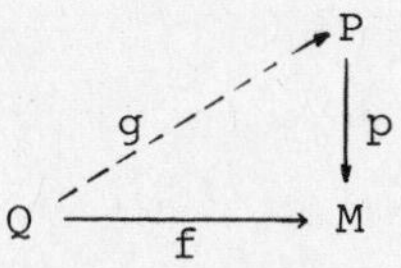

<u>Definition 1.</u> We call such a weak polytope P, or more precisely the map $p : P \to M$, a <u>partially complete core</u> of M.

Notice that it suffices to demand the universal property (9.0) for morphisms f with Q a polytope, since every weak polytope is an inductive limit of polytopes in the category Space(R). It is also clear, that any two partially complete hulls of M are isomorphic over M. Thus we may speak of "the" partially complete hull of M, if it exists.

How should P look like? Exploiting the universal property (9.0) for Q the one-point space we see that p must be bijective. Thus, if the partially complete hull $p : P \to M$ exists, we may assume that the spaces P and M have the same underlying set and p is the identity map of this set. In other words, P is the set M equipped with a finer space structure than the space M. {This means $\mathring{\mathcal{T}}(M) \subset \mathring{\mathcal{T}}(P)$, and $Cov_M(U) \subset Cov_P(U)$, $\mathcal{O}_M(U) \subset \mathcal{O}_M(P)$ for every $U \in \mathring{\mathcal{T}}(M)$.} Since id_M is a morphism from P to M, we also have $\gamma(P) \subset \gamma(M)$.

If $K \in \bar{\gamma}(P)$ then K is a polytope. Thus $p|K$ is a proper morphism from K as a subspace of P to K as a subspace of M. This must be an isomorphism. In other words, $K \in \gamma_c(M)$ and the subspace structures of K in M and P are the same.

Conversely, if $K \in \gamma_c(M)$ then we can apply the universal property (9.0) to the inclusion morphism $K \hookrightarrow M$ of the subspace K of M, and we learn that $K \in \bar{\gamma}(P)$. Thus $\bar{\gamma}(P) = \gamma_c(M)$.

Since the space P is weakly semialgebraic we now can describe the structure of P: The space P is the set M equipped with the inductive limit structure of the directed family of polytopes $\gamma_c(M)$ in the category Space(R). {Every element of $\gamma_c(M)$ is indexed by itself; the transition morphisms are the inclusions.}

More explicitly this means the following: A subset U of M is an element of $\mathring{\mathcal{T}}(P)$ iff $K \cap U \in \mathring{\gamma}(K)$ for every $K \in \gamma_c(M)$. A family $(U_\lambda | \lambda \in \Lambda)$ in $\mathring{\mathcal{T}}(P)$ is an element of Cov_P if, for every $K \in \gamma_c(M)$, the family $(K \cap U_\lambda | \lambda \in \Lambda)$ is an element of Cov_K. {This means that the intersection of K with $U := \cup(U_\lambda | \lambda \in \Lambda)$ is the union of finitely many sets $K \cap U_\lambda$.} A function $f : U \to R$ on some $U \in \mathring{\mathcal{T}}(P)$ is an element of $\mathcal{O}_P(U)$ iff $f|K \cap U \in \mathcal{O}_K(K \cap U)$ for every $K \in \gamma_c(M)$.

All this has been said under the assumption that a partially complete hull exists. We now make the following general definition.

<u>Definition 2.</u> P(M) denotes the inductive limit of the directed family of weak polytopes $\gamma_c(M)$ in the category Space(R), with underlying set M,

and $p_M : P(M) \to M$ is the morphism induced by the family of inclusion morphisms $K \hookrightarrow M$, with K running through $\gamma_c(M)$.

Notice that, on the set theoretical level, p_M is the identity map of the set M. By our discussion above the first half of the following proposition is already clear.

<u>Proposition 9.1.</u> i) If M admits a partially complete core then P(M) is a weak polytope and $p_M : P(M) \to M$ is a partially complete core of M. Also $\gamma_c(M) = \overline{\gamma}(P(M))$ and, for every $K \in \gamma_c(M)$, the space structures on M as a subspace of M and P(M) are the same.
ii) If the function ringed space P(M) is a weak polytope then $p_M : P(M) \to M$ is a partially complete core of M.

<u>Proof.</u> It remains to prove the second claim ii). If K is a complete semialgebraic subset of the space M then it is easily verified that K is closed semialgebraic in P(M) and that the subspace structure on the set K in M and P(M) are the same (cf. §1, Def. 5; for this we do not need that P(M) is weakly semialgebraic).

Let now a morphism $f : Q \to M$ be given with Q a polytope. Then $K := f(Q)$ is an element of $\gamma_c(M)$. We equip the set K with its subspace structure in M. We have a unique factorization $f = i \circ g$ with g a morphism from Q to K and i the inclusion morphism $K \hookrightarrow M$. As just has been proved, the space K is also a subspace of P(M). Denoting the inclusion morphism from K to P(M) by j we have the factorization $i = p_M \circ j$ of i, hence the factorization $f = p_M \circ g$ of f with $g := j \circ f$. Since p_M is bijective this is the only factorization of f through p_M, and we have verified the required universal property (9.0) of p_M. q.e.d.

<u>Definition 3.</u> We call the real closed field R <u>sequential</u>, if there

exists in R a sequence $(\varepsilon_k | k \in \mathbb{N})$ of positive elements converging to zero. Replacing in such a sequence every ε_k by the minimum of $\varepsilon_1, \varepsilon_2, \ldots, \varepsilon_k$ we may then assume that $\varepsilon_k \geq \varepsilon_{k+1}$, and omitting some ε_k, even that $\varepsilon_k > \varepsilon_{k+1}$ for every $k \in \mathbb{N}$.

Very many real closed fields are sequential. For example, every real closure of a field, which is finitely generated over $\mathbb{Q}$ or $\mathbb{R}$, is sequential, cf. [Du], [LSA, p. 44]. (These fields are even microbial, i.e. contain an element $\vartheta > 0$, such that $(\vartheta^n | n \in \mathbb{N})$ is a null sequence.)

Our main goal in this section is to prove

Theorem 9.2. Assume that R is sequential. Then M admits a partially complete core.

The proof of the theorem will be based on three easy lemmas, two of them stated in a slightly more general form than actually needed for that.

Lemma 9.3. Assume that M admits a partially complete core $q : Q \to M$. Let $A \in \overline{\mathfrak{T}}(M)$. Then the restriction $q_A : q^{-1}(A) \to A$ of q is a partially complete core of A. Thus P(A) is a closed subspace of the weak polytope P(M).

Indeed, it is evident from our subspace theory (Prop. 3.2) that q_A fulfills the universal property characterizing a partially complete hull of A.

Lemma 9.4. Assume that, for every $\alpha \in I$, the space $P(M_\alpha)$ is a weak polytope. Then P(M) is a weak polytope.

Proof. If $\beta < \alpha$ then $P(M_\beta)$ is a closed subspace of $P(M_\alpha)$ by Lemma 9.3. Theorem 7.1 applies to the set M and the family of spaces $(P(M_\alpha) | \alpha \in I)$. Thus there exists a structure of a weakly semialgebraic space Q on the set M such that, for every $\alpha \in I$, the space $P(M_\alpha)$ is a closed subspace of Q, and $(P(M_\alpha) | \alpha \in I)$ is an admissible covering of P(M). This space Q is a weak polytope. The identity map of M is a weakly semialgebraic map from Q to M which fulfills the universal property characterizing a partially complete core of M. We conclude from Proposition 9.1.i that Q = P(M). q.e.d.

Lemma 9.5. Assume that there exists a countable family $(P_k | k \in \mathbb{N})$ in $\mathcal{T}_c(M)$, with $P_k \subset P_{k+1}$ for every $k \in \mathbb{N}$, such that every $K \in \gamma_c(M)$ is contained in some P_k. Then P(M) is a weak polytope. {Thus, by 9.1.ii, p_M is a partially complete core of M.}

Proof. Every space P_k is the inductive limit of the family of spaces $\gamma_c(P_k)$ in the category Space(R). Thus the hypothesis of the lemma implies that the space P(M) is the inductive limit in Space(R) of the directed family of spaces $(P_k | k \in \mathbb{N})$. Theorem 7.1 now implies that P(M) is a weak polytope (and that $(P_k | k \in \mathbb{N})$ is an admissible covering of P(M)).

Remark. In the proof of Th. 9.2 this lemma will be only needed for M semialgebraic. In that case Th. 1.6 instead of Th. 7.1 suffices for the proof. But Th. 7.1 is needed in full strength to prove Lemma 9.4.

We now prove Theorem 9.2. By Lemma 9.4 it suffices to consider the case that M is semialgebraic.

Claim. There exists a family $(P_k | k \in \mathbb{N})$ in $\gamma_c(M)$, with $P_k \subset P_{k+1}$ for every $k \in \mathbb{N}$, such that every $K \in \gamma_c(M)$ is contained in some P_k.

Once we have proved this claim we are done by Lemma 9.5.

Choosing a triangulation of M we assume, without loss of generality, that M is a finite simplicial complex over R. It suffices to find, for every open simplex $\sigma \in \Sigma(M)$, a family $(P_k(\sigma) \mid k \in \mathbb{N})$ in $\gamma_c(\bar{\sigma} \cap M)$ which fulfills the claim for the closure $\bar{\sigma} \cap M$ of σ in M. Then the family of sets

$$P_k := \bigcup(P_k(\sigma) \mid \sigma \in \Sigma(M))$$

will fulfill the claim for M. Thus we retreat to the case that M is, for some $n \in \mathbb{N}_o$, a subcomplex of the closed standard n-simplex $\nabla(n)$ containing the open standard n-simplex $\mathring{\nabla}(n)$. We proceed by induction on the number t of open simplices in $\nabla(n) \setminus M$. If $t = 0$ then $M = \nabla(n)$ is complete and we are done. Let $t > 0$. We choose some open simplex σ in $\nabla(n) \setminus M$. Let $N := M \cup \sigma$. By induction hypothesis there exists a countable family $(Q_k \mid k \in \mathbb{N})$ in $\gamma_c(N)$ such that $Q_k \subset Q_{k+1}$ and every $L \in \gamma_c(N)$ is contained in some Q_k. We choose a numbering $e_o, e_1, \ldots, e_n$ of the vertices of $\nabla(n)$ such that $\sigma =]e_o, e_1, \ldots, e_r[$. Let $\lambda_o(x), \ldots, \lambda_n(x)$ denote the barycentric coordinates of a point $x \in \nabla(n) \{\lambda_i(x) \geq 0, \lambda_o(x) + \ldots + \lambda_n(x) = 1,\ \lambda_o(x)e_o + \ldots + \lambda_n(x)e_n = x\}$. We choose some null sequence $(\varepsilon_k \mid k \in \mathbb{N})$ of positive elements in R with $\varepsilon_k \geq \varepsilon_{k+1}$ for every $k \in \mathbb{N}$, and we define, for every $k \in \mathbb{N}$ the set*)

$$X_k := \left\{x \in \nabla(n) \;\middle|\; \sum_{j=r+1}^{n} \lambda_j(x) \geq \varepsilon_k \left(\prod_{i=0}^{r} \lambda_i(x)\right)^k \right\}.$$

X_k is closed semialgebraic in $\nabla(n)$, hence complete. We have $X_k \cap \sigma = \emptyset$ and $X_k \subset X_{k+1}$. The set $P_k := X_k \cap Q_k$ is complete semialgebraic and contained in M, and $P_k \subset P_{k+1}$. Let $K \in \gamma_c(M)$ be given. For every $x \in K$ the following holds, since $\sigma \cap M = \emptyset$:

---

*) I am indebted to Hans Delfs for this clever definition.

$$\sum_{j=r+1}^{n} \lambda_j(x) = 0 \Rightarrow \prod_{i=0}^{r} \lambda_i(x) = 0 .$$

By the inequality of Lojasiewicz (generalized to an arbitrary real closed field, cf. [D, p. 43], [BCR, Chap. 2, §6]) there exists some $k \in \mathbb{N}$ with

$$\sum_{j=r+1}^{n} \lambda_j(x) \geq \varepsilon_k \left(\prod_{i=0}^{r} \lambda_i(x)\right)^k$$

for every $x \in K$. Moreover, there exists some $s \geq k$ with $K \subset Q_s$. Thus $K \subset P_s$, and our claim is proved. This finishes the proof of Theorem 9.2.

Corollary 9.6. If R is sequential and M is of countable type then also P(M) is of countable type.

Proof. We start with an exhaustion $(M_n \mid n \in \mathbb{N})$ of M. Then $(P(M_n) \mid n \in \mathbb{N})$ is an admissible covering of P(M), as is evident from the proof of Lemma 9.4. As we have just seen in the proof of Theorem 9.2, there exists an admissible covering $(P_{n,k} \mid k \in \mathbb{N})$ of $P(M_n)$ by polytopes, for every $n \in \mathbb{N}$. Now $(P_{n,k} \mid (n,k) \in \mathbb{N} \times \mathbb{N})$ is an admissible covering of P(M) by polytopes with countable index set $\mathbb{N} \times \mathbb{N}$. This proves that P(M) is of countable type.

Proposition 9.7. Assume that M is locally semialgebraic (and, as always, weakly semialgebraic). Assume also that M is polytopic, i.e. locally complete (cf. 6.3). Then the function ringed space P(M) is the same as the locally semialgebraic space M_{loc} constructed in I, §7.

This is evident from I.7.8.

We now assume about our space M that P(M) is a weak polytope. We do not assume that the field R is sequential. On the contrary, we will prove this below except in the trivial case that M is a weak polytope.

We want to describe the semialgebraic subsets and the exhaustions of the space M. Since the identity of M is a morphism from P(M) to M we have $\gamma(P(M)) \subset \gamma(M)$ (as already observed above) and $\mathcal{T}(P(M)) \supset \mathcal{T}(M)$. By Proposition 9.1.i we have $\overline{\gamma}(P(M)) = \gamma_c(M)$, and for every $K \in \gamma_c(M)$ the subspace structures on K in M and P(M) are the same. The following proposition is now evident by Corollary 3.16.

Proposition 9.8. An ordered family $(K_\lambda | \lambda \in \Lambda)$ in $\gamma_c(M)$ is an exhaustion of P(M) iff the family fulfills E2-E5 and every $K \in \gamma_c(M)$ is contained in some K_λ.

By the universal property of p_M it is evident that a path in M is the same thing as a path in P(M). We conclude from Prop. 3.6 and the semialgebraic curve selection lemma ([BCR], [DK_2, §12], [DK_4, §2]) that, for every $X \in \gamma(M)$, the closures of X in M and in P(M) are equal. We denote this set unambiguously by $\overline{X}$.

Proposition 9.9. $\gamma(P(M))$ is the set of all $X \in \gamma(M)$ with $\overline{X} \in \gamma_c(M)$. For every such set X the subspace structures on X in the spaces M and P(M) are equal.

Proof. If $X \in \gamma(M)$ and $\overline{X} \in \gamma_c(M)$ then $X \in \mathcal{T}(P(M))$ and $\overline{X} \in \overline{\gamma}(P(M))$. This implies that $X \in \gamma(P(M))$. Conversely, if $X \in \gamma(P(M))$ then, by Prop. 3.6, $\overline{X} \in \overline{\gamma}(P(M)) = \gamma_c(M)$. The subspace structures on X with respect to M and P(M) both coincide with the subspace structure in the polytope $\overline{X}$.
q.e.d.

We now state a converse to Theorem 9.2.

Theorem 9.10. Assume that M is not a weak polytope but P(M) is a weak polytope. Then R is sequential.

Proof. Since M is not a weak polytope there exists an incomplete path $\gamma : [0,1[\to M$ which cannot be completed (Cor. 5.8). There exists some $c \in]0,1[$ such that $\gamma | [c,1[$ is an embedding (cf. the argument in [DK_4, p. 305f]). The set $A := \gamma([c,1[)$ is closed semialgebraic in M, since γ is proper (cf. II.9.9). The subspace A of M is isomorphic to $]0,1]$. By Lemma 9.3, $P(A)$ is a weak polytope. Thus $P(]0,1])$ is a weak polytope.

Let $(K_\alpha | \alpha \in I)$ be a faithful exhaustion of $P(]0,1])$. Every K_α is a complete semialgebraic subset of $]0,1]$. Omitting from the family $(K_\alpha | \alpha \in I)$ all sets which do not contain 1, we may assume that $1 \in K_\alpha$ for every $\alpha \in I$. The connected component L_α of 1 in K_α is a closed interval $[\varepsilon_\alpha, 1]$, with some $\varepsilon_\alpha \in]0,1]$. For any $c \in]0,1[$ there exists some $\alpha \in I$ with $\varepsilon_\alpha < c$. Indeed, by Prop. 9.8, there exists some $\alpha \in I$ with $[c,1] \subset K_\alpha$, and this implies $[c,1] \subset L_\alpha$.

For any $\alpha \in I$ let $m(\alpha)$ denote, as before, the number of elements $\beta \in I$ with $\beta < \alpha$. We claim that, for every non negative integer n, the set $\{\varepsilon_\alpha | \alpha \in I, m(\alpha) \leq n\}$ contains at most $n+1$ elements. We prove this claim by induction on n. For $n = 0$ there is nothing to prove since there exists only one index α with $m(\alpha) = 0$. Let $n > 0$. Assume that α and β are two different indices with $m(\alpha) = m(\beta) = n$. We have $K_\alpha \cap K_\beta = K_\gamma$ with some $\gamma \in I$ such that $\gamma < \alpha$, $\gamma < \beta$, $m(\gamma) \leq n-1$. From $K_\alpha \supset K_\gamma$ and $K_\beta \supset K_\gamma$ we conclude that $L_\alpha \supset L_\gamma$ and $L_\beta \supset L_\gamma$, hence $\varepsilon_\alpha \leq \varepsilon_\gamma$ and $\varepsilon_\beta \leq \varepsilon_\gamma$. Suppose that $\varepsilon_\alpha < \varepsilon_\gamma$, $\varepsilon_\beta < \varepsilon_\gamma$, and, say, $\varepsilon_\alpha \leq \varepsilon_\beta$. Then $[\varepsilon_\beta, 1]$ is contained in L_γ, contradicting the definition of ε_γ. Thus $\varepsilon_\alpha = \varepsilon_\gamma$ or $\varepsilon_\beta = \varepsilon_\gamma$. We see that the set $\{\varepsilon_\alpha | m(\alpha) \leq n\}$ contains at most one more element than $\{\varepsilon_\alpha | m(\alpha) \leq n-1\}$, and we conclude from the induction hypothesis that $\{\varepsilon_\alpha | m(\alpha) \leq n\}$ contains at most $n+1$ elements. Our claim is proved.

Since the set $\{\varepsilon_\alpha | m(\alpha) \leq n\}$ is finite it makes sense to define, for every $n \in \mathbb{N}_o$,

$$\varepsilon_n := \min \{\varepsilon_\alpha | m(\alpha) \leq n\} .$$

We have $\varepsilon_n > 0$ and $\varepsilon_n \geq \varepsilon_{n+1}$. For every $c \in]0,1[$ there exists some $\alpha \in I$ with $\varepsilon_\alpha < c$, as shown above. Then $\varepsilon_m < c$ for $m := m(\alpha)$. This proves that the sequence $(\varepsilon_n | n \in \mathbb{N}_o)$ converges to zero. R is sequential. q.e.d.

In the following we assume the field R to be sequential. We want to compare the strong topologies of the spaces M and P(M). We denote by M_{top} and $P(M)_{top}$ the set M equipped with the strong topologies of the spaces M and P(M) respectively (cf. §3, Def. 4). These are topological spaces in the classical sense. p_M is a continuous map from $P(M)_{top}$ to M_{top}, which means that the topology $P(M)_{top}$ is finer than M_{top}.

<u>Theorem 9.11.</u> The following are equivalent.

a) $P(M)_{top} = M_{top}$.

b) Every compact subset C of M_{top} is contained in some $K \in \gamma_c(M)$.

c) Every converging sequence $(x_n | n \in \mathbb{N})$ in M_{top} is contained in some $K \in \gamma_c(M)$.

d) M is polytopic.

d) ⇒ a): Let W be an open subset of $P(M)_{top}$. We have to verify that W is also open in M_{top}, which means that $W \cap M_\alpha$ is open in $(M_\alpha)_{top}$ for every $\alpha \in I$. Let some $\alpha \in I$ and a point $x \in W \cap M_\alpha$ be given. There exists a set $T \in \mathring{\gamma}(M_\alpha)$ with $x \in T$ and $\overline{T}$ complete. Here $\overline{T}$ is the closure of T in $(M_\alpha)_{top}$. The strong topology of $\overline{T}$ coincides with the subspace topology in M_{top} (Prop. 3.5.ii). Since $\overline{T} \in \gamma_c(M)$, it also coincides with the subspace topology in $P(M)_{top}$. Thus $W \cap \overline{T}$ is open in $(\overline{T})_{top}$. Intersecting with T we see that $W \cap T$ is open in T_{top}, hence open in

$(M_\alpha)_{top}$. Since $x \in W \cap T$ this proves that W is open in M_{top}.

a) ⇒ b): Let $(K_\lambda | \lambda \in \Lambda)$ be an exhaustion of P(M). Suppose that C is a compactum in $M_{top} = P(M)_{top}$ which is not contained in any $K \in \gamma_c(M)$. Then $C \not\subset K_\lambda$ for every $\lambda \in \Lambda$. Let $(K_\lambda^o | \lambda \in \Lambda^o)$ denote the set of patches of the exhaustion $(K_\lambda | \lambda \in \Lambda)$, cf. end of §1. The set J of primitive indices $\lambda \in \Lambda$ with $C \cap K_\lambda^o \neq \emptyset$ is infinite. For every $\lambda \in J$ we choose an element $x_\lambda \in C \cap K_\lambda^o$. The set $A := \{x_\lambda | \lambda \in J\}$ has finite intersection with K_μ for every $\mu \in \Lambda$. Thus $A \in \overline{\mathcal{T}}(P(M))$. Since $P(M)_{top} = M_{top}$, the set A is closed in M_{top}. The same goes for every subset of A. Thus A is also discrete in M_{top}. Since $A \subset C$, the set A is compact in M_{top}, hence finite. This absurdity proves the claim b).

b) ⇒ c) is trivial.

c) ⇒ d): Suppose M is not polytopic. Then there exists some $\alpha \in I$ such that the subspace M_α of M is not locally complete. M_α is isomorphic to a semialgebraic subset N of some R^p. Let $(K_n | n \in \mathbb{N})$ be an exhaustion of P(N) (cf. 9.6). It follows from the assumption c) and 9.8 that every sequence in N which converges in N is contained in some K_n. But now we construct "explicitly" a converging sequence for which this does not hold. This will be the desired contradiction.

We choose a point $x \in N$ which has no complete neighbourhood in N. (Recall that N is not locally complete.) We further choose a sequence $(\varepsilon_n | n \in \mathbb{N})$ of positive elements in R converging to zero. Let $\overline{B}(x, \varepsilon_n)$ denote the closed euclidean ball in R^p with center x and radius ε_n. For every $n \in \mathbb{N}$ the intersection $\overline{B}(x, \varepsilon_n) \cap N$ is not contained in K_n, since otherwise $\overline{B}(x, \varepsilon_n) \cap N$ would be complete. We choose a point $x_n \in \overline{B}(x, \varepsilon_n) \cap N$ with $x_n \notin K_n$. The sequence $(x_n | n \in \mathbb{N})$ converges to x but is not contained in K_m for any $m \in \mathbb{N}$. q.e.d.

Here is a somewhat typical example which illustrates the contents of Theorem 9.11.

Example 9.12. Let $R = \mathbb{R}$, and $M = \mathbb{R}^2 \setminus (]0,1] \times \{0\})$. We consider the map $\gamma :]0,1] \to M$, $\gamma(t) = (t, e^{-t^{-1}})$. With respect to M_{top}, this map is continuous and $\lim\limits_{t \to 0} \gamma(t) = (0,0)$. The set $A := \gamma(]0,1])$ is not closed in M_{top}. But for every polytope K in M the set $K \cap A$ is closed in K_{top} since the arc A ultimately "leaves" the polytope K. (A formal proof of this is possible by the inequality of Lojasiewicz.) Since $P(M)_{top}$ is the inductive limit of the spaces K_{top} with $K \in \gamma_c(M)$ we see that A is closed in $P(M)_{top}$. The set $C := A \cup \{(0,0)\}$ is compact in M_{top} but not contained in any $K \in \gamma_c(M)$.

We now discuss the functorial behaviour of partially complete cores. As before, we assume that R is sequential. A map $f : M \to N$ between spaces over R induces a unique map $P(f) : P(M) \to P(N)$ such that the diagram

$$\begin{array}{ccc} P(M) & \xrightarrow{P(f)} & P(N) \\ \downarrow{\scriptstyle p_M} & & \downarrow{\scriptstyle p_N} \\ M & \xrightarrow[f]{} & N \end{array}$$

commutes. This follows from the universal property of p_N. Thus we have a functor P from the category $WSA(R)$ to the full subcategory $\mathcal{P}(R)$ of $WSA(R)$ whose objects are the weak polytopes over R. The universal property of the maps p_M means that P is a reflector [Mt, p. 129] of $\mathcal{P}(R)$ in $WSA(R)$.

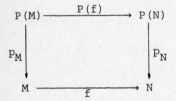

<u>Remarks 9.13.</u> i) Let $M \xrightarrow{f} S$ and $N \xrightarrow{g} S$ be maps between spaces. Let $M \times_S N$ denote the fibre product with respect to f and g, and let $P(M) \times_{P(S)} P(N)$ denote the fibre product with respect to P(f) and P(g). The natural map from $P(M \times_S N)$ to $P(M) \times_{P(S)} P(N)$ is an isomorphism. In short,

$$P(M \times_S N) = P(M) \times_{P(S)} P(N).$$

This is a formal consequence of the fact that the functor P is a reflector. In particular (S = one-point space), we have

$$P(M \times N) = P(M) \times P(N) \ .$$

ii) Let $(M_\lambda | \lambda \in \Lambda)$ be a family of spaces. Then the natural map from the direct sum of the family $(P(M_\lambda) | \lambda \in \Lambda)$ to the partially complete hull of $\sqcup (M_\lambda | \lambda \in \Lambda)$ is an isomorphism. In short,

$$P(\sqcup (M_\lambda | \lambda \in \Lambda)) = \sqcup (P(M_\lambda) | \lambda \in \Lambda) \ .$$

<u>Proposition 9.14.</u> If (M,A) is a closed pair of spaces and $f : A \rightarrow N$ is a map which is partially proper near $M \setminus A$ then the natural map from $P(M) \cup_{P(f)} P(N)$ to $P(M \cup_f N)$ {which on the set theoretical level is the identity of $M \cup_f N$} is an isomorphism. In short,

$$P(M \cup_f N) = P(M) \cup_{P(f)} P(N).$$

<u>Proof.</u> As in §8 let p denote the natural projection from the space $M \sqcup N$ to the space $M \cup_f N$. Then P(p) is a partially proper map from $P(M) \sqcup P(N)$ to $P(M \cup_f N)$. We will prove that P(p) is strongly surjective. Then we are done, since this implies that $P(M \cup_f N)$ is the strong quotient of $P(M) \sqcup P(N)$ by P(p), which is just $P(M) \cup_{P(f)} P(N)$.

Given a set $K \in \gamma_c(M \cup_f N)$ we have to find sets $C \in \gamma_c(M)$ and $D \in \gamma_c(N)$ with $p(C \cup D) \supset K$. We choose $D := K \cap N$, and we choose for C the closure of $K \cap (M \setminus A)$ in M. Then f yields a proper map $g : C \cap A \to D$ by restriction. We have $p(C \cup D) = K$. By 8.7.i we may identify $C \cup_g D$ with the subspace K of $M \cup_f N$, the natural projection $\pi : C \sqcup D \to C \cup_g D$ being obtained from p by restriction. Now π is proper (II, §10) and K is complete. Thus $C \sqcup D$ is complete. q.e.d.

Example 9.15 ($N = *$). If A is a closed subspace of M which is partially complete near $M \setminus A$ then

$$P(M/A) = P(M)/P(A).$$

Final remark 9.16. In the case that M is locally semialgebraic (and paracompact, as always) we defined "cores" of M already in [LSA]. Namely, if $\varphi : X \xrightarrow{\sim} M$ is a weak triangulation of M [LSA, p. 135], we defined the core of M with respect to φ as the image $\varphi(\mathrm{co}\,X)$ of the core co X of the simplicial complex X [LSA, p. 229]. The set of these cores $\varphi(\mathrm{co}\,X)$ with φ running through all triangulations (or all weak triangulations) is cofinal in the set of all partially complete subspaces of M, as follows from the triangulation theorem II.4.4. Thus the partially complete core P(M), as defined now, is the inductive limit of the family of all the cores in M defined in [LSA], partially ordered by inclusion. One can take as well the inductive limit with respect to suitable equivalence classes of triangulations or weak triangulations of M, partially ordered by a refinement relation.

In order to obtain P(M) as an inductive limit it suffices to use the cores with respect to good triangulations of M [LSA, p. 228f.]. These are strong deformation retracts of M [loc.cit.]. Thus we may expect that $p_M : P(M) \to M$ is a homotopy equivalence. We shall confirm this in the next chapter for all weakly semialgebraic spaces by other methods, cf. Th. V.4.13.

§10 - The spaces $P^A(M)$ and $P_f(M)$

In the last section we have seen how to turn a space into a polytope in a canonical way if R is sequential. In this section we give two canonical constructions which turn a closed pair (M,A) of spaces into a relative weak polytope without changing the subspace A and turn a map $f : M \to N$ between spaces into a partially proper map without changing the space N, again, if R is sequential. These constructions may be regarded as two generalizations of the construction $M \rightsquigarrow P(M)$.

For a short time we still work over an arbitrary real closed base field R.

Definition 1. A closed pair of spaces (M,A) is called a relative polytope if the closure $\overline{M \setminus A}$ of $M \setminus A$ in M is a polytope.

Clearly the relative polytopes are precisely those relative weak polytopes (M,A) for which $M \setminus A$ is semialgebraic. Alternatively we can say that a closed pair (M,A) is a relative polytope iff there exists some polytope L in M, i.e. $L \in \gamma_c(M)$, with $M = L \cup A$.

Henceforth let (M,A) be a fixed closed pair of spaces.

Definition 2. We denote the set of all subspaces K of M with $K \supset A$ and (K,A) a relative polytope by $\gamma_c(M,A)$. This is the set of all unions $L \cup A$ with L running through $\gamma_c(M)$.

Notice that the union and the intersection of any two sets in $\gamma_c(M,A)$ is again an element of $\gamma_c(M,A)$. In particular, $\gamma_c(M,A)$ is a directed family of spaces, the transition maps given by inclusions.

<u>Remark 10.1.</u> If $(X_\alpha \mid \alpha\in I)$ is an exhaustion of $M \setminus A$ then $(\overline{X}_\alpha \cup A \mid \alpha\in I)$ is an ordered family of subspaces of M fulfilling E2-E5. It is also an admissible covering of M. Thus, by Theorem 3.15, M is the inductive limit of this family of spaces in the category Space(R). If now (M,A) is a relative weak polytope then $(\overline{X}_\alpha \cup A \mid \alpha\in I)$ is a cofinal family in $\gamma_c(M,A)$. Thus, in this case, M is the inductive limit of the family of spaces $\gamma_c(M,A)$ in Space(R).

We now assume that the base field R is sequential.

<u>Definitions 3.</u> $P^A(M)$ denotes the space obtained by gluing P(M) to A along P(A) by the map $p_A : P(A) \to A$, and $p_M^A : P^A(M) \to M$ denotes the map induced by $p_M : P(M) \to M$ and the identity of A. Finally $p_{(M,A)}$ denotes the map (p_M^A, id_A) from $(P^A(M),A)$ to (M,A).

Notice that $(P^A(M),A)$ is a relative weak polytope. The map $p_{(M,A)}$ is universal in the following strong sense.

<u>Theorem 10.2.</u> Let $f : (Q,B) \to (M,A)$ be a map from a relative weak polytope (Q,B) to (M,A). Then there exists a unique map $\varphi : (Q,B) \to (P^A(M),A)$ such that the triangle

$$\begin{array}{ccc} & & (P^A(M),A) \\ & \overset{\varphi}{\nearrow} & \downarrow p_{(M,A)} \\ (Q,B) & \xrightarrow{\;f\;} & (M,A) \end{array}$$

commutes.

<u>Proof.</u> The pair (Q,B) is the inductive limit of the family of spaces $((K,B) \mid K \in \gamma_c(K,B))$ by the remark above. Thus, it suffices to prove the claim for the restrictions of f to all the pairs (Q,B). This

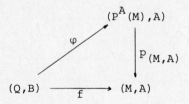

means that we may assume from the beginning that (Q,B) is a relative polytope.

We choose some $L \in \gamma_c(Q)$ with $Q = L \cup B$, say $L := \overline{Q \setminus B}$. Let $g : Q \to M$ and $h : B \to A$ be the restrictions of f to Q and B. By the universal property of p_M (cf. §9) there exists a unique map $\tilde{\psi} : L \to P(M)$ such that $p_M \circ \tilde{\psi} = g|L$. The maps $\tilde{\psi}$ and h fit together to a map $\psi : Q \to P^A(M)$. We have $p_M^A \circ \psi = g$. Clearly $\varphi := (\psi, h)$ is the unique map from (Q,B) to $(P^A(M),A)$ with $p_{(M,A)} \circ \varphi = f$. q.e.d.

The map $p := p_M^A$ from $P^A(M)$ to M is bijective. Let $K \in \gamma_c(M,A)$. Applying the theorem to the inclusion map from (K,A) to (M,A) we see that p maps $p^{-1}(K)$ isomorphically onto K. Conversely, if $(L,A) \in \gamma_c(P^A(M),A)$ then trivially $(p(L),A) \in \gamma_c(M,A)$. Identifying the underlying set of $P^A(M)$ with the underlying set of M by the bijection p we may say that the directed families of spaces $\gamma_c(M,A)$ and $\gamma_c(P^A(M),A)$ are equal. We now obtain from Remark 10.1 above

<u>Corollary 10.3.</u> The space structure $P^A(M)$ on the set M is the inductive limit of the family of spaces $\gamma_c(M,A)$.

Theorem 10.2 also implies a new criterion for a closed pair (M,A) to be a relative weak polytope.

<u>Corollary 10.4.</u> (M,A) is a relative weak polytope iff the diagram

$$\begin{array}{ccc} P(A) & \xrightarrow{\quad p_A \quad} & A \\ \downarrow & & \downarrow \\ P(M) & \xrightarrow{\quad p_M \quad} & M \,, \end{array}$$

with the inclusion mappings as vertical arrows, is cocartesian (= push out).

Indeed, that the diagram is cocartesian means that the map p_M^A from $P(M) \cup_{P(A)} A$ to M is an isomorphism. This happens to be iff the identity map of (M,A) also fulfills the universal property stated in Theorem 10.2.

We now come to our second construction. For a short time the base field R may again be any real closed field. We also abandon the convention that a "space" means automatically a weakly semialgebraic space over R. Instead we work in the larger category Space(R) of function ringed spaces over R.

Let $f : M \to N$ be a morphism (= weakly semialgebraic map) between weakly semialgebraic spaces. We want to turn f into a partially proper map in a universal way.

Definition 4. A partially proper core of f is a morphism $\pi : P \to M$ from a weakly semialgebraic space P to M with the following properties:

a) The map $f\circ\pi$ is partially proper

b) If $\alpha : Q \to M$ is a morphism from a weakly semialgebraic space Q to M such that $f\circ\alpha$ is partially proper, then there exists a unique morphism $\beta : Q \to P$ such that $\pi\circ\beta = \alpha$.

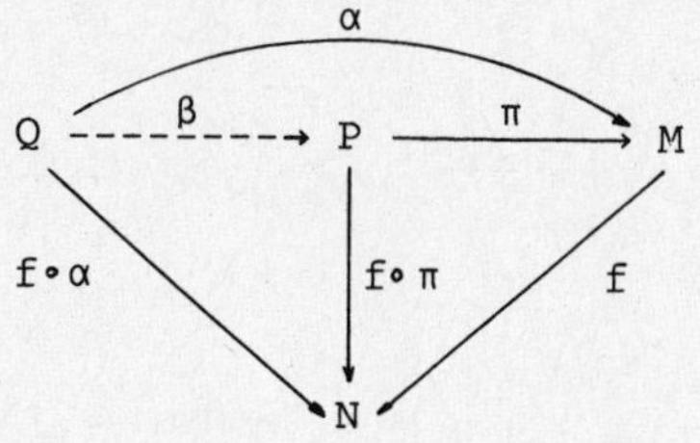

Notice that then the maps π, α, β are automatically partially proper. Of course, any two partially proper cores of f are isomorphic over M.

In the case that N is the one-point space and the field R is sequential we have found a partially proper core of f in §9, namely $p_M : P(M) \to M$. Looking for partially proper cores in general it is natural to study a set $\gamma_c(f)$ of subsets of M defined as follows.

Definition 5. We call a semialgebraic subset K of M proper over N, if the restriction $f|K : K \to N$ is a proper map. We denote the set of all $K \in \gamma(M)$ which are proper over N by $\gamma_c(f)$.

Remarks 10.5. i) $\gamma_c(f) \subset \overline{\gamma}(M)$. Indeed, let $K \in \gamma_c(f)$ and let i denote the inclusion from K to M. Then $f \circ i$ is proper, hence i is proper. This implies that $K = i(K)$ is closed in M.

ii) If $K \in \gamma_c(f)$ and L is a closed semialgebraic subset of K then $L \in \gamma_c(f)$.

iii) The subset $\gamma_c(f)$ of $\overline{\gamma}(M)$ is closed under finite unions. Thus $\gamma_c(f)$ is a directed family of closed semialgebraic subspaces of M (the transition maps being inclusions).

iv) The union of all sets in $\gamma_c(f)$ is the whole set M.

v) f is partially proper iff $\gamma_c(f) = \overline{\gamma}(M)$.

Definitions 6. a) $P_f(M)$ denotes the function ringed space over R with underlying set M which is the direct limit of the family $\gamma_c(f)$ in the category Space(R).

b) The identity map of the set M is a morphism from $P_f(M)$ to M. We denote this morphism by p_f.

c) We denote the morphism $f \circ p_f$ from $P_f(M)$ to N by $\tilde{f}$.

By arguments very similar (and including) those which gave us Proposition 9.1 one obtains

Proposition 10.6. i) If f admits a partially proper core then $P_f(M)$ is weakly semialgebraic and p_f is a partially proper core of f. Also $\gamma_c(f) = \overline{\gamma}(P_f(M))$ and, for every $K \in \gamma_c(f)$, the space structures on K as a subspace of M and of $P_f(M)$ are the same.
ii) If $P_f(M)$ is weakly semialgebraic and $\tilde{f}$ is partially proper then p_f is a partially proper core of f.

We now start out to prove

Theorem 10.7. If R is sequential then every weakly semialgebraic map $f : M \to N$ admits a partially proper core.

To prove this we first write down a chain of three lemmas which generalize the lemmas 9.3, 9.4, 9.5 and can be proved in the same way.

Lemma 10.8. Assume that f admits a partially proper core $q : Q \to M$. Let $A \in \overline{\mathcal{T}}(M)$. Then the restriction $q_A : q^{-1}(A) \to A$ of q is a partially proper core of the map $f|A$ from A to N. Thus $P_{f|A}(A)$ is a closed subspace of the weakly semialgebraic space $P_f(M)$.

Lemma 10.9. Let $(M_\alpha | \alpha \in I)$ be an exhaustion of M. Assume that, for every $\alpha \in I$, the map $f_\alpha = f|M_\alpha$ from M_α to N has a partially proper core. Then f has a partially proper core.

Lemma 10.10. Assume that there exists a countable family $(P_k | k \in \mathbb{N})$ in $\overline{\mathcal{T}}(M)$ with $P_k \subset P_{k+1}$ for every $k \in \mathbb{N}$, such that, for every $k \in \mathbb{N}$, the restriction $P_k \to N$ of f is partially proper, and every $K \in \gamma_c(f)$ is contained in some P_k. Then $P_f(M)$ is weakly semialgebraic and $\tilde{f} : P_f(M) \to N$ is partially proper. {Thus, by Prop. 10.6.ii, p_f is a partially proper core of f.}

By Lemma 10.9 it suffices to prove Theorem 10.7 in the case that the space M is semialgebraic. Of course, then replacing N by $\overline{f(M)}$ we may also assume that N is semialgebraic. By Lemma 10.10 it now suffices to find an increasing countable family $(P_k \mid k \in \mathbb{N})$ in $\gamma_c(f)$, such that every $K \in \gamma_c(f)$ is contained in some P_k.

In order to find such a family we choose a commuting triangle

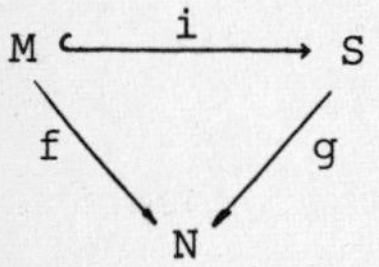

with i an embedding of M into a semialgebraic space S and g a proper map (cf. II.12.10; things are particularly easy here since M and N are semialgebraic). We regard M as a subspace of S with i the inclusion mapping. We further choose an embedding $S \hookrightarrow T$ of S into a complete (semialgebraic) space T which again we regard as an inclusion. Let $A := S \setminus M$. For X a subset of M we denote the closure of X in T by $\overline{X}^T$.

<u>Lemma 10.11.</u> Let $X \in \overline{\gamma}(M)$. Then $X \in \gamma_c(f)$ iff $A \cap \overline{X}^T = \emptyset$.

<u>Proof.</u> X is an element of $\gamma_c(f)$ iff the map $f|X = g|X$ is proper. Since g is proper this means that X is closed in S. We can express this by the equation $S \cap \overline{X}^T = X$ and also by $A \cap \overline{X}^T = \emptyset$. q.e.d.

From this lemma we conclude that $\gamma_c(f) = \{M \cap P \mid P \in \gamma_c(T \setminus A)\}$. Now we know from §9 (e.g. Cor. 9.6, to give a formal reference) that there exists a countable increasing family $(P_k \mid k \in \mathbb{N})$ in $\gamma_c(T \setminus A)$ such that every $P \in \gamma_c(T \setminus A)$ is contained in some set P_k. This implies that $(M \cap P_k \mid k \in \mathbb{N})$ is a countable increasing family in $\gamma_c(f)$ such that every $X \in \gamma_c(f)$ is contained in some set $M \cap P_k$. This completes the proof of Theorem 10.7.

Proposition 10.12 (This generalizes Prop. 9.7). Let $f : M \to N$ be a morphism between weakly semialgebraic spaces. Assume that M is locally semialgebraic and polytopic. Then the function ringes space $P_f(M)$ is locally semialgebraic. (Here R is not necessarily sequential.)

Proof. Given some $K \in \gamma_c(f)$ we shall prove that there exists some $U \in \mathring{\gamma}(M)$ with $K \subset U$ and $\overline{U} \in \gamma_c(f)$. Then we shall be done. We choose some $V \in \mathring{\gamma}(M)$ with $K \subset V$ and a commuting triangle

$$\begin{array}{ccc} V & \overset{i}{\hookrightarrow} & S \\ & {\scriptstyle f|V} \searrow \quad \swarrow {\scriptstyle g} & \\ & N & \end{array}$$

with i a dense embedding into a semialgebraic space S and g a proper map. We regard V as a subspace of S. Then V is open in S since V is locally complete, and K is closed in S since $f|K$ is proper. We finally choose U as an open semialgebraic neighborhood of K in V whose closure in S is contained in V. q.e.d.

We will write down some formal properties of the construction $f \mapsto P_f(M)$, always assuming that R is sequential. Now again a "space" means a weakly semialgebraic space over R and a "map" between spaces means a weakly semialgebraic map.

Notations 10.13. We sometimes think of a map $f : M \to N$ as a "space M over N", and then write $P_N(M)$ instead of $P_f(M)$, regarding $P_N(M)$ as a space over N via $\tilde{f}$. This is a particularly useful notation if somehow the map from M to N is given in a natural way starting from fixed data. If f is partially proper then we also say that M is "partially proper over N" (cf. Def. 5 above).

Remark 10.14. If $f : M \to N$ is any weakly semialgebraic map and $h : N \to S$ is a partially proper map then $\gamma_c(h \circ f) = \gamma_c(f)$, hence

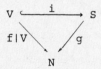

$P_{h\circ f}(M) = P_f(M)$, $p_{h\circ f} = p_f$, $(h\circ f)^{\sim} = h\circ\tilde{f}$.

Proposition 10.15 (Functoriality of $f \rightsquigarrow P_f(M)$).

i) Let

$$(*)\qquad \begin{array}{ccc} M' & \xrightarrow{\beta} & M \\ {\scriptstyle g}\downarrow & & \downarrow{\scriptstyle f} \\ N' & \xrightarrow[\alpha]{} & N \end{array}$$

be a commuting square of weakly semialgebraic maps and assume that α is partially proper. Then there exists a unique map γ from $P_g(M')$ to $P_f(M)$ such that $p_f\circ\gamma = \beta\circ p_g$.

ii) Assume in addition that the square (*) is cartesian. Then also the square

$$\begin{array}{ccc} P_g(M') & \xrightarrow{\gamma} & P_f(M) \\ \downarrow{\scriptstyle p_g} & & \downarrow{\scriptstyle p_f} \\ M' & \xrightarrow[\beta]{} & M \end{array}$$

is cartesian. In short, if N' is partially proper over N, then

$$P_{N'}(N' \times_N M) = N' \times_N P_N(M).$$

Proof. i) The map $f\circ\beta\circ p_g = \alpha\circ\tilde{g}$ from $P_g(M')$ to N is partially proper. Thus there exists a unique map γ from $P_g(M')$ to $P_f(M)$ with $p_f\circ\gamma = \beta\circ p_g$.

ii) We look at the following diagram consisting of two cartesian squares (solid arrows).

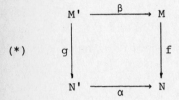

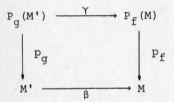

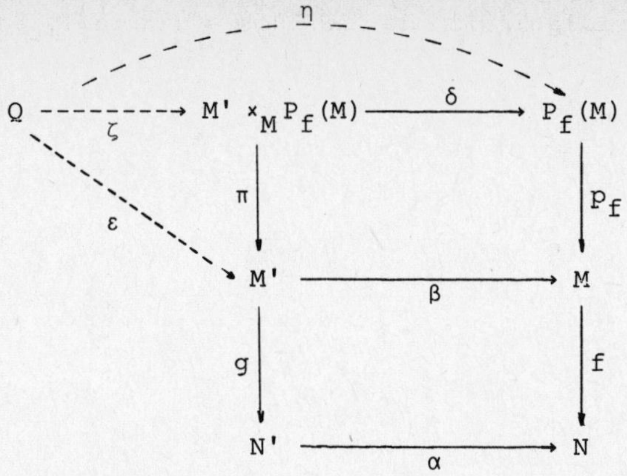

We have to verify that π is a partially proper core of g. The map π is bijective since p_f is bijective. The maps β and δ are partially proper since α is assumed to be partially proper. We conclude that $\alpha \circ g \circ \pi$ is partially proper, and then, that $g \circ \pi$ is partially proper.

It remains to prove that for a given map $\varepsilon : Q \to M'$ with $g \circ \varepsilon$ partially proper there exists a map ζ from Q to $M' \times_M P_f(M)$ with $\pi \circ \zeta = \varepsilon$. The proof is indicated by the dotted arrows in the commutative diagram above. η exists since $f \circ \beta \circ \varepsilon$ is partially proper. q.e.d.

From the second part of the proposition and Remark 10.14 we obtain

<u>Corollary 10.16.</u> If M is any space over N and N' is partially proper over N then

$$P_N(N' \times_N M) = N' \times_N P_N(M).$$

<u>Examples 10.17.</u> i) If L is a weak polytope and M is a space over N then

$$P_N(L \times M) = L \times P_N(M),$$

$L \times M$ being regarded as a space over N by the obvious map $L \times M \to M \to N$.

ii) If $M \to N$ is a weakly semialgebraic map and A is a partially complete (hence closed) subspace of N, then

$$\tilde{f}^{-1}(A) = P(f^{-1}(A)),$$

as follows from Remark 10.14 and Proposition 10.15.ii. In particular, for every $y \in N$,

$$\tilde{f}^{-1}(y) = P(f^{-1}(y))$$

<u>Proposition 10.18</u> ("Schachtelungssatz" for partially proper cores). Let $f : M \to N$ and $g : L \to M$ be weakly semialgebraic maps and $\pi := p_f$, $\pi' := p_g$. We form the cartesian square for π and $g \circ \pi'$.

$$\begin{array}{ccc}
P_N(M) \times_M P_M(L) & \xrightarrow{p_2} & P_M(L) \\
 & & \downarrow \pi' \\
\downarrow p_1 & & L \\
 & & \downarrow g \\
P_N(M) & \xrightarrow{\pi} & M
\end{array}$$

Then $\pi' \circ p_2$ is a partially proper core of $f \circ g$. In short,

$$P_N(M) \times_M P_M(L) = P_N(L).$$

<u>Proof.</u> $\pi' \circ p_2$ is bijective and $f \circ g \circ \pi' \circ p_2$ is partially proper. It remains to prove that, given a map $\alpha : Q \to L$ with $f \circ g \circ \alpha$ partially proper, there exists a map δ from Q to $P_N(M) \times_M P_M(L)$ such that $\pi' \circ p_2 \circ \delta = \alpha$. The proof is indicated by the following commuting diagram. Here β exists since $g \circ \alpha$ is partially proper and γ exists since $f \circ (g \circ \alpha)$ is partially proper.

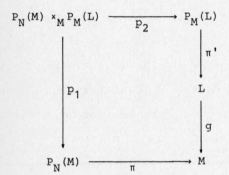

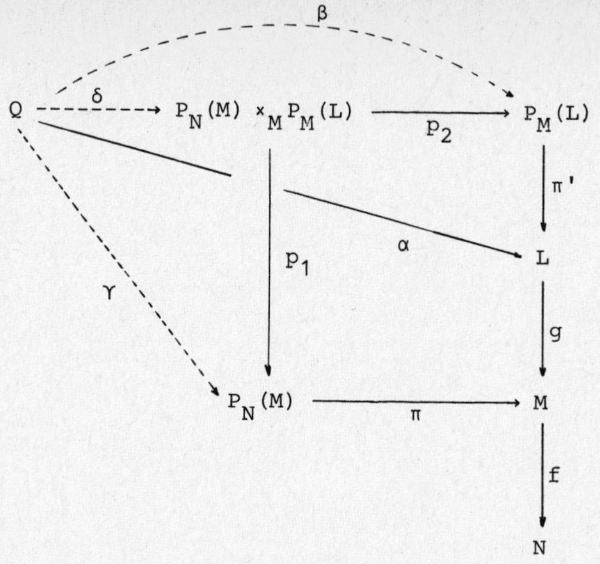

<u>Proposition 10.19</u> (Compatibility with fibre products). Assume that S is a space over T and M,N are spaces over S, hence also over T. Then we have the following equalities of spaces over T:

$$P_T(M \times_S N) = P_T(M) \times_S P_T(N) = P_T(M) \times_{P_T(S)} P_T(N) .$$

Here the second equality is obvious since the map $P_T(S) \to S$ is bijective. (Injectivity would suffice.) We leave the proof of the first equality to the reader. One verifies in a straightforward manner that the natural map from $P_T(M) \times_S P_T(N)$ to $M \times_S N$ has the universal property characterizing the partially proper core of $M \times_S N$ over T.

This proposition generalizes Corollary 10.16 above.

<u>Proposition 10.20</u> (Idempotency). Given weakly semialgebraic maps $f : M \to N$ and $g : N \to S$, let $L := P_f(M)$ and $\pi := p_f : L \to M$. Then $P_{gf\pi}(L) = P_{gf}(M)$ and $\pi \circ p_{gf\pi} = p_{gf}$. In short,

$$P_S(P_N(M)) = P_S(M).$$

<u>Proof.</u> One verifies in a straightforward way that $\pi \circ p_{gf\pi} : P_S(L) \to L \to M$ has the universal property required for p_{gf}. Since, on the set theoretical level, $\pi \circ p_{gf\pi}$ is the identity of M the assertion follows.

§11 - The quotient by a partially proper equivalence relation*)

Definition 1. a) An *equivalence relation* T *on a space* M is a weakly semialgebraic subset T of $M \times M$ which is an equivalence relation on M in the set theoretic sense. This means the following.

i) T contains the diagonal Diag M of $M \times M$.

ii) The switch automorphism $(x,y) \mapsto (y,x)$ of $M \times M$ maps T into (and hence onto) T.

iii) The natural projection $pr_{13} : M \times M \times M \to M \times M$ of $M \times M \times M$ to the first and the third factor M maps $(T \times M) \cap (M \times T)$ into (and hence onto) T.

b) If T is an equivalence relation on M then we denote the automorphism $(x,y) \mapsto (y,x)$ of the space T by τ_T and the natural projections $(x,y) \mapsto x$ and $(x,y) \mapsto y$ of T to M by p_1 and p_2. Notice that $p_2 = p_1 \circ \tau_T$. Notice also that p_1 is strongly surjective since p_1 has the section $x \mapsto (x,x)$.

c) The equivalence relation T is called *closed* if the set T is closed in $M \times M$.

Example 11.1. If $f : M \to N$ is a map between spaces then

$$E(f) := \{(x,y) \in M \times M \mid f(x) = f(y)\}$$

is a closed equivalence relation on M. We have the cartesian square

$$\begin{array}{ccc} E(f) & \xrightarrow{p_1} & M \\ {\scriptstyle p_1}\downarrow & & \downarrow{\scriptstyle f} \\ M & \xrightarrow[f]{} & N \end{array} \quad .$$

Definition 2. We call an equivalence relation T *partially proper* (semialgebraic, proper) if the map $p_1 : T \to M$ - and hence also p_2 -

*) This section may be skipped at first reading since it will not be used before Chapter VII and will have its main applications in the third volume [SFC].

is partially proper (semialgebraic, proper respectively). It is easily seen that a partially proper equivalence relation is closed.

Remark 11.2. If a map $f : M \to N$ between spaces is partially proper (semialgebraic, proper), then the equivalence relation $E(f)$ is partially proper (semialgebraic, proper). This follows from the cartesian square in 11.1.

We ask for the existence of a "strong quotient" or even a "partially proper quotient" of a space M by a given equivalence relation T. By this we mean the following.

Definition 3. A map $f : M \to N$ is a strong quotient of M by T if $T = E(f)$ and f is identifying (cf. §8). f is called a partially proper (resp. proper) quotient of M by T if $T = E(f)$ and f is strongly surjective and partially proper (resp. proper). Notice that in these cases f is certainly identifying (Th. 8.3).

Remarks 11.3. i) Let M/T denote the set of equivalence classes of the set M by the equivalence relation T, and let p_T denote the natural projection from M to M/T. We equip M/T with the function ringed space structure over R which is the strong quotient of M by p_T (§8, Def. 2). If there exists a strong quotient $f : M \to N$ of M by T in the sense of the definition above then f gives us a set theoretic bijection $\overline{f} : M/T \to N$ with $\overline{f} \circ p_T = f$. It is clear from §8 that f is an isomorphism of function ringed spaces over R. Thus M/T is a weakly semialgebraic space in this case and p_T is a strong quotient of M by T. If there exists a partially proper (resp. proper) quotient of M by T then we conclude that p_T is partially proper (resp. proper).
ii) If $f : M \to N$ is a strong quotient of M by T and $g : M \to L$ is a map to a space L with $E(g) \supset T$ then there exists a unique map $h : N \to L$ such that $h \circ f = g$. In Grothendieck's terminology [Gr, §2] the map f

is a strict epimorphism in the category WSA(R), in fact even in Space(R).

iii) If there exists a partially proper quotient of M by T then we know from 11.1 and 11.2 that T must be closed and partially proper.

Brumfiel has proved the important theorem [B_2] that if M is (affine) semialgebraic and T is a proper equivalence relation on M then indeed the proper quotient of M by T exists. This theorem extends readily to the following statement which we shall also call "Brumfiel's theorem" later on.

Theorem 11.4. If T is a partially proper equivalence relation on a space M then the function ringed space M/T (cf. 11.3) is weakly semialgebraic, and $p_T : M \to M/T$ is partially proper. If T happens to be proper then p_T is proper.

Proof. We choose an exhaustion $(M_\alpha | \alpha \in I)$ of M. For every $\alpha \in I$ the equivalence relation $T_\alpha := T \cap (M_\alpha \times M_\alpha)$ on M_α is proper. By Brumfiel's result the set M_α/T_α carries the structure of a semialgebraic space such that the natural projections $p_\alpha : M_\alpha \to M_\alpha/T_\alpha$ is proper. We regard the sets M_α/T_α as subsets of the set M/T. By Theorem 1.6 there exists on M/T the structure of a (weakly semialgebraic) space such that, for every α, M_α/T_α - in its given structure - is a closed subspace of M/T and $(M_\alpha/T_\alpha | \alpha \in I)$ is an exhaustion of M/T. Since the restrictions p_α of p_T are proper maps we conclude that p_T is a strongly surjective and partially proper (weakly semialgebraic) map. We now know that the present space structure on M/T coincides with the one defined above.

Assume that $p_1 : T \to M$ is proper. We want to prove that $p_T : M \to M/T$ is again proper. It suffices to prove that p_T is semialgebraic (cf. 5.6).

Let $X \in \gamma(M/T)$ be given. Since p_T is strongly surjective there exists some $Y \in \gamma(M)$ with $p_T(Y) = X$. Then $p_T^{-1}(X) = p_2(p_1^{-1}Y)$. This is indeed a semialgebraic set, since p_1 is semialgebraic. q.e.d.

Remark 11.5. The previous Theorem 8.6 on gluing of spaces is a special case of the present theorem 11.4, provided the gluing map f there is partially proper. We decided to do this special case earlier without using Brumfiel's result since the full strength of Theorem 11.4 will not be needed in the present volume in an essential way. Theorem 11.4 will turn out to be important in the third volume [SFC].

Remark 11.6. Brumfiel's theorem is a best possible result. Indeed, C. Scheiderer recently has proved the following "converse" to it [Schd]: Let M be a locally complete semialgebraic space, and let T be a closed equivalence relation on M such that the strong quotient of M by T exists. Then M contains a closed semialgebraic subset K such that the restriction of $p_1 : T \to M$ to $(M \times K) \cap T$ is a proper map onto M. Scheiderer makes essential use of abstract semialgebraic spaces (cf. [LSA, App. A]).

We now explicate what Theorem 11.4 gives us in the important case of group actions. We start with obvious definitions.

Definition 4. A weakly semialgebraic (locally semialgebraic, semialgebraic) group over R is a group object G in the category WSA(R) (resp. LSA(R), SA(R)). This means that G is a space (locally semialgebraic space, semialgebraic space respectively) over R and also an abstract group such that the multiplication map $m : G \times G \to G$, $(x,y) \mapsto xy$, and the map $i : G \to G$, $x \mapsto x^{-1}$, both are weakly semialgebraic maps.

Examples 11.7. i) If G is an algebraic group scheme over R (or over $C = R(\sqrt{-1})$) then G(R) (resp. G(C)) is a semialgebraic group over R.
ii) The orthogonal groups O(n,R) over R as well as the unitary groups U(n,R) and symplectic groups Sp(n,R) over R (cf. [Ch, Chap. I] with $\mathbb{R}$ there replaced by R) are semialgebraic groups over R. {We are just to use the letter R in our notation of unitary and symplectic groups although $R(\sqrt{-1})$ and the quaternions over R are used in the definition of these groups.}
iii) We embed O(n,R) into O(n+1,R) as usual by $A \longmapsto \begin{pmatrix} A & 0 \\ 0 & 1 \end{pmatrix}$. Then

$$O(\infty,R) := \lim_{n\to\infty} O(n,R) = \bigcup_n O(n,R)$$

is a weakly semialgebraic group which has the exhaustion $(O(n,R) \mid n \in \mathbb{N})$ by semialgebraic groups. Similarly we have weakly semialgebraic groups $U(\infty,R)$, $Sp(\infty,R)$. All these groups are partially complete.
iv) Assume that R is sequential. If G is a weakly semialgebraic group over R then the space P(G) with the same underlying abstract group G is a partially complete group over R. In particular, if $\mathcal{G}$ is any algebraic group scheme over R (or over C) then $\mathcal{G}(R)_{loc}$ (resp. $\mathcal{G}(C)_{loc}$) is a partially complete locally semialgebraic group over R.

We shall meet other examples of genuinely weakly semialgebraic groups in VII, §4 and VII, §9.

Definition 5. Let G be a weakly semialgebraic group (over R). A left operation of G on a space M is a left operation of the abstract group G on the set M such that the map $G \times M \to M$, $(g,x) \mapsto gx$ is weakly semialgebraic. We then call M a (left) G-space over R.

If M is a G-space and the group G is semialgebraic then the action of G gives us an equivalence relation

$$T(G) := \{(gx,x) \mid x \in M, g \in G\}$$

on M. Indeed, T(G) is the image of the semialgebraic map $G \times M \to M \times M$, $(g,x) \mapsto (gx,x)$, and hence a weakly semialgebraic subset of $M \times M$. This equivalence relation is semialgebraic (cf. Def. 2). The strong quotient M/T(G) - if it exists - will be denoted more briefly by $G\backslash M$. Its underlying set is the set of orbits of G on M.

Assume now that G is complete. Then $p_2 : T(G) \to M$ is proper. Thus Brumfiel's theorem gives us the following somewhat special result on orbit spaces.

Corollary 11.8. If G is a complete semialgebraic group then the proper quotient $G\backslash M$ exists for every G-space M.

Up to now we did not care for the question under which conditions the quotient of a locally semialgebraic (and, as always, weakly semialgebraic) space by an equivalence relation is again locally semialgebraic. Here is a rather special (and trivial) statement in this direction.

Remark 11.9. In the situation of Corollary 11.8, if M is locally semialgebraic, then also $G\backslash M$ is locally semialgebraic.

Proof. Let $(U_\alpha | \alpha \in I)$ be an admissible covering of M by open semialgebraic subsets. Then $(GU_\alpha | \alpha \in I)$ is again such a covering. Let p denote the projection from M to $G\backslash M$. We have $p^{-1}(pU_\alpha) = GU_\alpha$. Thus pU_α is open semialgebraic in $G\backslash M$ for every α, and $(pU_\alpha | \alpha \in I)$ is an admissible covering of $G\backslash M$. q.e.d.

Remark 11.10. Recently C. Scheiderer has proved a much better theorem [Schd]: Let M be a locally semialgebraic partially complete space and T a closed semialgebraic equivalence relation on M with the following property.

(*) If $U \in \mathring{\mathcal{E}}(M)$ then $p_2 p_1^{-1} U$ is open in M (hence $p_2 p_1^{-1} U \in \mathring{\mathcal{E}}(M)$).

Then the strong quotient M/T exists and is again locally semialgebraic. {Scheiderer only deals with the case that M is semialgebraic, but the statement above follows trivially from this case. Scheiderer also proves that (*) just means that the map p_1 is open.} If G is a semi-algebraic group acting (from the left) on a locally semialgebraic space M and if T(G) is closed then (*) is clearly fulfilled, hence the quotient G\M exists and is locally semialgebraic.

Chapter V - Patch complexes, and homotopies again

In this chapter a "space" means a weakly semialgebraic space over a fixed real closed field R and a "map" between spaces means a weakly semialgebraic map, if nothing else is said. All the homotopy theoretic notions and notations used in Chapter III retain their sense for these spaces and maps and will be used, starting from §2, without further explanation. The homotopy category of spaces over R is denoted by HWSA(R). Its objects are the spaces over R and its morphisms are the homotopy classes [f] of maps $f : M \to N$ between spaces.

§1 - Patch decompositions

In the following M is a fixed space.

Definition 1. A *semialgebraic partition* of M is a subset Σ of $\gamma(M)$ with the following properties.

PD1. If σ and τ are different elements of Σ then $\sigma \cap \tau = \emptyset$.

PD2. Every $X \in \gamma(M)$ is contained in the union of finitely many elements of Σ.

{The letters "PD" refer to "patch decomposition", see below.}

For example, if $(M_\alpha | \alpha \in I)$ is an exhaustion of M, then the set of patches $\{M_\alpha^o | \alpha \in I^o\}$ is a semialgebraic partition of M. Notice that PD2 just means that Σ is an admissible covering of M (IV, §3, Def. 7).

Definition 2. Let Σ be a semialgebraic partition of M, and let $\sigma \in \Sigma$.

i) The *boundary* of σ is the set $\partial\sigma := \overline{\sigma} \setminus \sigma$.

ii) An element τ of Σ is an *immediate face* of σ if $\tau \cap \partial\sigma \neq \emptyset$.

We then write $\tau \prec \sigma$.

Every $\sigma \in \Sigma$ has only finitely many immediate faces since the semialgebraic set $\partial\sigma$ is covered by finitely many elements of Σ.

Definition 3. a) A patch decomposition of the space M is a semialgebraic partition Σ of M fulfilling the following condition.

PD3. For every $\sigma \in \Sigma$ there exists a number $n \in \mathbb{N}_o$ such that every chain

$$\tau_r \prec \tau_{r-1} \prec \dots \prec \tau_o = \sigma$$ in Σ has length $r \leq n$.

The maximum of these lengths is then called the height $h(\sigma)$ of σ.

b) A patch complex (over R) is a pair $(X,\Sigma(X))$ consisting of a space X and a patch decomposition $\Sigma(X)$ of X. We often write the single letter X for the patch complex. The elements of $\Sigma(X)$ are called the patches of X.

Example 1.1. Let M be semialgebraic and let Σ be a finite set of pairwise disjoint semialgebraic subsets of M which cover M. Assume that there does not exist a cyclic chain

$$\sigma = \tau_r \prec \tau_{r-1} \prec \dots \prec \tau_o = \sigma$$

in Σ. Then (M,Σ) is a patch complex. We call these pairs (M,Σ) the finite patch complexes.

Example 1.2. Let M be any space and let Σ be a semialgebraic partition of M. Assume that for any two elements σ,τ of Σ with $\tau \prec \sigma$ the dimension of τ is strictly smaller than the dimension of σ. Then PD3 obviously holds. Thus (M,Σ) is a patch complex. We call these (M,Σ) the dimensional patch complexes.

Example 1.3. Again let M be any space and Σ a semialgebraic partition of M. Assume that the boundary $\partial\sigma$ of every $\sigma \in \Sigma$ is a union of elements of Σ. This means that every immediate face τ of σ is contained in $\partial\sigma$.

Then $\dim \tau < \dim \sigma$, and we see that (M,Σ) is a dimensional patch complex. We call these (M,Σ) normal patch complexes, in analogy to normal cell complexes in topology, cf. [LW, p.9]. Every simplicial complex is a normal patch complex (cf. IV.1.7 and IV.1.19).

Of course, a patch decomposition Σ of our space M is called finite (resp. dimensional, resp. normal) if the patch complex (M,Σ) is finite (dimensional, normal).

We are interested in patch complexes as a very weak substitute of simplicial complexes. Our spaces often cannot be triangulated (cf. IV.4.8), but they admit patch decompositions.

Proposition 1.4. Let $(M_\alpha | \alpha \in I)$ be an exhaustion of our space M. Then the set of patches $\{M_\alpha^o | \alpha \in I^o\}$, in the sense of IV, §1, is a patch decomposition of M.

Proof. If M_β^o is an immediate face of M_α^o then $M_\beta^o \cap M_\alpha \neq \emptyset$, hence $\beta \leq \alpha$ by IV.1.18. Certainly $\beta \neq \alpha$, hence $\beta < \alpha$, and the axiom PD3 is evident.

Example 1.5. Consider the exhaustion $(\mathbb{P}^n(R) | n \in \mathbb{N}_o)$ of the infinite dimensional projective space $\mathbb{P}^\infty(R)$ (IV, Ex. 1.9). The corresponding patch decomposition consists of the patches $\sigma_n := \mathbb{P}^n(R) \setminus \mathbb{P}^{n-1}(R) \cong R^n$ ($n \geq 0$, read $\mathbb{P}^{-1}(R) = \emptyset$). We have $\sigma_k \prec \sigma_l$ iff $k < l$. This patch decomposition is normal.

Example 1.6. We consider again the countable comb from IV.4.8 with the exhaustion $(M_n | n \in \mathbb{N}_o)$ indicated there. This time the patches are $\sigma_n := \{\frac{1}{n}\} \times]0,1]$ with $n \in \mathbb{N}$, and $\sigma_o := [0,1] \times \{0\}$. The only immediate face relations are $\sigma_o \prec \sigma_n$ with $n \in \mathbb{N}$. This patch decomposition is not dimensional.

<u>Definition 4.</u> Assume that Σ is a patch decomposition of M. A patch $\tau \in \Sigma$ is called a <u>face</u> of a patch $\sigma \in \Sigma$ if there exists a chain $\tau = \tau_r \prec \tau_{r-1} \prec \dots \prec \tau_o = \sigma$. We then write $\tau \leq \sigma$. If $r \geq 1$ then we call τ a <u>proper face</u> of σ and write $\tau < \sigma$.

It is evident from the axiom PD3 that the face relation is a partial ordering of Σ and that $\tau < \sigma$ just means $\tau \leq \sigma$, $\tau \neq \sigma$.

<u>Remark 1.7.</u> If Σ is a patch decomposition then every patch $\sigma \in \Sigma$ has only finitely many faces.

This follows by induction on the heigt $h(\sigma)$ from the fact that σ has only finitely many immediate faces.

<u>Remark 1.8.</u> If $\{M_\alpha^o \mid \alpha \in I^o\}$ is the patch decomposition of M coming from an exhaustion $(M_\alpha \mid \alpha \in I)$ then it is evident from the proof of Proposition 1.4 that the face relation $M_\beta^o \leq M_\alpha^o$ implies $\beta \leq \alpha$.

We need an obvious extension of Definition 4 in I, §1.

<u>Definition 5.</u> A family $(X_\lambda \mid \lambda \in \Lambda)$ of subsets of M is called <u>partially finite</u> (in M) if, for every semialgebraic subset Y of M, the set of all $\lambda \in \Lambda$ with $X_\lambda \cap Y \neq \emptyset$ is finite.

Of course, it suffices to check this property for Y running through the members of any admissible covering (IV, §3) of M, e.g. the patches of some patch decomposition of M.

Any semialgebraic partition Σ of M (cf. Def. 1), and in particular any patch decomposition of M, is a partially finite family of subsets of M (each set being indexed by itself).

Remark 1.9. Let $(A_\lambda | \lambda \in \Lambda)$ be a partially finite family in M such that every A_λ is weakly semialgebraic in M. Then, for every subset J of Λ, the set $B := \cup(A_\lambda | \lambda \in J)$ is again weakly semialgebraic in M since, for every $C \in \gamma(M)$, the intersection $B \cap C$ is clearly semialgebraic. The family $(A_\lambda | \lambda \in J)$ is partially finite in the space B.

Definition 6. A subset Y of the space X is called a subcomplex of the patch complex X if Y is the union of some patches of X. Then, by 1.9, Y is weakly semialgebraic in X, and the subspace Y of X together with the subset

$$\Sigma(Y) := \{\sigma \in \Sigma(X) | \sigma \subset Y\}$$

of $\gamma(Y)$ is a patch complex, which we also denote by Y.

Definition 7. A subcomplex Y of X is closed (resp. open) in X if the subset Y of the space X is closed (resp. open) in X.

Theorem 1.10. i) A subcomplex Y of X is closed in X iff for every $\sigma \in \Sigma(Y)$ all (immediate) faces of σ are contained in Y.
ii) The set H(X) of all finite closed subcomplexes of X is a faithful lattice exhaustion of the space X, and $\Sigma(X)$ is the set of patches of H(X).

Proof. a) Let Y be a closed subcomplex of X and $\sigma \in \Sigma(Y)$. Then $\partial\sigma \subset Y$. If τ is an immediate face of σ, then $\tau \cap Y \neq \emptyset$, hence $\tau \subset Y$. Of course, this implies that all faces of σ are contained in Y.
b) Let now Y be a finite subcomplex of X. Then $\overline{Y} = \cup(\overline{\sigma} | \sigma \in \Sigma(Y))$. If all immediate faces of all $\sigma \in \Sigma(Y)$ are contained in Y, then $\overline{\sigma} \subset Y$ for every $\sigma \in \Sigma(Y)$, and we conclude that $\overline{Y} = Y$, i.e. Y is closed in X.
c) It is now trivially checked that the set H(X) of finite closed subcomplexes of X fulfills the conditions E2', E3-E5, E7, E8 in IV, §1.

{Of course, we equip $H(X)$ with the ordering by inclusion, and regard $H(X)$ as a family of sets in $\overline{\mathfrak{F}}(X)$, each element of $H(X)$ being indexed by itself.} For every $\sigma \in \Sigma(X)$ the union $X(\sigma)$ of all faces of σ is an element of $H(X)$ by step b) of the proof. We have $\sigma \subset X(\sigma)$. Since $\Sigma(X)$ is an admissible covering of the space X, also $H(X)$ is an admissible covering of X. We conclude from IV, Cor. 3.16 that $H(X)$ is an exhaustion of the space X.

d) We look at the index function $\eta : X \to H(X)$ of this exhaustion (cf. end of IV, §1). Let $x \in X$ be given and let σ denote the unique patch which contains x. If Y is an element of $H(X)$ with $x \in Y$, then $\sigma \subset Y$ and hence $X(\sigma) \subset Y$ by step a) of the proof. Thus $\eta(x) = X(\sigma)$. Thus the $X(\sigma)$ with σ running through $\Sigma(X)$ are the primitive elements of $H(X)$. The patch corresponding to a given $X(\sigma)$ is the set

$$X(\sigma)^{O} = X(\sigma) \smallsetminus \bigcup(X(\tau) \mid X(\tau) \subsetneqq X(\sigma)) = X(\sigma) \smallsetminus \bigcup(X(\tau) \mid \tau < \sigma)$$

and this is just the set σ. Thus $\Sigma(X)$ is the set of patches of $H(X)$.

e) Let now Y be any subcomplex of X which has the property that Y contains all faces of all $\sigma \in \Sigma(Y)$. Then, for every $Z \in H(X)$, the subcomplex $Y \cap Z$ has the same property, and thus $Y \cap Z \in H(X)$ by step b). In particular, $Y \cap Z \in \overline{\mathfrak{F}}(Z)$. Since $H(X)$ is an exhaustion, we conclude that $Y \in \overline{\mathfrak{T}}(X)$. This concludes the proof of the theorem.

Notice that Proposition 1.4 and Theorem 1.10.ii together give us a new procedure how to enlarge a given faithful exhaustion $(M_\alpha \mid \alpha \in I)$ of a space M to a faithful lattice exhaustion H (Prop. IV.1.15). Namely we take for H the family of all closed finite subcomplexes of M with respect to the patch decomposition $(M_\alpha^{O} \mid \alpha \in I^{O})$. This procedure is more constructive than the one indicated in the proof of Prop. IV.1.15.

As a consequence of Theorem 1.10.i we obtain

Corollary 1.11. A subcomplex Y of a patch complex X is open in X iff for any $\tau \in \Sigma(Y)$ and $\sigma \in \Sigma(X)$ with $\tau \prec \sigma$ also $\sigma \in \Sigma(Y)$. In particular, for every $\sigma \in \Sigma(X)$ the union of all $\rho \in \Sigma(X)$ with $\sigma \leq \rho$ is the smallest open subcomplex of X which contains σ.

Definition 8. This open subcomplex of X is called the star of σ in X (or open star of σ in X) and denoted by $St_X(\sigma)$.

We now look for patch decompositions of a given space M with special properties. We need some more definitions.

Definition 9. A patch decomposition Σ of M is called special if, for every $\sigma \in \Sigma$, the pair $(\overline{\sigma},\sigma)$ of semialgebraic spaces is isomorphic to a pair $(N,\mathring{\nabla}(n))$ with N the union of the standard open n-simplex $\mathring{\nabla}(n)$ in R^{n+1} and some open faces of $\mathring{\nabla}(n)$. In this case (M,Σ) is called a special patch complex.

Notice that, if M is a weak polytope then, in this definition, N is forced to be the whole closed simplex $\nabla(n)$.

Definition 10. Let $(A_\lambda | \lambda \in \Lambda)$ be a family in $\mathcal{T}(M)$. A simultaneous patch decomposition of M and the family $(A_\lambda | \lambda \in \Lambda)$ is a patch decomposition of M such that every A_λ is a subcomplex of (M,Σ). Such a patch decomposition is called a simultaneous special patch decomposition of M and the family $(A_\lambda | \lambda \in \Lambda)$ if, in addition, M and every A_λ becomes a special patch complex by this decomposition.

Theorem 1.12. Let M be a space and $(A_\lambda | \lambda \in \Lambda)$ a partially finite family in $\mathcal{T}(M)$. Further let $(M_\alpha | \alpha \in I)$ be an exhaustion of M. Then there exists a simultaneous special patch decomposition of M and both families $(A_\lambda | \lambda \in \Lambda)$ and $(M_\alpha | \alpha \in I)$.

Proof. This is a straightforward consequence of the semialgebraic triangulation theorem I.2.13. Indeed, for any $\alpha \in I$ let $\Lambda(\alpha)$ denote the finite set of all $\lambda \in \Lambda$ with $M_\alpha \cap A_\lambda \neq \emptyset$. For every primitive index $\alpha \in I^o$ (cf. IV, §1, Def. 11) we choose a triangulation $\chi_\alpha : X_\alpha \xrightarrow{\sim} M_\alpha$ of M_α which simultaneously triangulates the finite families $(M_\alpha \cap A_\lambda \mid \lambda \in \Lambda(\alpha))$ and $(M_\beta \mid \beta \in I^o, \beta < \alpha)$. Let Σ^α denote the finite set of images $\chi_\alpha(S)$ of all open simplices S of X_α with $\chi_\alpha(S) \subset M_\alpha^o$. It is evident that the union Σ of all sets Σ^α with α running through I^o is a semialgebraic partition of M. If two elements $\sigma = \chi_\alpha(S)$, $\tau = \chi_\beta(T)$ of Σ are given then $\tau \prec \sigma$ implies that either $\beta < \alpha$ or $\beta = \alpha$ and T is a face of S. Thus it is clear that Σ also fulfills PD3, hence is a patch decomposition of M. If again $\sigma = \chi_\alpha(S)$ is an element of Σ^α then the pair $(\bar{\sigma}, \sigma)$ is isomorphic, via χ_α, to the pair $(\bar{S} \cap X_\alpha, S)$. {As usual, $\bar{S}$ denotes the union of S and all its faces, hence $\bar{S} \cap X_\alpha$ denotes the union of S and all faces contained in X_α.} Thus Σ is a special patch decomposition of M. Every M_α and every A_λ is a union of patches. q.e.d.

§2 - Some deformation retractions, and related homotopy equivalences

We want to extend some of the results of III, §1 to weakly semialgebraic spaces. This needs various preparations.

Definitions 1. a) A relative patch decomposition of a closed pair (M,A) of spaces (cf. IV, §8, Def. 8) is a patch decomposition Σ of the space $M \setminus A$. We then call the triple (M,A,Σ) a relative patch complex.

b) If σ is a patch of a relative patch complex (M,A,Σ), i.e. $\sigma \in \Sigma$, then $\bar{\sigma}$ denotes the closure of σ in M and $\partial\sigma$ denotes the set $\bar{\sigma} \setminus \sigma$, which is called the boundary of the patch (in M). Notions like "face of σ" and "height of σ" refer to the "absolute" patch complex $(M \setminus A,\Sigma)$. The same holds for notions derived from this like "dimensional" or "normal". We call the relative patch complex (M,A,Σ) finite if Σ is finite.

c) A subcomplex of the relative patch complex (M,A,Σ) is a subspace L of M with $A \subset L$ and $L \setminus A$ a union of patches. Then, of course, the triple (L,A,Σ') with Σ' the set of all $\sigma \in \Sigma$ contained in $L \setminus A$ is again a relative patch complex. Sometimes also this triple will be called a subcomplex of (M,A,Σ). The subspace L is closed in M iff

$$\sigma \in \Sigma', \ \tau < \sigma \Rightarrow \tau \in \Sigma' .$$

In this case we call L a closed subcomplex of (M,A).

d) The direct product $(M,A,\Sigma) \times (N,B,\Sigma')$ of two relative patch complexes (M,A,Σ) and (N,B,Σ') is defined as the closed pair of spaces $(M\times N,(M\times B) \cup (A\times N))$ equipped with the relative patch decomposition which has as patches the sets $\sigma\times\tau$ with $\sigma \in \Sigma$ and $\tau \in \Sigma'$. Notice that $\sigma_1 \times \tau_1 \le \sigma_2 \times \tau_2$ iff $\sigma_1 \le \sigma_2$ and $\tau_1 \le \tau_2$.

e) A relative patch decomposition Σ of a closed pair (M,A) and the corresponding patch complex (M,A,Σ) are called special iff, for every $\sigma \in \Sigma$, the pair $(\bar{\sigma},\sigma)$ is isomorphic to a pair $(N,\mathring{\nabla}(n))$ with N the union of the open standard simplex $\mathring{\nabla}(n)$ and some faces of $\mathring{\nabla}(n)$. (Of course, $n = \dim \sigma$.) Notice that if (M,A) is a relative weak polytope (IV, §8, Def. 8) this forces $N = \nabla(n)$.

Remark 2.1. We know from §1 that every closed pair (M,A) admits a simultaneous special patch decomposition (Th. 1.12). Then the patches in $M \setminus A$ give a special relative patch decomposition of (M,A). Conversely, if Σ' is a patch decomposition of A and Σ is a relative patch decomposition of (M,A) then $\Sigma \cup \Sigma'$ is a patch decomposition of M. If both Σ' and Σ are special then also $\Sigma \cup \Sigma'$ is special.

We could avoid below the use of the "relative" notions explicated in Definition 1, always working instead with simultaneous patch decompositions of pairs of spaces. But, whenever the patches in A do not play a significant role, it is conceptually simpler to use these relative notions.

In the following (M,A) is a closed pair of spaces. We equip (M,A) with a fixed relative patch decomposition Σ. We abusively denote the relative patch complex (M,A,Σ) again by (M,A). Sometimes we write $\Sigma(M,A)$ instead of Σ. (This will be common practice also later on.)

Notations 2.2. For any $n \in \mathbb{N}_0$ we denote the set of patches $\sigma \in \Sigma$ with height $h(\sigma) = n$ by $\Sigma(n)$. We denote the union of A and all patches with $h(\sigma) \leq n$ by M_n. Every M_n is a closed subcomplex of (M,A). We put $M_{-1} = A$ for convenience. We denote the direct sum (IV, 1.10) of the spaces $\overline{\sigma}$ with σ running through $\Sigma(n)$ by $M(n)$. Here $\overline{\sigma}$ carries, of course, the subspace structure in M. We further denote by $\partial M(n)$ the direct sum of the spaces $\partial\sigma$ with σ running through $\Sigma(n)$. This is a closed subspace of $M(n)$. We denote by $\psi_n : M(n) \to M_n$ the map whose restriction to any summand $\overline{\sigma}$ of $M(n)$ $(\sigma \in \Sigma(n))$ is the inclusion $\overline{\sigma} \hookrightarrow M_n$.

We finally denote by $\varphi_n : \partial M(n) \to M_{n-1}$ the map obtained from ψ_n by restrictions to $\partial M(n)$.

Both maps ψ_n and φ_n are partially proper. We have a commuting square

$$(*) \qquad \begin{array}{ccc} \partial M(n) & \xrightarrow{\varphi_n} & M_{n-1} \\ i \big\downarrow & & \big\downarrow j \\ M(n) & \xrightarrow{\psi_n} & M_n \end{array}$$

with inclusion maps i and j. The map $p_n : M(n) \sqcup M_{n-1} \to M_n$ which extends ψ_n and j is strongly surjective and partially proper. Thus, by the diagram (*),

$$M_n = M(n) \cup_{\varphi_n} M_{n-1} ,$$

i.e. we obtain M_n by gluing $M(n)$ to M_{n-1} along $\partial M(n)$ by the map φ_n (cf. IV, §8).

Definitions 2. We call M_n the *n-chunk* and $M(n)$ the *n-belt* of the relative patch complex (M,A). We further call $\partial M(n)$ the *boundary* of the n-belt, and $\varphi_n : \partial M(n) \to M_{n-1}$ the *attaching map* of the n-belt. Finally we call p_n the canonical projection from $M(n) \sqcup M_{n-1}$ to M_n.

Definition 3. An *admissible filtration* of a space X is an admissible covering $(X_n | n \in \mathbb{N})$ of X by closed weakly semialgebraic subsets with $X_n \subset X_{n+1}$ for every $n \in \mathbb{N}$. Instead of $\mathbb{N}$ we also allow shifted index sets like $\mathbb{N}_o$ or $\mathbb{N}_o \cup \{-1\}$.

In the present situation the family of chunks $(M_n | n \in \mathbb{N}_o)$ is an admissible filtration of M and thus M is the inductive limit of this family of spaces.

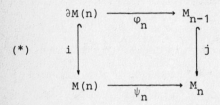

M can be "built up" starting from A by attaching one belt after the other using the maps φ_n. The whole space structure of M is given by the space A, the belts M(n), and the maps φ_n. Notice that the belts are spaces of rather simple type, namely direct sums of semialgebraic spaces. This way to analyze a space M will be very useful below and also later on. (We then often take $A = \emptyset$.)

Remark. If $M \setminus A$ would be of countable type (IV, §4), then we could find an admissible filtration $(N_n | n \geq -1)$ of M such that $N_{-1} = A$ and N_n is obtained from N_{n-1} by gluing a semialgebraic space to N_{n-1}. Our procedure in the general case is only slightly more complicated. Instead of semialgebraic spaces we use direct sums of semialgebraic spaces.

In the following we denote the unit interval [0,1] of R by I, as in previous chapters (but not Chapter IV where I has been an index set).

We study the following problem.

Problem 2.3. We are given a set $U \in \mathring{\mathfrak{T}}(A)$ and a set $D \in \mathring{\mathfrak{T}}(M)$ with $D \supset U$. Find a set $V \in \mathring{\mathfrak{T}}(D)$ such that $V \cap A = U$ and U is a strong deformation retract of V!

Lemma 2.4. There exists a set $X \in \mathring{\mathfrak{T}}(D \cap M_o)$ such that $X \cap A = U$ and U is a strong deformation retract of X. {N.B. M_o is the 0-chunk of M.}

Proof. Let $F := A \setminus U$. We have $U \subset D \setminus F$ and $(D \setminus F) \cap A = U$. Also, $(M \setminus F) \cap A = U$ and $(M \setminus F) \setminus U = M \setminus A$. Thus $(M \setminus F, U)$ is again a relative patch complex. Replacing M,A,D by $M \setminus F$, U, $D \setminus F$ respectively we assume without loss of generality that U = A.

For every $\sigma \in \Sigma(0)$ there exists a neighbourhood $W_\sigma \in \mathring{\mathcal{T}}(\bar{\sigma})$ of $\partial\sigma$ in $D \cap \bar{\sigma}$ together with a strong deformation retraction $G_\sigma : W_\sigma \times I \to W_\sigma$ from W_σ to $\partial\sigma$ (cf. III, 1.1 or already [DK_5, §2]). The G_σ combine to a strong deformation retraction $G : W\times I \to W$ from the direct sum W of all W_σ to $\partial M(o)$. We have $W \subset \psi_o^{-1}(D \cap M_o)$. Let $p : M(o) \sqcup A \to M_o$ be the canonical projection from $M(o) \sqcup A$ to M_o (i.e. $p = p_o$). Let $X := p(W \sqcup A)$. We have $A \subset X \subset D \cap M_o$. Moreover $p^{-1}(X) = W \sqcup A$. Since p is identifying (IV.8.3) this implies that $X \in \mathring{\mathcal{T}}(M_o)$. The restriction $\pi : W \sqcup A \to X$ of p is strongly surjective and partially proper since p is so. Let $pr_1 : A\times I \to A$ denote the projection of $A\times I$ to the first factor. There exists a unique set theoretic map $H : X\times I \to X$ such that the diagram

$$\begin{array}{ccc} (W\times I) \sqcup (A\times I) & \xrightarrow{G \sqcup pr_1} & W \sqcup A \\ \downarrow{\scriptstyle \pi\times id_I} & & \downarrow{\scriptstyle \pi} \\ X\times I & \xrightarrow[H]{} & X \end{array}$$

commutes. The reason is that $G(x,t) = x$ for every $(x,t) \in \partial M(o)\times I$. The map $\pi\times id_I$ is again partially proper and strongly surjective, hence identifying. Thus the map H is weakly semialgebraic. H is a strong deformation retraction from X to A. q.e.d.

<u>Theorem 2.5.</u> Problem 2.3 can always be solved. More precisely, there exists a set $V \in \mathring{\mathcal{T}}(D)$ with $V \cap A = U$ and a strong deformation retraction $F : V\times I \to V$ from V to U such that F maps $(V \cap M_n)\times I$ to $V \cap M_n$ for every $n \in \mathbb{N}_o$.

<u>Proof.</u> We apply Lemma 2.4 to the relative patch complexes (M,M_n) for all $n \geq -1$ and obtain successively for $n = -1,0,1,2,\ldots$ sets $V_n \in \mathring{\mathcal{T}}(D \cap M_n)$, with $V_{-1} = U$ and $V_n \cap M_{n-1} = V_{n-1}$ for $n \geq 0$, together with strong deformation retractions $H_n : V_n\times I \to V_n$ from V_n to V_{n-1} $(n \geq 0)$. Let V denote the union of all V_n. Then $V \cap M_n = V_n$ for

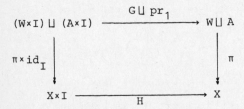

every $n \geq -1$. Thus $V \in \mathring{\mathcal{T}}(M)$, $V \subset D$, and $V \cap A = U$. We choose an infinite sequence

$$s_{-1} = 1 > s_o > s_1 > s_2 > \ldots$$

of positive elements in R (say, $s_n = 2^{-n-1}$). Let $\rho_n : V_n \to V_{n-1}$ denote the retraction $H_n(-,1)$ from V_n to V_{n-1}. If $n > m$, let

$$r_{n,m} := \rho_{m+1} \circ \rho_{m+2} \circ \ldots \circ \rho_n \ .$$

This is a retraction from V_n to V_m. We put $r_{n,n} := id_{V_n}$. We define a map $F_n : V_n \times I \to V_n$ as follows. Let $x \in V_n$. If $t \in [0,s_n]$ then $F_n(x,t) := x$. If $t \in [s_k, s_{k-1}]$ and $1 \leq k \leq n$, then

$$F_n(x,t) := H_k(r_{n,k}(x), (s_{k-1}-s_k)^{-1}(t-s_k)) \ .$$

F_n is a strong deformation retraction from V_n to U. For every $(x,t) \in V_n \times I$ we have $F_{n+1}(x,t) = F_n(x,t)$ since the homotopy H_{n+1} does not move the points in V_n. Thus the F_n fit together to a set theoretic map $F : V \times I \to V$. This map is weakly semialgebraic since $(V_n | n \in \mathbb{N})$ is an admissible covering of V by closed subspaces (IV.3.15.c). F is a strong deformation retraction from V to A. q.e.d.

Theorem 2.5 covers in particular (U = A) the following generalization of a part of Theorem III.1.1.

Corollary 2.6. If (M,A) is a closed pair of spaces then there exists an open weakly semialgebraic neighbourhood V of A in M such that A is a strong deformation retract of V.

We now can state a generalization of Proposition III.1.2. The proof is exactly the same as for locally semialgebraic spaces.

Proposition 2.7 (Extension of maps to a neighbourhood). Let M,A,V be as in Corollary 2.6. Any map $f : A \to Z$ into some space Z extends to a map $g : V \to Z$. If g and h are two extensions of f to V then there exists a homotopy $G : V\times I \to Z$ with $G_o = g$, $G_1 = h$ and $G_t|A = f$ for every $t \in I$.

Also Theorem III.1.3 extends to weakly semialgebraic spaces, but this needs a new proof.

Theorem 2.8. If (M,A) is a closed pair of spaces then $(M\times O) \cup (A\times I)$ is a strong deformation retract of $M\times I$.

Proof. We choose a relative patch decomposition of (M,A). We regard the pair $(I,\{O\})$ as a relative patch complex with a unique patch $]O,1]$. We form the direct product of these two relative patch complexes (Def. 1.d). This is the closed pair

$$(N,B) := (M\times I,(M\times O) \cup (A\times I))$$

together with the patch decomposition

$$\Sigma(N,B) := \{\sigma\times]O,1] \mid \sigma\in\Sigma(M,A)\}$$

of $N\setminus B$. For any two patches σ,τ of (M,A) we have $\tau\times]O,1] \prec \sigma\times]O,1]$ iff $\tau \prec \sigma$. Thus the patch $\sigma\times]O,1]$ of (N,B) has the same height as σ.

We use the notations (2.2) for both relative patch complexes (M,A) and (N,B). We have $N(n) = M(n)\times I$,

$$\partial N(n) = (M(n)\times O) \cup (\partial M(n)\times I),$$
$$N_n = (M\times O) \cup (M_n\times I) \quad .$$

By Theorem III.1.3 (or already [DK_5, Th. 5.1]) there exists, for every $n \geq O$, a strong deformation retraction $G_n : N(n)\times I \to N(n)$ from

$N(n)$ to $\partial N(n)$. From these G_n we obtain, in exactly the same way as in the proof of Lemma 2.4, strong deformation retractions $H_n : N_n \times I \to N_n$ from N_n to N_{n-1}. "Composing" all the H_n, as in the proof of Theorem 2.5, we obtain a strong deformation retraction $F : N \times I \to N$ from N to B. q.e.d.

Using only the fact that $(M \times O) \cup (A \times I)$ is a retract of $M \times I$ we obtain immediately, as in III, §1, the following extension of Cor. III.1.4 to weakly semialgebraic spaces.

Corollary 2.9 (Homotopy extension theorem). Let A be a closed subspace of a space M. Given a map $g : M \to Z$ into some space Z and a homotopy $F : A \times I \to Z$ with $F_o = g|A$ there exists a homotopy $G : M \times I \to Z$ with $G_o = g$ and $G|A \times I = F$.

It seems more difficult to generalize the second half of III,§1 somehow to weakly semialgebraic spaces. We will say something about this in §4.

We conclude this section by writing down several somewhat formal consequences of the homotopy extension theorem 2.9. With two exceptions we do not give the proofs, albeit some of them are tricky, since they run exactly as in the topological setting.

Proposition 2.10 (cf. [DKP, 2.9]). Let

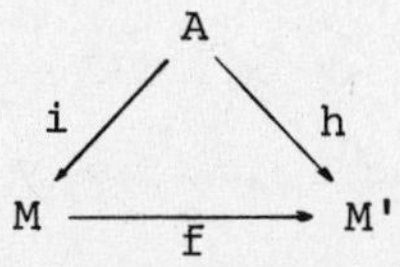

be a triangle of maps which commutes up to homotopy ($fi \simeq h$). Assume that i is a closed embedding, i.e. an isomorphism of A onto a closed subspace of M. Then there exists a map $g : M \to M'$ such that $g \simeq f$ and $g \circ i = h$.

The proof is almost trivial. We choose a homotopy $H : A \times I \to M'$ from $f \circ i$ to h. By Cor. 2.9 there exists a homotopy $F : M \times I \to M'$ such that $F(-,0) = f$ and $F \circ (i \times id_I) = H$. The map $g := F(-,1)$ has the required properties.

Definition 4 (also for later use). a) Let C be a space over R. A space under C is a map $\alpha : C \to M$ into a space M over R. A map $f : \alpha \to \beta$ from α to a space $\beta : C \to N$ under C is a map $f : M \to N$ such that $f \circ \alpha = \beta$. We also call f a map from M to N under C, if there is no doubt which "structural maps" α, β are under consideration, and we write $f : M \xrightarrow{C} N$. The category of spaces and maps under C is denoted by $WSA(R)^C$.
b) A homotopy H from a map $f : \alpha \to \beta$ to a map $g : \alpha \to \beta$ is a homotopy $H : M \times I \to N$ in the usual sense with $H(\alpha(z),t) = \beta(z)$ for all $z \in C, t \in I$. We then also say that H is a homotopy under C and we write $H : M \times I \xrightarrow{C} N$. If there exists a homotopy H under C from f to g then we write $f \simeq^C g$.
c) The homotopy class under C of a map $f : M \xrightarrow{C} N$ is denoted by $[f]^C$. The category whose objects are the spaces under C and whose morphisms are the homotopy classes under C is denoted by $HWSA(R)^C$. A map f under C is called a homotopy equivalence under C if $[f]^C$ is an isomorphism in $HWSA(R)^C$.

Theorem 2.11 (Dold [Do , 3.6], [DKP, 2.18]). Assume that $f : M \xrightarrow{C} N$ is a map under C and that the structural maps $\alpha : C \to M$, $\beta : C \to N$ are closed embeddings. Assume further that f is a homotopy equivalence (forgetting C). Then f is a homotopy equivalence under C.

Proposition 2.10 and Theorem 2.11 imply the following nice results about retractions, cf. [Do , 3.5, 3.7], [DKP, 2.27].

Corollary 2.12. Let (M,A) be a closed pair of spaces and let $i : A \to M$ denote the inclusion mapping.

i) Assume there exists a map $r : M \to A$ with $r \circ i \simeq id_A$. Then there exists a retraction $\rho : M \to A$ which is homotopic to r.

ii) If i is a homotopy equivalence then A is a strong deformation retract of M.

Definition 5. As in [LSA] we mean by a system of spaces a finite family of spaces $(M_o, M_1, \ldots, M_r)$ such that every M_i is a subspace of M_o. We call the system decreasing if $M_{i+1} \subset M_i$, and hence M_{i+1} is a subspace of M_i, for $1 \leq i \leq r-1$. We call the system closed if every M_i is closed in M. A map from $(M_o, M_1, \ldots, M_r)$ to a second system $(N_o, N_1, \ldots, N_r)$ is a family $(f_o, \ldots, f_r)$ of maps $f_i : M_i \to N_i$ such that every f_i is a restriction of $f_o : M_o \to N_o$.

Theorem 2.13. Let $(f_o, f_1, \ldots, f_r) : (M_o, M_1, \ldots, M_r) \to (N_o, N_1, \ldots, N_r)$ be a map between closed decreasing systems of spaces. Assume that every component $f_i : M_i \to N_i$ of this map is a homotopy equivalence. Then $(f_o, f_1, \ldots, f_r)$ is a homotopy equivalence.

For $r = 1$ a proof can be read off from [DKP, p. 64]. The arguments there and a straightforward induction on r give the proof in general.

From the case $r = 1$ of 2.13 one easily deduces the following interesting corollary, cf. [BD, 5.13].

Corollary 2.14 [BD, 5.13]. Let $f : M \to N$ be a map between spaces. Let M_1, M_2 be closed subspaces of M and N_1, N_2 closed subspaces of N such that $M = M_1 \cup M_2$, $N = N_1 \cup N_2$, $f(M_1) \subset N_1$, $f(M_2) \subset N_2$. Assume that the restrictions $M_1 \to N_1$, $M_2 \to N_2$, $M_1 \cap M_2 \to N_1 \cap N_2$ of f are homotopy equivalences. Then f is a homotopy equivalence.

Proposition 2.15 (cf. e.g. [DKP, 2.36]). Assume that A is a contractible closed subspace of the space M and that A is partially complete near $M \setminus A$. Then the natural projection $p : M \to M/A$ is a homotopy equivalence. Thus, by 2.13, also the map between pairs $p : (M,A) \to (M/A, A/A)$ is a homotopy equivalence.

For the proof one needs that $p \times id : M \times I \to (M/A) \times I$ is identifying, cf. Remark IV.8.7.ii.

The following modest generalization and easy consequence of Corollary 2.12 will be needed only much later (VII, §8) and thus may now be skipped by the reader.

Proposition 2.16. Let $(M_o, \ldots, M_r)$ be a closed decreasing system of spaces and let C be a closed subspace of M_o. Assume that, for every $k \in \{0, \ldots, r\}$, the inclusion map from $C \cap M_k$ to M_k is a homotopy equivalence. Then the system $(C, C \cap M_1, \ldots, C \cap M_r)$ is a strong deformation retract of $(M_o, \ldots, M_r)$.

Proof. The case $r = 0$ is the Corollary 2.12 above. We study the case $r = 1$. By 2.12 there exist strong deformation retractions $D_k : M_k \times I \to M_k$ from M_k to $C \cap M_k$ for $k = 0$ and $k = 1$. By the homotopy extension theorem (Cor. 2.9) there exists a homotopy $E : M_o \times I \to M_o$ relative C which starts with the identity of M_o and extends D_1. Now the map $G : M_o \times I \to M_o$, defined by $G(x,t) := E(x,2t)$ if $0 \leq t \leq \frac{1}{2}$, and $G(x,t) := D_o(E(x,1), 2t-1)$ if $\frac{1}{2} \leq t \leq 1$, is a strong deformation retraction from M_o to C. It maps $M_1 \times I$ to M_1 and gives by restriction a strong deformation retraction of M_1 to $C \cap M_1$. For $r \geq 2$ the proof runs by the same argument and induction on r. q.e.d.

§3 - Partially finite open coverings

We will prove the existence of partially finite open coverings of a space M which have special properties, some of them analogous to common properties of open coverings of paracompact topological spaces. These properties are important for sheaf cohomology on M. They will also turn out to be useful later in the theory of fibre bundles and covering maps but will play no role in the rest of the present chapter.

If $(U_\lambda | \lambda\in\Lambda)$ is a partially finite open covering of M, i.e. a partially finite family in $\mathring{\mathcal{T}}(M)$ with union M, then certainly $(U_\lambda | \lambda\in\Lambda) \in \mathrm{Cov}_M(M)$. We have to live with the fact that, in general, the U_λ cannot be semialgebraic sets, since, in general, M is not locally semialgebraic.

We choose a patch decomposition Σ of our space M. Abusively we denote the patch complex (M,Σ) also by M.

Proposition 3.1. The family of stars $(\mathrm{St}_M(\sigma) | \sigma\in\Sigma)$ is a partially finite open covering of the space M.

Proof. Let ρ be a given patch of M. Then for any $\sigma \in \Sigma$, the intersection $\mathrm{St}_M(\sigma) \cap \rho$ is not empty iff $\rho \subset \mathrm{St}_M(\sigma)$, and this means that σ is a face of ρ. But ρ has only finitely many faces (1.7). q.e.d.

If M is a simplicial complex then the stars $\mathrm{St}_M(\sigma)$ are contractible. In general we cannot expect this. Nevertheless we will construct, starting from the patch decomposition, a partially finite covering of M by contractible open weakly semialgebraic subsets {but, in contrast to the case of a simplicial complex, the intersections of finitely many sets of the covering will, in general, not be contractible}. In our construction Theorem 2.5 from the preceding section will play a crucial role.

Theorem 3.2. Let $(D_\lambda | \lambda \in \Lambda)$ be an admissible open covering of the space M. Then there exists a partially finite covering $(U_\alpha | \alpha \in J)$ of M by contractible open weakly semialgebraic sets which refines $(D_\lambda | \lambda \in \Lambda)$.

Proof. For every $\sigma \in \Sigma(M)$ we choose a finite subset $\Lambda(\sigma)$ of Λ such that σ is contained in the union of the D_λ with $\lambda \in \Lambda(\sigma)$. Then we choose a finite covering $(U_{\sigma,\alpha} | \alpha \in K_\sigma)$ of the space σ by open contractible semialgebraic sets $U_{\sigma,\alpha}$ which refines the covering $(\sigma \cap D_\lambda | \lambda \in \Lambda(\sigma))$ of σ. This can be done by triangulating σ and the sets $\sigma \cap D_\lambda$ with $\lambda \in \Lambda(\sigma)$ simultaneously and taking as sets $U_{\sigma,\alpha}$ open stars of this triangulation.

We choose, for every $\alpha \in K_\sigma$, some index $\lambda(\sigma,\alpha) \in \Lambda$ with $U_{\sigma,\alpha} \subset \sigma \cap D_{\lambda(\sigma,\alpha)}$. Let J denote the set of all pairs (σ,α) with $\sigma \in \Sigma(M)$ and $\alpha \in K_\sigma$. By Theorem 2.5, applied to the closed pairs $(St_M(\sigma),\sigma)$, there exists, for every $(\sigma,\alpha) \in J$, a set $V_{\sigma,\alpha} \in \mathring{\mathcal{J}}(M)$ such that

$$U_{\sigma,\alpha} \subset V_{\sigma,\alpha} \subset St_M(\sigma) \cap D_{\lambda(\sigma,\alpha)}$$

and $U_{\sigma,\alpha}$ is a strong deformation retract of $V_{\sigma,\alpha}$. Since $U_{\sigma,\alpha}$ is contractible also $V_{\sigma,\alpha}$ is contractible. Since the family $(St_M(\sigma) | \sigma \in \Sigma(M))$ is partially finite (Prop. 3.1) and the sets K_σ are finite we conclude that the family $(V_{\sigma,\alpha} | (\sigma,\alpha) \in J)$ is partially finite. This family covers M since every $\sigma \in \Sigma(M)$ is covered by the subfamily $(V_{\sigma,\alpha} | \alpha \in K_\sigma)$. q.e.d.

As a by-product we obtain a second proof of a part of Proposition IV.4.16.

Corollary 3.3. Every locally semialgebraic space M which is also weakly semialgebraic is paracompact.

Proof. There exists, by the very definition of locally semialgebraic spaces, an admissible covering $(D_\lambda|\lambda\in\Lambda)$ of M by open semialgebraic sets. By the theorem this covering has a partially finite refinement $(U_\alpha|\alpha\in J)$ with $U_\alpha \in \mathring{\mathfrak{T}}(M)$. Since M is locally semialgebraic the family $(U_\alpha|\alpha\in J)$ is locally finite in the sense of [LSA, p. 6]. Every U_α is contained in some D_λ, hence $U_\alpha \in \mathring{\mathfrak{T}}(M)$. Thus we have found a locally finite covering of M by open semialgebraics, which proves M to be paracompact (I, §4, Def. 2). q.e.d.

Proposition 3.4 ("Shrinking lemma"). Let $(D_\lambda|\lambda\in\Lambda) \in \mathrm{Cov}_M(M)$ be partially finite. Then there exist families $(V_\lambda|\lambda\in\Lambda)$ in $\mathring{\mathfrak{T}}(M)$ and $(A_\lambda|\lambda\in\Lambda)$ in $\bar{\mathfrak{T}}(M)$ such that M is the union of all V_λ and $V_\lambda \subset A_\lambda \subset D_\lambda$. {In particular, $(V_\lambda|\lambda\in\Lambda)$ is again a partially finite, hence admissible, open covering of M.}

This can be proved by use of Zorn's lemma, as in the topological setting (e.g. [Q, p. 79f]), once we have verified the following

Claim. Let Γ be a subset of Λ and α an index in $\Lambda\setminus\Gamma$. Let $\Gamma' := \Gamma\cup\{\alpha\}$. Assume we are given families $(V_\lambda|\lambda\in\Lambda)$ in $\mathring{\mathfrak{T}}(M)$ and $(A_\lambda|\lambda\in\Lambda)$ in $\bar{\mathfrak{T}}(M)$ such that $V_\lambda \subset A_\lambda \subset D_\lambda$ for every $\lambda\in\Gamma$, and

$$(*) \qquad \bigcup(V_\lambda|\lambda\in\Gamma) \cup \bigcup(D_\mu|\mu\in\Lambda\setminus\Gamma) = M .$$

Then there exist two sets $V_\alpha \in \mathring{\mathfrak{T}}(M)$ and $A_\alpha \in \bar{\mathfrak{T}}(M)$ such that $V_\alpha \subset A_\alpha \subset D_\alpha$ and

$$\bigcup(V_\lambda|\lambda\in\Gamma') \cup \bigcup(D_\mu|\mu\in\Lambda\setminus\Gamma') = M .$$

In order to verify the claim we look at the set

$$X_\alpha := M\setminus[\bigcup(V_\lambda|\lambda\in\Gamma) \cup \bigcup(D_\mu|\mu\in\Lambda\setminus\Gamma')] .$$

This set X_α is closed and weakly semialgebraic in M since the families $(V_\lambda|\lambda\in\Gamma)$ and $(D_\mu|\mu\in\Lambda\setminus\Gamma')$ are partially finite. It follows from (*) that $X_\alpha \subset D_\alpha$. By "Urysohn's lemma" (IV.3.12) there exists a weakly semialgebraic function $\varphi : M \to I$ with $\varphi^{-1}(0) \supset M\setminus D_\alpha$ and $\varphi^{-1}(1) \supset X_\alpha$. We define

$$V_\alpha := \varphi^{-1}(]\tfrac{1}{2},1]),\ A_\alpha := \varphi^{-1}([\tfrac{1}{2},1]).$$

These sets fulfill the claim.

Definition 1. i) A partition of unity of a space M is a family $(\varphi_\alpha|\alpha\in J)$ of weakly semialgebraic functions $\varphi_\alpha : M \to [0,1]$ such that the family $(\varphi_\alpha^{-1}(]0,1])|\alpha\in J)$ is partially finite in M and, for every $x \in M$ the sum of the values $\varphi_\alpha(x)$, with α running through J, is 1.
ii) Let $(D_\lambda|\lambda\in\Lambda)$ be a partially finite covering of M by weakly semialgebraic sets. A partition of unity subordinate to this covering (resp. strictly subordinate to this covering) is a partition of unity $(\varphi_\lambda|\lambda\in\Lambda)$ such that, for every $\lambda \in \Lambda$, $\varphi_\lambda^{-1}(]0,1]) \subset D_\lambda$ (resp. $\varphi_\lambda^{-1}(]0,1]) \subset A_\lambda \subset D_\lambda$ with some set $A_\lambda \in \bar{\mathfrak{I}}(M)$).

N.B. We have been a little cautious in writing down the last condition since the closure of $\varphi_\lambda^{-1}(]0,1])$ is not necessarily weakly semialgebraic.

Theorem 3.5. Given a partially finite covering $(D_\lambda|\lambda\in\Lambda)$ of a space M by open weakly semialgebraic sets there exists a partition of unity $(\varphi_\lambda|\lambda\in\Lambda)$ which is strictly subordinate to $(D_\lambda|\lambda\in\Lambda)$.

Proof. By the last proposition there exist families $(V_\lambda|\lambda\in\Lambda)$ in $\mathring{\mathfrak{I}}(M)$ and $(A_\lambda|\lambda\in\Lambda)$, $(B_\lambda|\lambda\in\Lambda)$ in $\bar{\mathfrak{I}}(M)$ respectively with $\bigcup(B_\lambda|\lambda\in\Lambda) = M$ and $B_\lambda \subset V_\lambda \subset A_\lambda \subset D_\lambda$ for every $\lambda \in \Lambda$. We choose weakly semialgebraic functions $\psi_\lambda : M \to [0,\infty[$ with $B_\lambda \subset \psi_\lambda^{-1}(]0,\infty[) \subset V_\lambda$ (cf. IV.3.12). Then $\psi := \sum_{\lambda\in\Lambda} \psi_\lambda$ is a well defined weakly semialgebraic function on M which is positive everywhere. The functions $\varphi_\lambda := \psi_\lambda/\psi$ and the sets A_λ have the properties required in Definition 1.ii above. q.e.d.

In Corollary 3.3 we have obtained, as a by-product, a result on locally semialgebraic spaces. One may ask for a "combinatorial" criterion that a given space M is locally semialgebraic, starting from a patch decomposition of M. In general this seems to be difficult since the face relation may be very loose from a geometric view point. We can provide a good answer if the patch complex M is normal (cf. 1.3).

Proposition 3.6. Let M be a normal patch complex. The space M is locally semialgebraic iff, for every patch σ of M, the complex $St_M(\sigma)$ is finite. Of course, it suffices to check this for the patches of height zero.

Proof. If all the stars are finite complexes then $(St_M(\sigma) \mid \sigma \in \Sigma(M))$ is a covering of M by open semialgebraic sets. This covering is locally finite (Prop. 3.1), hence admissible. Thus M is locally semialgebraic.

Assume now that M is locally semialgebraic. Let σ be a patch of M. There exists some $U \in \mathring{\gamma}(M)$ which contains σ. Let ρ be a patch in $St_M(\sigma)$, i.e. with $\sigma \leq \rho$. If $\sigma \neq \rho$ then $\sigma \prec \rho$, since the patch complex M is normal. This implies $\partial\rho \cap U \neq \emptyset$ and then $\rho \cap U \neq \emptyset$. But U meets only finitely many patches since U is semialgebraic. Thus the complex $St_M(\sigma)$ is finite. q.e.d.

Sometimes it may be advisable to work with open coverings which are "locally finite", as to be defined now, instead of just partially finite. It also may be useful, or at least comfortable, to have an open covering with countable index set at disposal. We now will construct such coverings out of a given partially finite open covering.

<u>Definition 2.</u> A family $(X_\lambda \mid \lambda\in\Lambda)$ of subsets of M is called <u>locally finite</u> (in M) if there exists an admissible open covering $(U_\alpha \mid \alpha\in I)$ of M such that, for every $\alpha \in I$, the set of all $\lambda \in \Lambda$ with $X_\lambda \cap U_\alpha \neq \emptyset$ is finite.

Notice that every such family $(X_\lambda \mid \lambda\in\Lambda)$ is partially finite, since any semialgebraic subset of M is contained in the union of finitely many U_α. If the space M is locally semialgebraic then the terms "partially finite", "locally finite", as defined now, and "locally finite", as defined in [LSA, p. 6], all mean the same.

<u>Lemma 3.7.</u> Given a partially finite covering $(U_n \mid n\in\mathbb{N})$ of M by open weakly semialgebraic sets with countable index set there exists a locally finite covering $(W_n \mid n\in\mathbb{N})$ of M by open weakly semialgebraic sets such that $W_n \subset U_n$ for every $n \in \mathbb{N}$.

<u>Proof.</u> We choose a partition of unity $(u_n \mid n\in\mathbb{N})$ subordinate to $(U_n \mid n\in\mathbb{N})$. For every $n \in \mathbb{N}$ we define a new weakly semialgebraic function $w_n : M \to [0,1]$,

$$w_n(x) := \max(0, u_n(x) - \sum_{i<n} u_i(x)).$$

Let $W_n := w_n^{-1}(]0,1])$. We have $W_n \subset U_n$. We claim that $(W_n \mid n\in\mathbb{N})$ is a locally finite covering of M.

Given a point x in M let m be the smallest index with $u_m(x) \neq 0$. Then $w_m(x) = u_m(x) > 0$, which means that $x \in W_m$. Thus the union of the sets W_n is the whole space M.

We define an increasing family $(N_k \mid k\in\mathbb{N})$ in $\mathring{\mathcal{T}}(M)$ by

$$N_k := \{x \in M \mid \sum_{i\leq k} u_i(x) > \tfrac{1}{2}\} .$$

Given a semialgebraic subset X of M there exists a largest index l in $\mathbb{N}$ such that u_l does not vanish everywhere on X. For every $x \in X$ we have $\sum_{i<l} u_i(x) = 1$. Thus $X \subset N_l$. This proves that $(N_k | k \in \mathbb{N})$ is an admissible open covering of M. If $n > k$ then $W_n \cap N_k = \emptyset$. Thus $(W_n | n \in \mathbb{N})$ is indeed a locally finite open covering of M. q.e.d.

Theorem 3.8. Given an admissible open covering $(U_\alpha | \alpha \in I)$ of a space M there exists a locally finite open covering $(W_n | n \in \mathbb{N})$ of M such that every connected component of every set W_n is contained in U_α for some $\alpha \in I$.

Proof (cf. [tD, p. 397]). By Theorem 3.2 we may already assume that the covering $(U_\alpha | \alpha \in I)$ is partially finite. We choose a partition of unity $(u_\alpha | \alpha \in I)$ subordinate to this covering (Th. 3.5). We may assume that I is infinite. Let Λ denote the set of finite non empty subsets of I. For every $E \in \Lambda$ we define a weakly semialgebraic function $q_E : M \to R$ by

$$q_E(x) := \min_{\alpha \in E} u_\alpha(x) - \max_{\beta \in I \setminus E} u_\beta(x) .$$

Let $V_E := \{x \in M | q_E(x) > 0\}$. If E and F are different finite subsets of I with the same cardinality $|E| = |F| \neq 0$ then $V_E \cap V_F = \emptyset$. Indeed, assume there exists some point x in $V_E \cap V_F$. Choosing indices $k \in F \setminus E$ and $l \in E \setminus F$ we have $u_k(x) < u_l(x)$ since $x \in V_E$, and $u_l(x) < u_k(x)$ since $x \in V_F$, a contradiction.

Every set V_E is contained in $\cap(u_\alpha^{-1}(]0,1]) | \alpha \in E)$. In particular, $V_E \subset U_\alpha$ for any $\alpha \in E$. Thus $(V_E | E \in \Lambda)$ is certainly a partially finite family in $\mathring{\mathcal{T}}(M)$. This implies that, for every $n \in \mathbb{N}$, the set

$$V_n := \cup(V_E | E \in \Lambda, |E| = n)$$

is (open and) weakly semialgebraic, and that the family $(V_n | n \in \mathbb{N})$ is again partially finite. Also $(V_E | E \in \Lambda, |E| = n)$ is an admissible open covering of V_n. Since the sets in this family are pairwise disjoint the space V_n is the direct sum of this family of spaces. Thus certainly every connected component of V_n is contained in some set U_α.

Given a point $x \in M$ the set E consisting of all $\alpha \in I$ with $u_\alpha(x) \neq 0$ is finite and non empty. We have $x \in V_E$, hence $x \in V_n$ with $|E| = n$. Thus the union of the sets V_n is the whole space M, and we conclude that $(V_n | n \in \mathbb{N})$ is a partially finite open covering of M.

By the preceding lemma there exists a locally finite open covering $(W_n | n \in \mathbb{N})$ of M with $W_n \subset V_n$ for every n. This covering fulfills the requirements of the theorem. q.e.d.

§4 - Approximation of spaces by weak polytopes

As before we consider spaces over a fixed real closed field R. If a space M is locally semialgebraic, then there exists a weak polytope P in M which is a strong deformation retract of M. Indeed, the core P of M with respect to the first barycentric subdivision of any triangulation of M will do (III, §1). Thus the inclusion map $P \hookrightarrow M$ is a homotopy equivalence.

This result and its refinements in III, §1 have proved to be very useful in the development of homotopy theory of locally semialgebraic spaces (Chapter III) and also of homology theory [D], [DK_3], [D_2]. For the spaces studied now triangulations are not available. We nevertheless look for some generalization of the result above.

<u>Definition 1.</u> A <u>WP-approximation</u> of a space M is a homotopy equivalence $\varphi : P \to M$ with P a weak polytope.

The goal of this section is to prove

<u>Theorem 4.1.</u> Every space M has a WP-approximation.

This will be done by a string of four lemmas which we label by capital letters A,B,C,D. We will use mapping cylinders and related more complicated constructions ("telescopes").

<u>Definition 2.</u> Let $f : M \to N$ be a partially proper map. The <u>mapping cylinder</u> Z(f) of f is the space (cf. IV, §8)

$$Z(f) := (M\times I) \cup_{M\times 1, f} N.$$

Here we have identified $M\times 1$ with M in the obvious way in order to glue M×1 to N by f.

Mapping cylinders had been constructed in II, §12 for proper maps within the category LSA(R). Although we now are much better off having mapping cylinders for partially proper maps it will cause us sometimes trouble (in later sections) that we do not have them for all maps, as one has in topology.

If f is partially proper then we have a canonical factorization of f by a closed embedding i and a homotopy equivalence p.

$$
(4.2)\qquad
\begin{array}{ccc}
M & \xrightarrow{\;i\;} & Z(f) \\
 & {\scriptstyle f}\searrow & \downarrow{\scriptstyle p} \\
 & & N
\end{array}
$$

Here i and p are defined by $i(x) = (x,0)$ and $p(x,t) = f(x)$, $p(y) = y$, for $x \in M$, $t \in [0,1]$, $y \in N$. As usual, we regard $M\times[0,1[$ and N as subspaces of Z(f), cf. IV, §8. We identify M with the subspace $M\times 0$ of Z(f).

<u>Lemma A.</u> Let $f : A \to M$ be any map between spaces which have WP-approximations. Given a WP-approximation $\psi : Q \to A$ of A there exists a commuting square

$$
\begin{array}{ccc}
Q & \overset{i}{\hookrightarrow} & P \\
{\scriptstyle \psi}\downarrow & & \downarrow{\scriptstyle \varphi} \\
A & \xrightarrow[f]{} & M
\end{array}
$$

with φ a WP-approximation of M and i a closed embedding (i.e. isomorphism of Q onto a closed subspace of P).

<u>Proof.</u> We start with any WP-approximation $\varphi_1 : P_1 \to M$ of M. Then we have a square

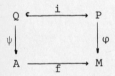

$$\begin{array}{ccc} Q & \xrightarrow{\ g\ } & P_1 \\ {\scriptstyle\psi}\downarrow & & \downarrow{\scriptstyle\varphi_1} \\ A & \xrightarrow[\ f\ ]{} & M \end{array}$$

which commutes up to homotopy. Using the factorization

$$g : Q \xrightarrow[i]{} Z(g) \xrightarrow[p]{} P_1$$

of g via its mapping cylinder (which exists since Q is a weak polytope) we obtain a square

$$\begin{array}{ccc} Q & \xrightarrow{\ i\ } & P \\ {\scriptstyle\psi}\downarrow & & \downarrow{\scriptstyle\varphi_2} \\ A & \xrightarrow[\ f\ ]{} & M \end{array}$$

with $P := Z(g)$ and $\varphi_2 := \varphi_1 \circ p$, which is again a WP-approximation of M. Moreover we have a homotopy $F : Q \times I \to M$ with $F(-,0) = \varphi_2 \circ i$ and $F(-,1) = f \circ \psi$. We extend F to a homotopy $G : P \times I \to M$ with $G(-,0) = \varphi_2$ (cf. 2.9). Then $\varphi := G(-,1)$ is a WP-approximation of M with $\varphi \circ i = f \circ \psi$. q.e.d.

<u>Lemma 4.3.</u> Let A be a closed subspace of a space M and let $f : A \to N$ be a partially proper map. The natural map

$$q : M \cup_A Z(f) \to M \cup_f N,$$

which is induced by the identity $id_M : M \to M$ and the natural projection $p : Z(f) \to N$, is a homotopy equivalence.

<u>Proof.</u> We consider the space $L := M \times I \cup_{A \times 1, f} N$. It is evident that $M \cup_f N = M \times 1 \cup_{A \times 1, f} N$ is a strong deformation retract of L, the retraction map $r : L \to M \cup_f N$ being given by $r[x,t] = [x,1]$, $r(y) = y$ $\{x \in M, t \in I, y \in N\}$. We now prove that also the subspace $M \cup_A Z(f) = M \times 0 \cup_{A \times 0} Z(f)$

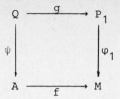

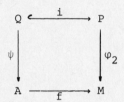

of L is a strong deformation retract of L. Then we will be done since $q = r \circ j$ with $j : M \cup_A Z(f) \to L$ the inclusion map.

We have a strong deformation retraction $F : M \times I \times I \to M \times I$ of $M \times I$ to $(M \times O) \cup (A \times I)$, cf. Theorem 2.8. Since F is constant on $A \times I$, this homotopy extends to a strong deformation retraction

$$G : L \times I = (M \times I \times I) \cup_{A \times 1 \times I, f \times id} N \times I \to L$$

from L to $M \cup_A Z(f)$. q.e.d.

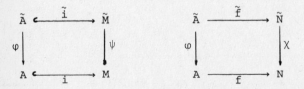

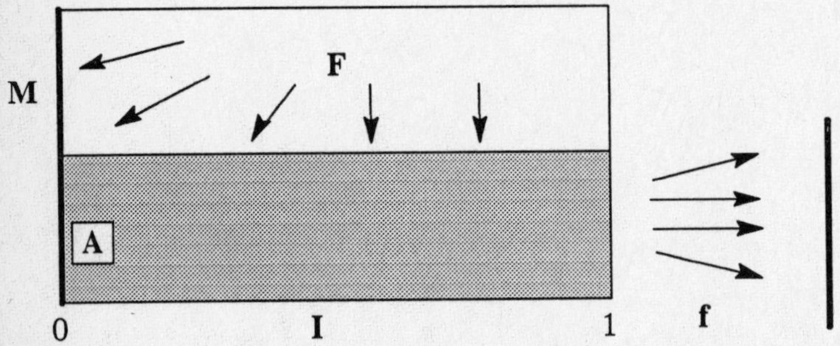

Lemma B. Let A be a closed subspace of M and $f : A \to N$ be a partially proper map. Assume that M,A, and N have WP-approximations. Then also $M \cup_f N$ has a WP-approximation.

Proof. Let i denote the inclusion $A \hookrightarrow M$. By the preceding lemma it suffices to find a WP-approximation of the space $M \cup_A Z(f)$. We choose a WP-approximation $\varphi : \tilde{A} \to A$ of A. By Lemma A there exist commuting squares

$$\begin{array}{ccc} \tilde{A} & \overset{\tilde{i}}{\hookrightarrow} & \tilde{M} \\ \varphi \downarrow & & \downarrow \psi \\ A & \underset{i}{\hookrightarrow} & M \end{array} \qquad \begin{array}{ccc} \tilde{A} & \overset{\tilde{f}}{\longrightarrow} & \tilde{N} \\ \varphi \downarrow & & \downarrow \chi \\ A & \underset{f}{\longrightarrow} & N \end{array}$$

with WP-approximations ψ, χ and closed embeddings $\tilde{i}, \tilde{f}$. (We will only need that $\tilde{i}$ is a closed embedding.) φ and χ together induce a map $\Phi : Z(\tilde{f}) \to Z(f)$. The diagram

$$\begin{array}{ccc} Z(\tilde{f}) & \xrightarrow{\tilde{p}} & \tilde{N} \\ \Phi \downarrow & & \downarrow \chi \\ Z(f) & \xrightarrow{p} & N \end{array}$$

with natural projections p and $\tilde{p}$ commutes.
Since $p, \tilde{p}, \chi$ are homotopy equivalences also Φ is a homotopy equivalence.
The maps ψ and Φ combine to a map

$$\psi : \tilde{M} \cup_{\tilde{A}} Z(\tilde{f}) \to M \cup_A Z(f) \ .$$

We conclude from Cor. 2.14 that ψ is again a homotopy equivalence.
This is the desired WP-approximation of $M \cup_A Z(f)$. q.e.d.

In the following we choose a patch decomposition of a given space M and use the notations 2.2. In particular $(M_n \mid n \in \mathbb{N})$ is the family of chunks of M.

Lemma C. Every space M_n has a WP-approximation.

We prove this by induction on n. M_o is a direct sum of semialgebraic spaces and thus has a WP-approximation. Let $n > 0$. Then M_n is obtained by gluing $M(n)$ to M_{n-1} along the subspace $\partial M(n)$ by a partially proper map $\varphi_n : \partial M(n) \to M_{n-1}$ (cf. §2). The spaces $M(n)$ and $\partial M(n)$ are direct sums of semialgebraic spaces and thus have WP-approximations. By induction hypothesis also M_{n-1} has a WP-approximation. We conclude from Lemma B that M_n has a WP-approximation.

The proof of Theorem 4.1 will be completed if we verify the following

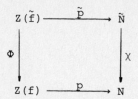

Lemma D. Let $(M_n \mid n \in \mathbb{N})$ be an admissible filtration of a space M. If every M_n has a WP-approximation then M has a WP-approximation.

This lemma will follow from Lemma A and some elementary facts about "telescopes".

Definition 3. Let $\mathfrak{a}$ be an infinite sequence of partially proper maps

$$\mathfrak{a} : A_1 \xrightarrow[\alpha_1]{} A_2 \xrightarrow[\alpha_2]{} A_3 \to \dots \quad .$$

The telescope $\mathrm{Tel}(\mathfrak{a})$ is the inductive limit of a family of spaces $(\mathrm{Tel}_n(\mathfrak{a}) \mid n \in \mathbb{N})$ which is defined inductively as follows.

$$\mathrm{Tel}_1(\mathfrak{a}) := Z(\alpha_1)$$
$$\mathrm{Tel}_n(\mathfrak{a}) := \mathrm{Tel}_{n-1}(\mathfrak{a}) \cup_{A_n} Z(\alpha_n) \quad (n > 1) \quad .$$

Here we regard $\mathrm{Tel}_n(\mathfrak{a})$ as a closed subspace of $\mathrm{Tel}_m(\mathfrak{a})$ for $n \leq m$ in the obvious way.

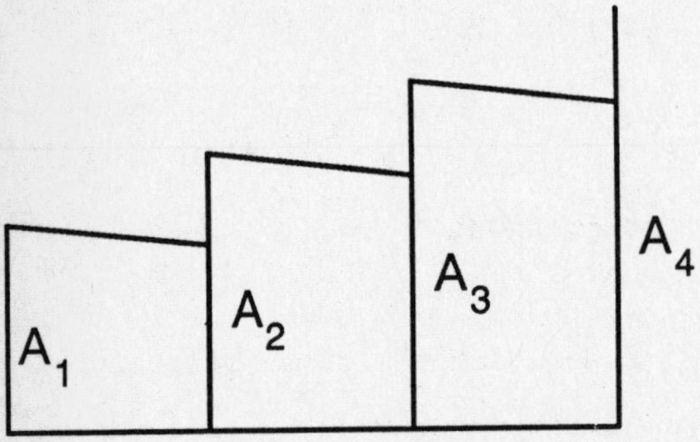

Sketch of $\mathrm{Tel}_3(\mathfrak{a})$.

This inductive limit exists by IV, §7. Every $\mathrm{Tel}_n(\mathfrak{a})$ is a closed subspace of $\mathrm{Tel}(\mathfrak{a})$ and $(\mathrm{Tel}_n(\mathfrak{a}) \mid n \in \mathbb{N})$ is an admissible filtration of the space $\mathrm{Tel}(\mathfrak{a})$.

The telescope $\mathrm{Tel}(\mathfrak{a})$ comes close to an inductive limit of the sequence $\mathfrak{a}$ in the category HWSA(R). Indeed, we have obvious inclusion maps

$j_n : A_n \to \mathrm{Tel}(\mathfrak{A})$. Clearly $j_n \simeq j_{n+1} \circ \alpha_n$. Given a family of maps $\varphi_n : A_n \to N$ into a fixed space N with $\varphi_n \simeq \varphi_{n+1} \circ \alpha_n$ for every $n \in \mathbb{N}$ it is easy to find a map $\varphi : \mathrm{Tel}(\mathfrak{A}) \to N$ such that $\varphi \circ j_n \simeq \varphi_n$ for every n. But the homotopy class $[\varphi]$ is not uniquely determined by the classes $[\varphi_n]$ since the construction of φ depends on the choice of homotopies from φ_n to $\varphi_{n+1} \circ \alpha_n$ for every n. Thus $\mathrm{Tel}(\mathfrak{A})$ is not a genuine inductive limit of the sequence $\mathfrak{A}$ in HWSA(R).

Lemma 4.4. Let

$$\begin{array}{ccccccc} A_1 & \xrightarrow{\alpha_1} & A_2 & \xrightarrow{\alpha_2} & A_3 & \longrightarrow & \cdots\cdots \\ \downarrow{\scriptstyle \varphi_1} & & \downarrow{\scriptstyle \varphi_2} & & \downarrow{\scriptstyle \varphi_3} & & \\ B_1 & \xrightarrow[\beta_1]{} & B_2 & \xrightarrow[\beta_2]{} & B_3 & \longrightarrow & \cdots\cdots \end{array}$$

be a commuting ladder of maps between spaces. We regard this as a morphism from the upper horizontal sequence $\mathfrak{A}$ to the lower horizontal sequence $\mathfrak{B}$. Assume that every φ_n is a homotopy equivalence and that the α_i and β_j are partially proper. Then the induced map $\Phi : \mathrm{Tel}(\mathfrak{A}) \to \mathrm{Tel}(\mathfrak{B})$ is again a homotopy equivalence.

This can be proved as in the topological setting, cf. [BD, p. 59ff] or [Pu, p. 314, Hfs. 7].

N.B. The presentation in [BD] is very lucid. We have to replace the function $e(t) = \exp(-t)$ there on p. 59 by a semialgebraic isomorphism $e : [0,\infty[\xrightarrow{\sim}]0,1]$. This does not destroy the argument. A statement similar to 4.4 holds for a ladder which only commutes up to homotopy (loc.cit.). We do not need this more general fact at present.

For our purpose in this section it suffices to consider a special kind of telescopes.

<u>Definition 4.</u> The telescope $\mathrm{Tel}(\mathfrak{M})$ of an admissible filtration $\mathfrak{M} = (M_n | n \in \mathbb{N})$ of a space M is defined as the telescope of the sequence

$$M_1 \xrightarrow[i_1]{} M_2 \xrightarrow[i_2]{} M_3 \hookrightarrow \ldots\ldots$$

with the inclusion maps i_n.

Let I_∞ denote the inductive limit of the intervals $[0,n]$. This is a locally semialgebraic weak polytope which has as underlying set the non negative elements in the smallest real valuation ring $\mathfrak{o}$ of R (cf. [KW], [P, §7], [L, §5]).

It is convenient to visualize $\mathrm{Tel}(\mathfrak{M})$ as a closed subspace of $M \times I_\infty$. This may be done as follows. We regard $\mathrm{Tel}_n(\mathfrak{M})$ as the closed subspace

$$(M_1 \times [0,1]) \cup (M_2 \times [1,2]) \cup \ldots \cup (M_n \times [n-1,n]) \cup (M_{n+1} \times \{n\})$$

of $M \times I_\infty$ and $\mathrm{Tel}(\mathfrak{M})$ as the union of these subspaces. (The upper lines in the picture of Tel_3 above can now be drawn horizontally.)

The projection of $M \times I_\infty$ to the first factor restricts to a natural projection $p_{\mathfrak{M}} : \mathrm{Tel}(\mathfrak{M}) \to M$.

<u>Lemma 4.5.</u> $p_{\mathfrak{M}}$ is a homotopy equivalence.

This follows from the fact that $\mathrm{Tel}(\mathfrak{M})$ is a strong deformation retract of $M \times I_\infty$, which can be proved by a rather obvious inductive procedure using Theorem 2.8, cf. [BD, p. 63f].

We now start out to prove Lemma D. By the lemma just proved it suffices to find a WP-approximation of the telescope $\mathrm{Tel}(\mathfrak{M})$ of the given admissible filtration. Using Lemma A repeatedly we obtain an infinite commutative ladder

$$\begin{array}{ccccccc}
\tilde{M}_1 & \xhookrightarrow{\tilde{i}_1} & \tilde{M}_2 & \xhookrightarrow{\tilde{i}_2} & \tilde{M}_3 & \hookrightarrow & \cdots\cdots \\
\downarrow \varphi_1 & & \downarrow \varphi_2 & & \downarrow \varphi_3 & & \\
M_1 & \xhookrightarrow[i_1]{} & M_2 & \xhookrightarrow[i_2]{} & M_3 & \hookrightarrow & \cdots\cdots
\end{array}$$

with WP-approximations φ_n and closed embeddings $\tilde{i}_n$. Let $\tilde{\mathfrak{M}}$ denote the upper sequence. It may be regarded as an admissible filtration of a weak polytope $\tilde{M}$. By Lemma 4.4 the map $\Phi : \mathrm{Tel}(\tilde{\mathfrak{M}}) \to \mathrm{Tel}(\mathfrak{M})$ induced by the φ_n is a homotopy equivalence. Clearly $\mathrm{Tel}(\tilde{\mathfrak{M}})$ is a weak polytope. Thus we have found a WP-approximation of $\mathrm{Tel}(\mathfrak{M})$ as desired. (Using again Lemma 4.5 it is clear that the inductive limit $\varphi : \tilde{M} \to M$ of the maps φ_n is also a homotopy equivalence, hence a WP-approximation of M.) This finishes the proof of Lemma D and of Theorem 4.1.

Looking again at Lemma A we draw from Theorem 4.1 the following consequence

Corollary 4.6. Given a map $f : M \to N$ between spaces and a WP-approximation $\varphi : P \to M$ there exists a commuting square

$$\begin{array}{ccc}
P & \xhookrightarrow{i} & Q \\
\downarrow \varphi & & \downarrow \psi \\
M & \xrightarrow[f]{} & N
\end{array}$$

with ψ a WP-approximation and i a closed embedding.

Theorem 4.7. Assume that the field R is sequential. Then, for every space M over R, the partially complete core $p_M : P(M) \to M$ (cf. IV, §9) is a homotopy equivalence.

Thus, in this case, every space has a canonical WP-approximation.

In order to prove this we start with a very general observation. If $f,g : X \rightrightarrows Y$ are maps between spaces over R, and $F : X\times I \to Y$ is a homotopy from f to g, then we have $P(X\times I) = P(X)\times I$ (cf. (IV.9.13.i), and $P(F)$ is a homotopy from $P(f)$ to $P(g)$. We conclude that, if $h : X \to Y$ is a homotopy equivalence, then $P(h) : P(X) \to P(Y)$ is again a homotopy equivalence.

Let now M be any space over the sequential field R. By Theorem 4.1 there exists a homotopy equivalence $\varphi : Q \to M$ with Q a weak polytope. Then $P(\varphi)$ is a homotopy equivalence from Q to $P(M)$ and $p_M \circ P(\varphi) = \varphi$. Since both maps φ and $P(\varphi)$ are homotopy equivalences also p_M is a homotopy equivalence.

The method in the proof of Lemma B yields also a very general homotopy result on gluing of spaces which will turn out to be useful later on (cf. §7 and proof of 4.11).

<u>Theorem 4.8.</u> Let $(\varphi,\psi) : (M,A) \to (M',A')$ be a homotopy equivalence between closed pairs of spaces. (N.B. By 2.13 it suffices to know for this that $\varphi : M \to M'$ and $\psi : A \to A'$ are homotopy equivalences.) Let

$$\begin{array}{ccc} A & \xrightarrow{\ f\ } & N \\ {\scriptstyle\psi}\downarrow & & \downarrow{\scriptstyle\chi} \\ A' & \xrightarrow[\ f'\ ]{} & N' \end{array}$$

be a commuting square with f and f' partially proper and (ψ and) χ a homotopy equivalence. Then the map

$$\Phi : M \cup_f N \to M' \cup_{f'} N'$$

induced by φ,ψ,χ is again a homotopy equivalence.

In order to prove this one replaces $M \cup_f N$ and $M' \cup_{f'} N'$ by $M \cup_A Z(f)$ and $M' \cup_{A'} Z(f')$ respectively. This is justified by Lemma 4.3. One then applies Corollary 2.14.

<u>Corollary 4.9.</u> Let (M,A) be a closed pair of spaces with A a weak polytope.

i) If $f,g : A \rightrightarrows N$ are homotopic maps into a space N, then $M \cup_f N$ is homotopy equivalent to $M \cup_g N$ under N (cf. §2, Def. 4).

ii) If $f : A \to N$ is any map and $h : N \to L$ is a homotopy equivalence then the map $u : M \cup_f N \to M \cup_{h \circ f} L$ induced by h and the identity map of M is again a homotopy equivalence.

<u>Proof.</u> The second claim is just a special case of Theorem 4.8. In the situation of i) let $F : A \times I \to N$ be a homotopy from f to g. Since A is is a weak polytope, all three maps f,g,F are partially proper. Consider the commuting diagram

$$\begin{array}{ccccc} M & \longleftarrow & A & \xrightarrow{\ f\ } & N \\ {\scriptstyle i}\downarrow & & {\scriptstyle j}\downarrow & & \downarrow{\scriptstyle id_N} \\ M\times I & \longleftarrow & A\times I & \xrightarrow[\ F\ ]{} & N \end{array}$$

with $i : M \hookrightarrow M \times I$ the injection $x \mapsto (x,0)$ and $j := i|A$. The three vertical arrows are homotopy equivalences. Thus, by Theorem 4.8 and 2.11, the space $(M\times I) \cup_F N$ is homotopy equivalent to $M \cup_f N$ under N. For the same reason $(M\times I) \cup_F N$ is homotopy equivalent to $M \cup_g N$ under N. This proves the first claim q.e.d.

We now discuss WP-approximations of decreasing systems of spaces (§2, Def. 5). These will turn out to be important in the next chapter.

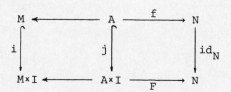

Definition 5. i) A system of weak polytopes, or WP-system for short, is a finite family $(P_o,P_1,\dots,P_r)$ of weak polytopes with P_i a subspace of P_o for $1 \le i \le r$. WP-systems form a special class of closed systems of spaces. In the cases $r = 1$, $r = 2,\dots$ we speak of pairs, triples, ... of weak polytopes, or WP-pairs, WP-triples, ..., for short.
ii) A WP-approximation of a decreasing system of spaces $(M_o,M_1,\dots,M_r)$ is a map

$$\varphi : (P_o,P_1,\dots,P_r) \to (M_o,M_1,\dots,M_r)$$

with $(P_o,\dots,P_r)$ a decreasing WP-system and every component $\varphi_i : P_i \to M_i$ of φ a homotopy equivalence.

If the system $(M_o,\dots,M_r)$ is closed then it follows from Theorem 2.13 that any WP-approximation of $(M_o,\dots,M_r)$ is a homotopy equivalence of systems.

Theorem 4.10. i) Every decreasing system $(M_o,\dots,M_r)$ of spaces has a WP-approximation $\varphi : (P_o,\dots,P_r) \to (M_o,\dots,M_r)$. If M_r is contractible then P_r can be chosen as a one-point space.
ii) If $\psi : (Q_o,\dots,Q_r) \to (M_o,\dots,M_r)$ is a second WP-approximation then there exists a map $f : (P_o,\dots,P_r) \to (Q_o,\dots,Q_r)$, unique up to homotopy, such that $\psi \circ f \simeq \varphi$. This map f is a homotopy equivalence.

The first part of the theorem follows from Theorem 4.1 and Cor. 4.6, applied several times. The second part can be proved easily by use of the homotopy extension theorem 2.9 and Theorem 2.13.

We will have to say more about WP-approximations of decreasing systems at the end of §6. Already in the next section we will need "relative WP-approximations". These are defined as follows.

Definition 6. Let (M,A) be a closed pair of spaces. A relative WP-approximation of (M,A) consists of a closed embedding $j : A \hookrightarrow P$ with $(P, j(A))$ a relative weak polytope and a commuting triangle

$$\begin{array}{ccc} & A & \\ {}^{j}\swarrow & & \searrow^{i} \\ P & \xrightarrow{\varphi} & M \end{array}$$

with i the inclusion from A to M and φ a homotopy equivalence. Notice that then, by Dold's theorem 2.11, φ is a homotopy equivalence under A.

Identifying A with j(A) we alternatively may regard a relative WP-approximation of (M,A) as a homotopy equivalence $\varphi : (P,A) \to (M,A)$ between pairs such that (P,A) is a relative weak polytope and φ is the identity on A.

Theorem 4.11. Let

$$\begin{array}{ccc} Q & \xhookrightarrow{k} & P \\ {\scriptstyle\chi}\downarrow & & \downarrow{\scriptstyle\psi} \\ A & \xhookrightarrow[i]{} & M \end{array}$$

be a commuting square with k a closed embedding and χ, ψ WP-approximations of M and A respectively. {N.B. Such a square exists for every closed pair (M,A), cf. 4.6.} Then the map

$$\varphi : (P \cup_{\chi} A, A) \to (M \cup_A A, A) = (M,A)$$

induced by the commuting diagram

$$\begin{array}{ccccc} P & \xleftarrow{k} & Q & \xrightarrow{\chi} & A \\ {\scriptstyle\psi}\downarrow & & \downarrow{\scriptstyle\chi} & & \downarrow{\scriptstyle id_A} \\ M & \xleftarrow[i]{} & A & \xrightarrow[id_A]{} & A \end{array}$$

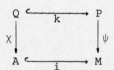

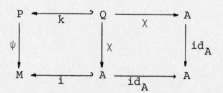

is a relative WP-approximation of (M,A).

<u>Proof.</u> It is obvious that $(P \cup_\chi A, A)$ is a relative weak polytope, and it follows from Theorem 4.8 that $\varphi : P \cup_\chi A \to M$ is a homotopy equivalence.

<u>Example 4.12.</u> Assume that the field R is sequential. For every closed pair (M,A) of spaces over R the map

$$p_{(M,A)} = (p_M^A, id) : (P^A(M), A) \to (M,A)$$

(cf. IV, §10) is a relative WP-approximation of (M,A), as is clear by the preceding theorem and Theorem 4.7. This relative WP-approximation is especially nice since it is canonical and $(P^A(M),A)$ has the same dimension as (M,A).

Also the other construction $f \longmapsto P_f(M)$ in IV, §10 gives us a homotopy equivalence which will be useful later.

<u>Theorem 4.13.</u> Assume again that the field R is sequential. Let $f : M \to N$ be any weakly semialgebraic map over R. The partially proper core $p_f : P_f(M) \to M$ of f is a homotopy equivalence.

<u>Proof.</u> Let $L := P_f(M)$. By IV.10.20 we have $P(L) = P(M)$ and $p_f \circ p_L = p_M$. We know from Theorem 4.7 that p_M and p_L are homotopy equivalences, and conclude that p_f is a homotopy equivalence. q.e.d.

§5 - The two main theorems on homotopy sets

If M is a space, x is a point of M, and A is a weakly semialgebraic subset of M containing x, then we define the absolute homotopy groups $\pi_n(M,x)$ $(n \geq 1)$ and the relative homotopy groups $\pi_n(M,A,x)$ $(n \geq 2)$ and also the pointed sets $\pi_o(M,x)$ and $\pi_1(M,A,x)$ in the same way as has been done in III, §6 for locally semialgebraic spaces. The more formal properties of these groups and sets stated there on pp. 265-270 (before Theorem III.6.3) remain in force in the present setting, together with their proofs.

If $(M_\alpha | \alpha \in I)$ is an exhaustion of M such that every M_α contains the point x then clearly

$$\pi_n(M,x) = \underset{\longrightarrow}{\lim}\, \pi_n(M_\alpha,x)$$

$$\pi_n(M,A,x) = \underset{\longrightarrow}{\lim}\, \pi_n(M_\alpha,A\cap M_\alpha,x).$$

From this fact it is evident that also the first and second main theorem for homotopy groups (III.6.3, 6.4) remain true for weakly semialgebraic spaces.

More generally, the main theorems on homotopy sets III.4.2 and III.5.1 can be generalized to weakly semialgebraic spaces. In order to prove this, and also for later use, we make explicit an easy lemma on the "composition" of infinitely many homotopies. (This lemma has already been used in the proof of III.4.2.)

Lemma 5.1. Let $(C_n | n \in \mathbb{N})$ be an admissible filtration (cf. §2, Def. 3) of a space M. Assume that $(G_n : M\times I \rightarrow N | n \in \mathbb{N})$ is a family of homotopies such that $G_n(-,1) = G_{n+1}(-,0)$ and G_n is constant on C_n. Let

$$s_o = 0 < s_1 < s_2 < \dots$$

be an infinite strictly increasing sequence in $[0,1[$. Then there exists a homotopy $F : M\times I \to N$ such that

$$F(x,t) = G_{k+1}(x,(s_{k+1}-s_k)^{-1}(t-s_k))$$

if $(x,t) \in C_n\times[s_k,s_{k+1}]$, $0 \leq k \leq n-2$, and

$$F(x,t) = G_n(x,0)$$

if $(x,t) \in C_n\times[s_{n-1},1]$.

Proof. Using these formulas we obtain a well defined homotopy $F_n : C_n\times I \to N$ from $G_1(-,0)|C_n$ to $G_n(-,0)|C_n$. We have $F_n|C_{n-1}\times I = F_{n-1}$ since G_{n-1} is constant on C_{n-1}. Thus the F_n fit together to the desired homotopy F.

Definition 1. We call F the composite of the family of homotopies $(G_n|n\in\mathbb{N})$, along the sequence $(s_n|n\in\mathbb{N}_0)$.

Let $(M,A_1,\ldots,A_r)$ and $(N,B_1,\ldots,B_r)$ be two systems of spaces over R with every A_i closed in M. We fix a map $h : C \to N$ on a closed subspace C of M with $h(C\cap A_i) \subset B_i$ for $1 \leq i \leq r$. We use the notations about relative homotopy sets established in III, §4 and III, §5.

Theorem 5.2. i) (First main theorem). Let S be a real closed field containing R. Then the natural map

$$\kappa : [(M,A_1,\ldots,A_r),(N,B_1,\ldots,B_r)]^h \to [(M,A_1,\ldots,A_r),(N,B_1,\ldots,B_r)]^h(S)$$

is bijective.

ii) (Second main theorem). Assume that $R = \mathbb{R}$. Then the natural map

$$\lambda : [(M,A_1,\ldots,A_r),(N,B_1,\ldots,B_r)]^h \to [(M,A_1,\ldots,A_r),(N,B_1,\ldots,B_r)]^h_{top}$$

is bijective.

The two claims can be proved in exactly the same way starting from Theorem III.4.2 and III.5.1 respectively. We give the proof of i).

As in III, §4 we see that it suffices to prove the surjectivity of κ, and as there we retreat to the case $r=0$ using the homotopy extension theorem 2.9. We are given a map $f : M(S) \to N(S)$ extending $h_S : C(S) \to N(S)$, and we look for a map $g : M \to N$ extending h with $g_S \simeq f$ rel. $C(S)$.

We choose a relative patch decomposition of (M,C) and we use the standard notations 2.2 from §2 (the letter A there being replaced by C, of course). We want to construct maps $h_n : M_n \to N$, $f_n : M(S) \to N(S)$ for every $n \geq -1$, and a homotopy $H_n : M(S) \times I(S) \to N(S)$ relative $M_{n-1}(S)$ for $n \geq 0$, such that the following holds: $h_{-1} = h$, $h_n | M_{n-1} = h_{n-1}$, $f_{-1} = f$, $f_n | M_n(S) = (h_n)_S$, $H_n(-,0) = f_{n-1}$, $H_n(-,1) = f_n$. Once we have accomplished this we are done. We have a map $g : M \to N$ with $g | M_n = h_n$ for every n. Composing the family $(H_n | n \geq 0)$ of homotopies along some strictly increasing sequence $(s_n | n \geq -1)$ in $[0,1[_R$ with $s_{-1} = 0$ (cf. Lemma 5.1, the indices are shifted by -1) we obtain a homotopy $G : M(S) \times I(S) \to N(S)$ relative $C(S)$ from f to g_S, as desired.

In order to construct h_n, f_n, H_n we proceed by induction on n. We start with $h_{-1} = h$ and $f_{-1} = f$, of course. Assume that the h_i, f_i, H_i are given for $i < n$. We use the pushout diagram over R (cf. 2.2)

$$(*) \qquad \begin{array}{ccc} \partial M(n) & \xrightarrow{\varphi_n} & M_{n-1} \\ {\scriptstyle i}\downarrow & & \downarrow{\scriptstyle j} \\ M(n) & \xrightarrow[\psi_n]{} & M_n \end{array}$$

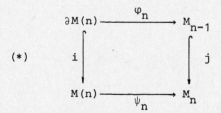

By base field extension, we obtain from (*) a pushout diagram $(*)_S$ over S. We introduce the maps

$$k_n := h_{n-1} \circ \varphi_n : \partial M(n) \to N ,$$

$$u_n := (f_{n-1} | M_n(S)) \circ (\psi_n)_S : M(n)(S) \to N(S) .$$

Notice that u_n extends $(k_n)_S$. The space M(n) is a direct sum of semi-algebraic spaces. By Theorem III.4.2 there exists a map $v_n : M(n) \to N$ extending k_n together with a homotopy

$$F_n : M(n)(S) \times I(S) \longrightarrow N(S)$$

relative $\partial M(n)(S)$ from u_n to $(v_n)_S$. By the pushout property of the diagram (*) the maps v_n and h_{n-1} combine to a map $h_n : M_n \to N$ with $h_n \circ \psi_n = v_n$ and $h_n | M_{n-1} = h_{n-1}$. "Multiplying" the diagram $(*)_S$ with the unit interval I(S) we obtain again a pushout diagram (cf. IV.8.7.ii). By this diagram the map F_n and the map $(x,t) \mapsto (h_{n-1})_S(x)$ from $M_{n-1}(S) \times I(S)$ to N(S) combine to a homotopy $\tilde{H}_n : M_n(S) \times I(S) \to N(S)$ relative $M_{n-1}(S)$ with $\tilde{H}_n(-,0) = f_{n-1} | M_n(S)$ and $\tilde{H}_n(-,1) = (h_n)_S$. We extend $\tilde{H}_n$ to a homotopy $H_n : M(S) \times I(S) \to N(S)$ with $H_n(-,0) = f_{n-1}$. We put $f_n := H_n(-,1)$. This finishes the induction step and the whole proof of Theorem 5.2.i.

We are interested in Theorem 5.2 mainly in the "absolute case" $C = \emptyset$. But for the proof (as already in III, §4-§5) it is necessary to work with relative homotopies.

Remark 5.3. Theorem 5.2 remains true for locally finite systems $(M,(A_\lambda | \lambda \in \Lambda)),(N,(B_\lambda | \lambda \in \Lambda))$ of spaces instead of finite ones, with all A_λ closed in M, of course. This can be proved in a similar way as above with more notational effort.

It will be more difficult here than in Chapter III to apply the transfer principles given by the main theorems on homotopy sets, since we do not know whether a given system of spaces $(M,A_1,\dots,A_r)$ over R is isomorphic to $(N(R),B_1(R),\dots,B_r(R))$ for some system of spaces $(N,B_1,\dots,B_r)$ over the field R_o of real algebraic numbers.*) This will make the homotopy theory in WSA(R) more laborious than in LSA(R).

*) Recall that R_o embeds in a unique way into any other real closed field and thus is the "prime field" in real algebraic geometry.

§6 - Compressions and n-equivalences

In this section we present some theorems which can be proved essentially by well known classical "compression arguments" mostly due to J.H.C. Whitehead. For some steps, which are important to attain the right level of generality (steps b)-d) in the proof of Theorem 6.8 below), we will need the spaces $P^A(M)$ and $\dot{P}_f(M)$, introduced in IV, §10, and the first main theorem on homotopy sets 5.2.i.

Definition 1. Let $f : (M,A) \to (N,B)$ be a map between pairs of spaces. Let further L be a weakly semialgebraic subset of N containing B. We say that f can be compressed to L, if f is homotopic relative A to a map $g : (M,A) \to (N,B)$ with $g(M) \subset L$. We then call a homotopy $F : (M\times I, A\times I) \to (N,B)$ from f to g a compression of f to L. We say that f is compressible if f can be compressed to B.

Definition 2. Let $n \in \mathbb{N}_0 \cup \infty$, i.e. either $n = 0$ or n is a natural number or $n = \infty$. We call a pair of spaces (M,A) n-connected if every map $f : (E^k, S^{k-1}) \to (M,A)$ with $k \in \mathbb{N}_0$ and $k \leq n$ is compressible. Here, as usual, E^k denotes the closed unit ball in R^k and S^{k-1} its boundary.

Proposition 6.1. Let (M,A) be a special relative patch complex (cf. §2, Def. 1). Assume that the pair of spaces (M,A) is a relative weak polytope and that, for a fixed $m \in \mathbb{N}$, every patch $\sigma \in \Sigma(M,A)$ has dimension at least m. Then (M,A) is $(m-1)$-connected.

This can be proved by a very basic "cell by cell" argument, which here is even simpler than in the topological setting, say, for simplicial complexes. We are given a map $f : (E^k, S^{k-1}) \to (M,A)$ with $k < m$. We have to compress f to A. There exists a finite closed subcomplex (M',A) of (M,A) with $f(E^k) \subset M'$. Replacing M by M' we may assume that the relative complex (M,A) is finite. If σ is a patch of dimension d, then $(\bar{\sigma},\partial\sigma) \cong (E^d,S^{d-1})$, since M is partially complete relative A. We choose a maximal patch in $M \setminus A$, i.e. a patch σ which is not a face of another patch. Then $(M \setminus \sigma, A)$ is a closed subcomplex of (M,A). Since $\dim f(E^k) < \dim \sigma$ there exists a point $p \in \sigma$ which is not contained in the image of f. The set $\partial\sigma$ is a strong deformation retract of $\bar{\sigma} \setminus \{p\}$. Thus $M \setminus \sigma$ is a strong deformation retract of $M \setminus \{p\}$, and f can be compressed to $M \setminus \sigma$. The proposition now follows by induction on the number of patches in $M \setminus A$.

Quite generally, n-connectedness can be expressed in terms of relative homotopy groups $\pi_k(M,A,x)$ (resp. sets if k = 1).

Proposition 6.2. A pair of spaces (M,A) is n-connected iff the following two properties hold.

a) The natural map $\pi_o(A) \to \pi_o(M)$ is surjective.

b) $\pi_k(M,A,x) = 0$ for every $x \in A$ and every natural number $k \leq n$.

Of course, it suffices to demand (b) for one point x in each connected component of A since the fundamental groupoid $\Pi(A)$ operates on the system $(\pi_k(M,A,x) \mid x \in A)$.

Proposition 6.2 is an immediate consequence of the following more general lemma.

Lemma 6.3 (Compression lemma, cf. [W, p. 70f]). Let $f : (M,A) \to (N,B)$ be a map from a closed pair of spaces (M,A) to a pair of spaces (N,B). Assume that f is homotopic to a map $g : (M,A) \to (N,B)$ with $g(M) \subset L$ for some $L \in \mathcal{T}(N)$ containing B. Then f can be compressed to L.

This can be proved, as in the topological theory, by applying the homotopy extension theorem 2.9 to the pairs (M,A) and $(M\times I, (A\times I) \cup (M\times \partial I))$, cf. [W, p. 71].

Having finished the preliminaries on compressibility and n-connectedness we head for our main result on compression of maps, which will be exploited afterwards.

Definition 3. The dimension $\dim X$ of a space X is the supremum of the dimensions $\dim X_\alpha$ of the semialgebraic spaces X_α for some exhaustion $(X_\alpha | \alpha \in J)$ of X. We have $\dim X \in \mathbb{N}_0 \cup \infty$. Of course, $\dim X$ is independent of the chosen exhaustion of X. If (M,A) is a pair of spaces then we set $\dim(M,A) := \dim(M \setminus A)$.

Notice that, if $(Y_\lambda | \lambda \in \Lambda)$ is any admissible covering of X by weakly semialgebraic sets (cf. IV, §3, Def. 7), then $\dim X = \sup(\dim Y_\lambda | \lambda \in \Lambda)$.

Theorem 6.4 (Compression theorem). Let $d \in \mathbb{N}_0 \cup \infty$. Assume that (M,A) is a relative weak polytope of dimension d, and that (N,B) is a d-connected pair of spaces. Then every map $f : (M,A) \to (N,B)$ is compressible.

Example 6.5 ($f = \mathrm{id}_{(M,A)}$). If (M,A) is a d-dimensional relative weak polytope which is also d-connected, then A is a strong deformation retract of M.

We start out to prove Theorem 6.4. We choose a special relative patch decomposition of the pair (M,A), and we use the standard notations 2.2. We want to construct inductively a family of homotopies $(H_n : M\times I \to N \mid n \in \mathbb{N}_0)$ with the following properties: i) $H_n(-,0) = H_{n-1}(-,1)$.
ii) $H_n(x,t) = H_{n-1}(x,1)$ for $x \in M_{n-1}$, $t \in I$.
iii) $H_n(M_n \times 1) \subset B$.
Here (for $n = 0$) $H_{-1} : M\times I \to N$ means the constant homotopy $H_{-1}(x,t) =$

$f(x)$. Once this is done we can apply Lemma 5.1 with $C_n = M_{n-1}$, $G_n = H_{n-1}$, and thus can compose the H_n to a homotopy $F : M\times I \to N$ relative A which obviously compresses f to B. The sequence $(s_n \mid n \in \mathbb{N}_0)$ can be chosen here in any way.

Suppose that for some $n \geq 0$ the homotopy H_{n-1} is already given (H_{-1} as above if $n = 0$). We want to construct H_n. We consider the map $u_n : (M_n, M_{n-1}) \to (N,B)$ defined by $u_n(x) = H_{n-1}(x,1)$. Composing u_n with the standard map $\psi_n : (M(n), \partial M(n)) \to (M_n, M_{n-1})$ we obtain a map $g_n : (M(n), \partial M(n)) \to (N,B)$. Now $(M(n), \partial M(n))$ is just a direct sum of pairs which are isomorphic to pairs (E^k, S^{k-1}) with - perhaps varying - natural numbers $k \leq d$. Since (N,B) is d-connected there exists a compression

$$G_n : (M(n)\times I, \partial M(n)\times I) \to (N,B)$$

of g_n to B. The diagram

$$\begin{array}{ccc} \partial M(n)\times I & \xrightarrow{\varphi_n \times id_I} & M_{n-1}\times I \\ \Big\downarrow{\scriptstyle i\times id_I} & & \Big\downarrow{\scriptstyle j\times id_I} \\ M(n)\times I & \xrightarrow[\psi_n \times id_I]{} & M_n \times I \end{array}$$

obtained from the cocartesian square (*) in (2.2) by multiplying with I is again cocartesian (cf. IV.8.7.ii). Thus there exists a unique homotopy $U_n : M_n \times I \to N$ with $U_n \circ (\psi_n \times id) = G_n$ and $U_n(x,t) = u_n(x)$ for $(x,t) \in M_{n-1}\times I$. Using the homotopy extension theorem we extend U_n to a homotopy $H_n : M\times I \to N$ with $H_n(-,0) = H_{n-1}(-,1)$. Clearly $H_n(M_n \times 1) = U_n(M_n\times 1) \subset B$. This finishes the proof of Theorem 6.4.

<u>Definition 5.</u> Let $d \in \mathbb{N}_0 \cup \infty$. A map $f : M \to N$ between spaces is called a <u>d-equivalence</u> if, for every $x \in M$, the induced homomorphism

$f_* : \pi_q(M,x) \to \pi_q(N,f(x))$ is bijective for $0 \leq q < d$ and, in case $d \neq \infty$, surjective for $q = d$. If $d = \infty$ then we also say that f is a <u>weak homotopy equivalence</u> if this holds.

<u>Example 6.6.</u> If (M,A) is a pair of spaces then the inclusion map $A \hookrightarrow M$ is a d-equivalence iff (M,A) is d-connected. This follows from the long exact homotopy sequence III.6.1.

<u>Remark 6.7.</u> A partially proper map $f : M \to N$ is a d-equivalence iff the closed pair (Z(f),M), with Z(f) the mapping cylinder of f (cf. §4) is d-connected. This follows from the factorization (4.2) of f via Z(f) since the projection $p : Z(f) \to N$ is a homotopy equivalence.

Starting from this remark we are able to transform the compression theorem 6.4 into a very general theorem about lifting of maps up to homotopy.

<u>Theorem 6.8</u> (cf. e.g. [Spa, p. 404]). Let $f : M \to N$ be a d-equivalence for some $d \in \mathbb{N}_0 \cup \infty$. Let (L,C) be a closed pair of spaces of dimension at most d and let h be a map from C to M. Then the map

$$f_* : [L,M]^h \to [L,N]^{f\circ h}$$

between relative homotopy sets (cf. notations in III, §4) induced by f is surjective.

<u>Remark.</u> More explicitly this means the following. Let (L,C) be a closed pair of spaces of dimension at most d. Assume there is a given commuting square of maps (solid arrows)

(*) 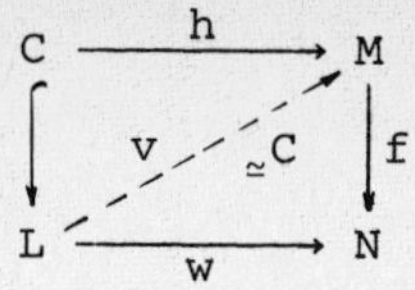

with f a d-equivalence and the left vertical arrow the inclusion of C into L. Then there exists a map $v : L \to M$ (dotted arrow) such that $v|C = h$ and $f \circ v$ is homotopic to w relative C.

Proof. a) We first deal with the case that (L,C) is a relative weak polytope and f is partially proper. In this case Z(f) exists and the classical argument [Spa, p. 404f] remains valid. We reproduce this argument for the convenience of the reader.

Consider the square (solid arrows)

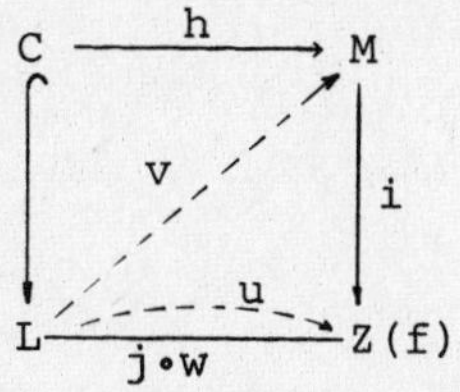

with $j : N \hookrightarrow Z(f)$ the inclusion of N into Z(f). This square does not commute, but there exists a homotopy $H : L \times I \to Z(f)$ from $j \circ w$ to a map $u : L \to Z(f)$ which extends the obvious homotopy from $j \circ w|C$ to $i \circ h$. The square commutes after replacing $j \circ w$ by u. According to Remark 6.7 and Theorem 6.4 we can compress $u : (L,C) \to (Z(f),M)$ to M. Thus we have a map $v : L \to M$ extending h with $i \circ v \simeq u$ rel. C. Composing with the canonical projection $p : Z(f) \to N$ we obtain $f \circ v \simeq p \circ u$ rel. C. By the special nature of the homotopy $H|C \times I$ we have $w = p \circ j \circ w \simeq p \circ u$ rel. C. Thus $f \circ v \simeq w$ rel. C.

b) We now prove the theorem under the additional assumptions that the field R is sequential and L is a weak polytope, which implies that also C is a weak polytope. But we do not assume that f is partially proper. We will use the partially proper core $p_f : P_f(M) \to M$ introduced in IV, §10. Starting from the given commuting square (*) of solid arrows we obtain a commuting diagram of solid arrows with $p_f \circ \tilde{h} = h$,

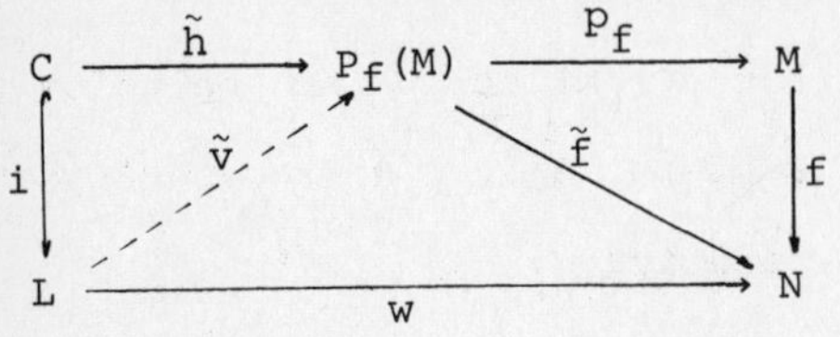

since the map $w \circ i$ from C to N is certainly partially proper. (Of course, i denotes the inclusion from C to L). The map $\tilde{f}$ is partially proper and p_f is a homotopy equivalence (Th. 4.13). Thus $\tilde{f}$ is again a d-equivalence. By step a) of the proof there exists a map $\tilde{v}$ from L to $P_f(M)$ (dotted arrow) with $\tilde{v} \circ i = \tilde{h}$ and $\tilde{f} \circ \tilde{v} \simeq w$ rel. C. The map $v := p_f \circ \tilde{v}$ from L to M has the required properties.

c) We now prove the theorem under the weaker assumption that R is se-sequential and (L,C) is a relative weak polytope. Starting again from the square (*) of solid arrows we look at the commuting diagram of solid arrows

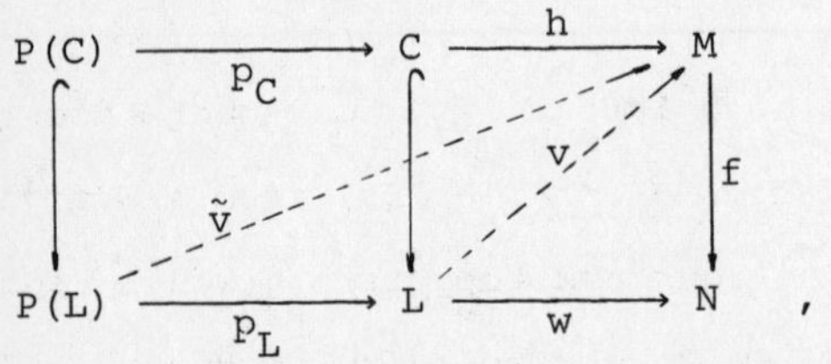

with p_L and p_C the partially complete cores of L and C (cf. IV, §9). Since (L,C) is a relative weak polytope the left hand square is cocartesian (Cor. IV.10.4). The pairs (P(L),P(C)) and (L,C) have the same relative dimension $\le d$. By step b) of the proof there exists a map $\tilde{v}$ from P(L) to M (dotted arrow above) extending $h \circ p_C$ such that $f \circ \tilde{v} \simeq w \circ p_L$ relative P(C). We obtain from $\tilde{v}$ a map $v : L \to M$ extending h such that $f \circ v \circ p_L = f \circ \tilde{v}$. We know that $f \circ v \circ p_L \simeq w \circ p_L$ rel. P(C) and conclude from this that $f \circ v \simeq w$ rel. C, since also the square

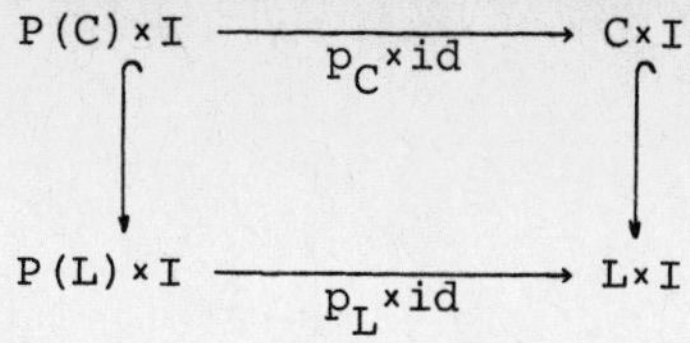

is cocartesian (IV.8.7.ii).

d) We still assume that R is sequential, but now (L,C) may be any closed pair of dimension at most d. We now use the relative WP-approximation $p_L^C : P^C(L) \to L$ described in 4.12. We denote this map by $\varphi : P \to L$. Thus (P,C) is a relative weak polytope of dimension at most d and φ is a homotopy equivalence under C. Let $\psi : L \to P$ denote a homotopy inverse of φ under C. We look at the commuting diagram (solid arrows)

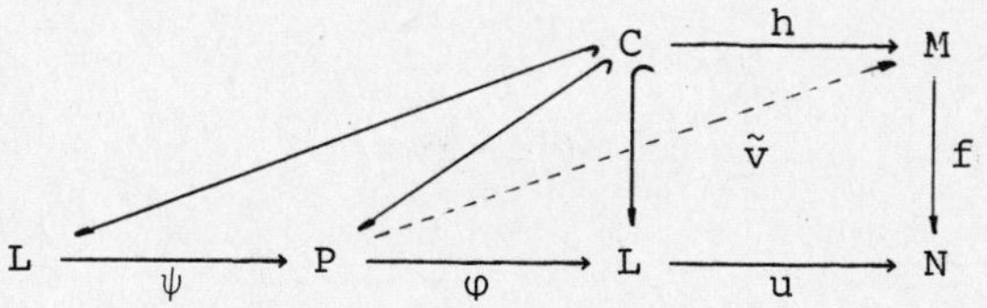

By the preceding step c) of the proof there exists a map $\tilde{v} : P \to M$ under C such that $f\circ\tilde{v} \simeq^C u\circ\varphi$. Then $v := \tilde{v}\circ\psi : L \to M$ is again a map under C and

$$f \circ v \simeq^C u\circ\varphi\circ\psi \simeq^C u \ ,$$

as desired.

e) We finally prove the theorem in the case that R is not sequential. Then we choose a real closed overfield S of R which is sequential. For example we can take for S the real closure of the rational function field R(t) over R (one indeterminate t) with respect to the ordering of R(t) in which t is positive and infinitely small over R. Alternatively we can take for S the real closure of R((t)) with respect to the unique ordering in which t is positive. In both cases $(t^n \mid n\in\mathbb{N})$ is a null sequence in S. We have a commuting square

$$\begin{array}{ccc} [L,M]^h & \xrightarrow{f_*} & [L,N]^{f\circ h} \\ \kappa \downarrow \cong & & \cong \downarrow \kappa \\ [L,M]^h(S) & \xrightarrow[(f_S)_*]{} & [L,N]^{f\circ h}(S) \end{array}$$

where the vertical arrows are the natural maps "base extension from R to S". They are bijective by Theorem 5.2.i. The lower horizontal map $(f_S)_*$ is bijective by step d) of the proof. Thus f_* is bijective. q.e.d.

Applying Theorem 6.8 to the product $(L,C) \times (I,\partial I) = (L\times I,(L\times\partial I) \cup (C\times I))$ instead of (L,C) we obtain, in the usual way,

<u>Corollary 6.9.</u> In the situation of Theorem 6.8 assume that even $\dim(L,C) \leq d-1$. Then the map f_* above is bijective.

The most interesting case here is that $C = \emptyset$ and $d = \infty$. Then $f : M \to N$ yields a bijection $f_* : [L,M] \to [L,N]$ for every space L over R. This means that f is a homotopy equivalence. Thus we have proved

<u>Theorem 6.10</u> ("Whitehead's theorem"). Every weak homotopy equivalence is a homotopy equivalence.

<u>Corollary 6.11.</u> Let $(M_n | n\in\mathbb{N})$ and $(N_n | n\in\mathbb{N})$ be admissible filtrations of spaces M and N. Let $f : M \to N$ be a map with $f(M_n) \subset N_n$ for every $n \in \mathbb{N}$. Assume that the restrictions $f_n : M_n \to N_n$ of f all are homotopy equivalences. Then f is a homotopy equivalence.

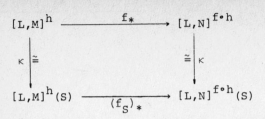

Indeed, f clearly is a weak homotopy equivalence, hence by the theorem a genuine homotopy equivalence. {The result 6.11 could also have been

gained by the telescopic methods of §4.}

We now look for a generalization of Theorem 6.8 to decreasing systems of spaces (cf. §2, Def. 5).

Definition 7. Let $d \in \mathbb{N}_o \cup \infty$. We call a map $f : (M_o,\dots,M_r) \to (N_o,\dots,N_r)$ between decreasing systems of spaces a d-equivalence (or weak homotopy equivalence if $d = \infty$) if all the components $f_i : M_i \to N_i$ of f are d-equivalences.

Remarks 6.12. i) If $r = 2$ and M_r, N_r are one point spaces $\{x\}$, $\{y\}$ this implies that the induced homomorphism $f_* : \pi_q(M_o,M_1,x) \to \pi_q(N_o,N_1,y)$ is bijective if $1 \leq q < d$ and surjective if $d \in \mathbb{N}$ and $q = d$.

ii) By Theorems 6.10 and 2.13 every weak homotopy equivalence between closed decreasing systems of spaces is a homotopy equivalence.

iii) In the present terminology a WP-approximation of a decreasing system of spaces $(M_o,\dots,M_r)$ (cf. §4, Def. 5) is just a weak homotopy equivalence $\varphi : (P_o,\dots,P_r) \to (M_o,\dots,M_r)$ with $(P_o,\dots,P_r)$ a decreasing system of weak polytopes. If the system $(M_o,\dots,M_r)$ is closed then φ is a homotopy equivalence. We conclude from Theorem 4.10.i that every closed decreasing system of spaces is homotopy equivalent to a decreasing WP-system.

Theorem 6.13. Let $f : (M_o,\dots,M_r) \to (N_o,\dots,N_r)$ be a d-equivalence between decreasing systems of spaces ($d \in \mathbb{N}_o \cup \infty$). Let further $(P_o,\dots,P_r)$ be a closed decreasing system. Let finally C be a closed subspace of P_o and h be a map from $(C, C \cap P_1,\dots,C \cap P_r)$ to $(M_o,\dots,M_r)$.

i) If $\dim P_o \leq d$ then the induced map

$$f_* : [(P_o,\dots,P_r),(M_o,\dots,M_r)]^h \to [(P_o,\dots,P_r),(N_o,\dots,N_r)]^{f\circ h}$$

between relative homotopy sets is surjective.

ii) If $\dim P_o \leq d-1$ then f_* is bijective.

Proof. Once we have proved the first claim i) we obtain the second claim ii) by applying i) to $(P_o\times I,\dots,P_r\times I)$ and $(C\times I)\cup(P_o\times\partial I)$ instead of $(P_o,\dots,P_r)$ and C.

We prove the first claim by inductions on r.
The case $r = 0$ is covered by Theorem 6.8. Let $r \geq 1$, and let $u : (P_o,\dots,P_r) \to (N_o,\dots,N_r)$ be a map extending $f\circ h$. In order to prove the first claim we have to find a map $v : (P_o,\dots,P_r) \to (M_o,\dots,M_r)$ extending h such that $f\circ v \simeq u$ rel. C. We denote the components of u,v,f,h by u_i,v_i,f_i,h_i respectively ($0 \leq i \leq r$). We first consider the commuting square of solid arrows

$$\begin{array}{ccc}
(C\cap P_1,\dots,C\cap P_r) & \xrightarrow{\;h'\;} & (M_1,\dots,M_r) \\
\big\downarrow & \overset{v'}{\nearrow} & \big\downarrow f' \\
(P_1,\dots,P_r) & \xrightarrow[\;u'\;]{} & (N_1,\dots,N_r)
\end{array}$$

with $h' = (h_1,\dots,h_r)$, $f' = (f_1,\dots,f_r)$, $u' = (u_1,\dots,u_r)$, and the left vertical arrow an inclusion map. By induction hypothesis there exists a map (dotted arrow)

$$v' = (v_1,\dots,v_r) : (P_1,\dots,P_r) \to (M_1,\dots,M_r)$$

extending h' and a homotopy

$$F' = (F_1,\dots,F_r) : (P_1\times I,\dots,P_r\times I) \to (N_1,\dots,N_r)$$

with $F'(-,0) = u'$, $F'(-,1) = f'\circ v'$, and $F_1(c,t) = f\circ h_1(c)$ for every $(c,t) \in (C\cap P_1)\times I$. We extend F_1 to a homotopy $F_o : P_o\times I \to N_o$ with $F_o(-,0) = u_o$ and $F_o(c,t) = f\circ h_o(c)$ for every $(c,t) \in C\times I$ (cf. 2.9). This means we have a homotopy

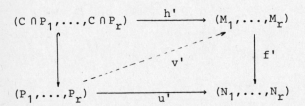

$F = (F_o,\ldots,F_r) : (P_o\times I,\ldots,P_r\times I) \to (N_o,\ldots,N_r)$ with $F(-,0) = u$ and F constant on C. Let $w := (w_o,\ldots,w_r) := F(-,1)$ and $w' := (w_1,\ldots,w_r)$. Then $w' = f'\circ v'$. Now consider the commuting square of solid arrows

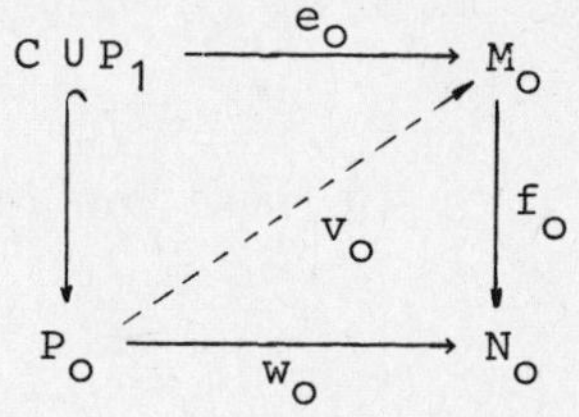

with e_o obtained by gluing h_o and v_1. By Theorem 6.8 (or by induction hypothesis) there exists a map $v_o : P_o \to M_o$ (dotted arrow) such that $v_o|C \cup P_1 = e_o$ and $f_o\circ v_o \simeq w_o$ rel. $C \cup P_1$. Let $v := (v_o,v') : (P_o,\ldots,P_r) \to (M_o,\ldots,M_r)$. Then $f\circ v \simeq w$ rel. $C \cup P_1$. On the other hand $w \simeq u$ rel. C. Thus $f\circ v \simeq u$ rel. C, as desired.

q.e.d.

<u>Remark.</u> As the proof shows it suffices for the first claim i) to assume that $\dim(P_r, C\cap P_r) \le d$ and $\dim(P_i,(C\cap P_i)\cup P_{i+1}) \le d$ for $0 \le i \le r-1$.

The special case $d = \infty$, C empty, of the theorem implies a characterization of WP-approximations by a universal property.

<u>Corollary 6.14.</u> Let $\varphi : (P_o,\ldots,P_r) \to (M_o,\ldots,M_r)$ be a WP-approximation of a decreasing system of spaces $(M_o,\ldots,M_r)$. Let f be any map from a decreasing WP-system $(Q_o,\ldots,Q_r)$ to $(M_o,\ldots,M_r)$. Then there exists a map g from $(Q_o,\ldots,Q_r)$ to $(P_o,\ldots,P_r)$, unique up to homotopy, such that $\varphi\circ g \simeq f$.

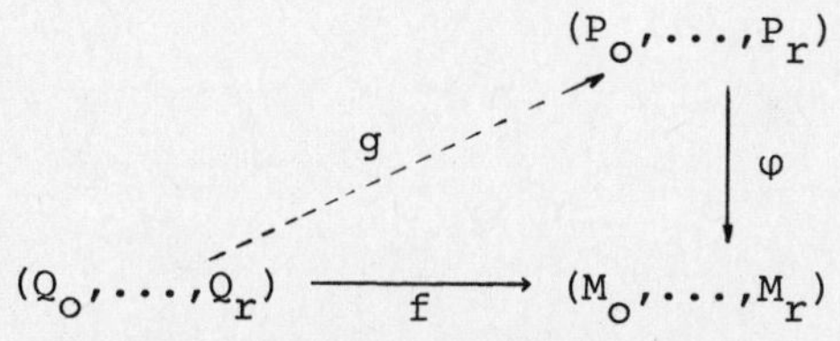

This universal property implies functoriality of WP-approximations.

Corollary 6.15. Let $\varphi : (P_o,\dots,P_r) \to (M_o,\dots,M_r)$ and $\psi : (Q_o,\dots,Q_r) \to (N_o,\dots,N_r)$ be WP-approximations of decreasing systems of spaces. For every map f from $(M_o,\dots,M_r)$ to $(N_o,\dots,N_r)$ there exists a map g from $(P_o,\dots,P_r)$ to $(Q_o,\dots,Q_r)$, unique up to homotopy, such that $\psi\circ g \simeq f\circ\varphi$.

$$\begin{array}{ccc}
(P_o,\dots,P_r) & \xrightarrow{\ \varphi\ } & (M_o,\dots,M_r) \\
{\scriptstyle g}\,\big\downarrow & & \big\downarrow\,{\scriptstyle f} \\
(Q_o,\dots,Q_r) & \xrightarrow[\ \psi\ ]{} & (N_o,\dots,N_r)
\end{array}$$

If f is a weak homotopy equivalence then g is a homotopy equivalence.

The last statement here is already clear by Theorem 4.10.ii.

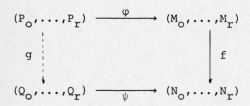

§7 - CW-complexes

Definition 1. a) A relative CW-complex (M,A) of R is a relative patch complex (M,A) with the following two additional properties.

CW1. The patch decomposition of $M \setminus A$ is dimensional (i.e. if $\tau \prec \sigma$ then $\dim \tau < \dim \sigma$, cf. §1).

CW2. For every patch $\sigma \in \Sigma(M,A)$ there exists a map $\chi_\sigma : E^n \to \overline{\sigma}$ from the closed n-ball ($n = \dim \sigma$) onto $\overline{\sigma}$ which maps $\mathring{E}^n$ isomorphically onto σ and S^{n-1} onto $\partial\sigma$.

b) If (M,A) is a relative CW-complex and A is empty then we call M a CW-complex (or absolute CW-complex) over R.

Notice that, as a consequence of CW2, the underlying pair of spaces of a relative CW-complex is a relative weak polytope.

If we are working over $R = \mathbb{R}$ we will call, if necessary, the (absolute or relative) CW-complexes defined here more precisely semialgebraic CW-complexes, in contrast to the classical CW-complexes which we will call topological CW-complexes.

Examples 7.1. i) If (M,A) is a dimensional special relative patch complex (§2, Def. 1) and if (M,A) is also a relative weak polytope then (M,A) is a relative CW-complex. These are precisely those relative CW-complexes where the maps χ_σ in CW2 can be chosen as isomorphisms. (They should be called regular CW-complexes, as in the topological theory [LW, p. 78].)

ii) Every closed simplicial complex (cf. IV.1.7) is a CW-complex.

iii) If L is a closed subcomplex of a relative CW-complex (M,A) (cf. §2, Def. 1.c) then (L,A) is again a relative CW-complex.

iv) The direct product (§2, Def. 1.d) of two relative CW-complexes is again a relative CW-complex. In particular, if (M,A) is a relative

CW-complex, then also

$$(M,A)\times(I,\partial I) = (M\times I,(A\times I)\cup(M\times\partial I))$$

is a relative CW-complex. The patches of $(M,A)\times(I,\partial I)$ are the sets $\sigma\times]0,1[$ with σ running through $\Sigma(M,A)$.

v) If (M,A) is a relative CW-complex, then any map $f : A \to N$ to a space N is partially proper near $M\setminus A$, and the pair $(M\cup_f N,N)$ is again a relative CW-complex with the same patches as (M,A). {As usual we identify $M\setminus A$ with the open subspace $(M\cup_f N)\setminus N$ of $M\cup_f N$.} In particular (N = one point space), M/A is an absolute CW-complex. The cells of M/A are those of (M,A) and one further zero-dimensional cell A/A.

We extend the terminology which is common for topological CW-complexes to the semialgebraic setting in the obvious way. Thus, if (M,A) is a relative CW-complex over R, we call the $\sigma\in\Sigma(M,A)$ the (open) <u>cells</u> of (M,A). For a given cell σ we call a map χ_σ as above a <u>characteristic map</u> for σ. Some authors incorporate a fixed choice of the characteristic maps into the structure of a CW-complex. We do not follow this convention. Our CW-complexes are just patch complexes of a special type.

The <u>n-skeleton</u> M^n of (M,A) is defined as the union of A and all cells of dimension $\le n$. We put $M^{-1} = A$. We should denote the n-skeleton of (M,A) more precisely by $(M,A)^n$, but most often this is not necessary. Notice that M^n is a subcomplex of (M,A) and that $(M^n \mid n\ge 1)$ is an admissible filtration (§2, Def. 3) of the space M.

For CW-complexes it is more natural to work with the skeletons than with the chunks. Thus we need a modification of the notations 2.2.

<u>Notations 7.2.</u> Let (M,A) be a relative CW-complex. $\Sigma_n(M,A)$, or more briefly Σ_n, denotes the set of cells of dimension n of (M,A). We choose

for every $\sigma \in \Sigma_n$ a characteristic map $\chi_\sigma : E^n \to \overline{\sigma}$. The composite of χ_σ with the inclusion $\overline{\sigma} \hookrightarrow M^n$ is denoted by ψ_σ. Abusively we also call $\psi_\sigma : E^n \to M^n$ a characteristic map for σ. Its restriction $\varphi_\sigma : S^{n-1} \to M^{n-1}$ is called an <u>attaching map</u> of σ. We have a commuting square of partially proper maps

$$(*) \qquad \begin{array}{ccc} \Sigma_n \times S^{n-1} & \xrightarrow{\varphi_n} & M^{n-1} \\ i \downarrow & & \downarrow j \\ \Sigma_n \times E^n & \xrightarrow[\psi_n]{} & M^n \end{array}$$

defined as follows. The set Σ_n is regarded as a discrete space. Thus $\Sigma_n \times E^n$ is the direct sum of copies of E^n indexed by the set Σ_n, and $\Sigma_n \times S^{n-1}$ is the closed subspace of $\Sigma_n \times E^n$ consisting of the boundaries of these copies of E^n. The map ψ_n, called the <u>big characteristic map in dimension</u> n, is defined by $\psi_n(\sigma,x) := \psi_\sigma(x)$. The map φ_n, called the <u>big attaching map in dimension</u> n, is obtained from ψ_n by restriction. i and j are inclusions. The maps ψ_n and j combine into a partially proper strongly surjective map

$$p_n : (\Sigma_n \times E^n) \sqcup M^{n-1} \to M^n.$$

Thus the square (*) is cocartesian, which means

$$M^n = (\Sigma_n \times E^n) \cup_{\varphi_n} M^{n-1} \quad .$$

We spell out the standard facts about "cellular approximation" of maps. There is no difference to the topological theory here. In the following (M,A) and (N,B) are relative CW-complexes over R.

<u>Proposition 7.3.</u> Assume that, for a fixed $m \in \mathbb{N}$, every cell of (M,A) has dimension at least m. Then (M,A) is (m-1)-connected.

The proof is very similar to the proof of Proposition 6.1.

<u>Definition 2.</u> a) A map f from the pair of spaces (M,A) to the pair (N,B) is called <u>cellular</u>, if $f(M^n) \subset N^n$ for every $n \geq 0$.

b) Consequently, a homotopy $H : (M\times I, A\times I) \to (N,B)$ is called cellular if H maps the n-skeleton $(M^n \times \partial I) \cup (M^{n-1} \times I)$ of $(M\times I, A\times I)$ into N^n for every $n \geq 0$. This forces the maps $H(-,0)$ and $H(-,1)$ from (M,A) to (N,B) to be cellular.

Applying Proposition 7.3 and the compression theorem 6.4 successively to all the relative CW-complexes (M^n, M^{n-1}) and (N, N^n), and using finally Lemma 5.1, we obtain, as in the topological theory,

<u>Proposition 7.4.</u> Every map $f : (M,A) \to (N,B)$ is homotopic relative A to a cellular map $g : (M,A) \to (N,B)$.

We call such a map g a <u>cellular approximation</u> of f. Cellular approximations are a very rough substitute of the simplicial approximations which we have used in Chapter III but which now are out of bounds.

Applying Proposition 7.4 to the direct product $(M\times I, (A\times I) \cup (M\times \partial I))$ of the relative CW-complexes (M,A) and $(I, \partial I)$ we obtain

<u>Corollary 7.5.</u> If two cellular maps $f,g : (M,A) \rightrightarrows (N,B)$ are homotopic then there exists a cellular homotopy from f to g.

Cellularity of the attaching map is a necessary assumption if one wants to glue a CW-complex to another CW-complex along a closed subcomplex without destroying the cell structure. It is also sufficient, as is stated by the following rather obvious proposition.

Proposition 7.6. Let L be a closed subcomplex of (M,A) and let $f : (L,A) \to (N,B)$ be a cellular map. Then the pair $(M \cup_f N,B)$ together with the patch decomposition

$$\Sigma(M \cup_f N,B) = \Sigma(M,L) \cup \Sigma(N,B)$$

is a relative CW-complex. {Notice that f is partially proper near $M \setminus L$ since (M,L) is a relative weak polytope.} N is a closed subcomplex of $(M \cup_f N,B)$. The natural map from (M,A) to $(M \cup_f N,B)$ extending f is again cellular.

We now start out to prove that every weak polytope over R is homotopy equivalent to a CW-complex which can be defined over the field R_o of real algebraic numbers, and also, that every topological CW-complex is topologically equivalent to a semialgebraic CW-complex over $\mathbb{R}$. Actually we will prove much more refined theorems (7.8, 7.10, 7.11). In these theorems we will establish homotopy equivalences between relative patch complexes or CW-complexes which induce homotopy equivalences between many subcomplexes. We need a definition to make this precise.

Definition 3. Let (M,A) and (N,B) be relative patch complexes over R. A patch watching homotopy equivalence from (M,A) to (N,B) is a pair (f,α) consisting of a map $f : (M,A) \to (N,B)$ between pairs of spaces and a bijection $\alpha : \Sigma(M,A) \to \Sigma(N,B)$ such that the following three axioms PW1-PW3 hold. For any $\sigma \in \Sigma(M,A)$ we write ${}^{\alpha}\sigma$ instead of $\alpha(\sigma)$. For any subcomplex L of (M,A) let

$${}^{\alpha}L := B \cup \bigcup({}^{\alpha}\sigma \mid \sigma\in\Sigma(L,A)).$$

This is a subcomplex of (N,B).

PW1. If σ and τ are patches of (M,A) and ${}^{\alpha}\tau$ is a face of ${}^{\alpha}\sigma$ then τ is a face of σ. Thus, if L is a closed subcomplex of (M,A) then ${}^{\alpha}L$

is a closed subcomplex of (N,B).

PW2. $f(L) \subset {}^{\alpha}L$ for every closed subcomplex L of (M,A). (Of course, it suffices to check this for L the union of A and the faces of a patch σ of (M,A).)

PW3. For every closed subcomplex L of (M,A) the map from (L,A) to $({}^{\alpha}L,B)$ obtained by restricting f to L is a homotopy equivalence. In particular, f itself is a homotopy equivalence.

For some time we will still work over a fixed real closed field R. The following lemma is obvious by Corollary 6.11 and Theorem 2.13. It indicates that patch watching homotopy equivalences can be constructed by an inductive procedure.

Lemma 7.7. Let (M,A) and (N,B) be relative patch complexes and let $(M_n | n \in \mathbb{N})$, $(N_n | n \in \mathbb{N})$ be (automatically admissible) filtrations of M and N by closed subcomplexes M_n and N_n of (M,A) and (N,B) respectively. Assume that for every $n \in \mathbb{N}$ there is given a patch watching homotopy equivalence (f_n, α_n) from (M_n, A) to (N_n, B) such that f_n extends f_{n-1} and α_n extends α_{n-1}. Let $f : (M,A) \to (N,B)$ be the map between pairs of spaces extending all the maps f_n, and let $\alpha : \Sigma(M,A) \to \Sigma(N,B)$ be the map between sets extending all the maps α_n. Then (f,α) is a patch watching homotopy equivalence from (M,A) to (N,B).

Theorem 7.8. Let (M,C) be a relative patch complex and A a closed subcomplex of (M,C) such that the relative patch complex (M,A) is special. {Notice that such a relative patch decomposition can be chosen on any decreasing closed triple of spaces (M,A,C), cf. 2.1.} Assume that (M,A) is a relative weak polytope. Let (g,β) be a patch watching homotopy equivalence from (A,C) to a relative CW-complex (B,D). Then there exists a relative CW-complex (N,D) containing B as a closed subcomplex (with the pregiven cell decomposition of (B,D)) and a patch watching homotopy equivalence (f,α) from (M,C) to (N,D) such that f extends g and α extend β. In particular ($A = \emptyset$, $B = \emptyset$), every weak polytope is homotopy equivalent to a CW-complex.

Proof. Let M_n denote the n-chunk of the special relative patch complex (M,A). We will construct inductively a relative CW-complex (N_n,D) and a patch watching homotopy equivalence (f_n,α_n) from (M_n,C) to (N_n,D) for every $n \geq -1$, starting with $N_{-1} = B$, $f_{-1} = g$, $\alpha_{-1} = \beta$, such that N_{n-1} is a closed subcomplex of (N_n,B), f_n extends f_{n-1}, and α_n extends α_{n-1}. Then we will be done: We equip the union N of all the sets N_n with the inductive limit space structure of the family of spaces $(N_n | n \geq -1)$. This space N is weakly semialgebraic. Every N_n is a closed subspace of N, and $(N_n | n \geq -1)$ is an admissible filtration of N (IV, Th. 7.1). Clearly the union Σ of all the sets of cells $\Sigma(N_n,D)$ is a relative patch decomposition of (N,D) which makes (N,D) a relative CW-complex containing the N_n as closed subcomplexes. By Lemma 7.7 the pairs (f_n,α_n) fit together to a patch watching homotopy equivalence (f,α) from (M,A) to (N,B), as desired.

Assume that, for some $n \geq 0$, the patch watching homotopy equivalence (f_{n-1},α_{n-1}) from (M_{n-1},C) to a relative CW-complex (N_{n-1},D) is already given. Then we construct (N_n,D) and (f_n,α_n) as follows. We have

$$M_n = M(n) \cup_{\varphi_n} M_{n-1}$$

with $M(n)$ the n-belt of (M,A) and $\varphi_n : \partial M(n) \to M_{n-1}$ the attaching map of the n-belt (cf. 2.2). Let $u : \partial M(n) \to N_{n-1}$ denote the composite $f_{n-1} \circ \varphi_n$. By Corollary 4.9.ii the homotopy equivalence $f_{n-1} : M_{n-1} \to N_{n-1}$ extends to a homotopy equivalence

$$h : M_n \to M(n) \cup_u N_{n-1}$$

in a canonical way.

Now $(M(n),\partial M(n))$ is the direct sum of the pairs of spaces $(\overline{\sigma},\partial\sigma)$ with σ running through the set $\Sigma(n) = \Sigma(M_n,M_{n-1})$ of patches of height n of (M,A). We fix some $\sigma \in \Sigma(n)$. Let $\varphi_\sigma : \partial\sigma \to M_{n-1}$ denote the restriction of φ_n to $\partial\sigma$. This is just the inclusion mapping from $\partial\sigma$ to M_{n-1}. Let $L(\sigma)$ denote the smallest closed subcomplex of (M_{n-1},C) containing $\partial\sigma$. We write $\lambda := \alpha_{n-1}$. Since (f_{n-1},λ) is patch watching the space $\partial\sigma$ is mapped by $u_\sigma = f_{n-1}\circ\varphi_\sigma$ into the subcomplex ${}^{\lambda}L(\sigma)$ of (N_{n-1},D). The pair $(\overline{\sigma},\partial\sigma)$ is isomorphic to (E^d,S^{d-1}) with $d := \dim\sigma$. By Proposition 7.3 the restriction $u_\sigma : \partial\sigma \to {}^{\lambda}L(\sigma)$ of u is homotopic to a map $v_\sigma : \partial\sigma \to {}^{\lambda}L(\sigma)$ with image in the (d-1)-skeleton ${}^{\lambda}L(\sigma) \cap N^{d-1}$ of $({}^{\lambda}L(\sigma),D)$.

All the v_σ with $\sigma \in \Sigma(n)$ together form a map $v : \partial M(n) \to N_{n-1}$ which is homotopic to u. We define

$$(N_n,D) := (M(n) \cup_v N_{n-1},D)$$

and equip this pair with the relative patch decomposition given by $\Sigma(N_{n-1},D)$ and the obvious relative patch decomposition of $(M(n),\partial M(n))$ (cf. 2.1). This makes (N_n,D) a relative CW-complex containing N_{n-1} as a closed subcomplex.

We have a tautological bijection γ from $\Sigma(M_n,M_{n-1}) = \Sigma(n)$ to $\Sigma(N_n,N_{n-1})$. Combining γ with the bijection $\alpha_{n-1} = \lambda$ from $\Sigma(M_{n-1},C)$ to $\Sigma(N_{n-1},D)$ we obtain a bijection α_n from $\Sigma(M_n,C)$ to $\Sigma(N_n,D)$. By construction α_n obeys the axiom PW1 and extends α_{n-1}. In the following we will write $\mu := \alpha_n$ in order to get rid of indices.

We fix again some $\sigma \in \Sigma(n)$. Let $Q(\sigma)$ denote the smallest closed subcomplex of (M,C) containing σ. We have $Q(\sigma) = \sigma \cup L(\sigma)$. The homotopy equivalence h from above restricts to a homotopy equivalence

$$h_\sigma : Q(\sigma) \to \overline{\sigma} \cup_u {}^{\lambda}L(\sigma)$$

which extends the homotopy equivalence $f_{n-1}|L(\sigma)$ from $L(\sigma)$ to ${}^{\lambda}L(\sigma)$. By Corollary 4.9.i there exists a homotopy equivalence

$$k_\sigma : \bar{\sigma} \cup_{u_\sigma} {}^{\lambda}L(\sigma) \to \bar{\sigma} \cup_{v_\sigma} {}^{\lambda}L(\sigma) = {}^{\mu}Q(\sigma)$$

under ${}^{\lambda}L(\sigma)$. The composite $f_\sigma := k_\sigma \circ h_\sigma$ is a homotopy equivalence from $Q(\sigma)$ to ${}^{\mu}Q(\sigma)$ which extends the homotopy equivalence $f_{n-1}|L(\sigma)$ from $L(\sigma)$ to ${}^{\lambda}L(\sigma)$.

All the maps f_σ fit together to a map $f_n : M_n \to N_n$ which extends f_{n-1}. By construction the pair (f_n,μ) obeys the axiom PW2. If L is a <u>finite</u> closed subcomplex of (M_n,C) we obtain from Corollary 2.14 by induction on the number of patches in $L \setminus (L \cap M_{n-1})$ that the restriction $f_n|L$ is a homotopy equivalence from L to ${}^{\mu}L$. By Theorem 6.10 this also holds for L infinite. The map $f_n|L$ extends the homotopy equivalence g between the closed subspaces C and D of L and ${}^{\mu}L$ respectively. Thus (Th. 2.13) $f_n|L$ is a homotopy equivalence from the pair (L,C) to $({}^{\mu}L,D)$. We have verified that (f_n,μ) obeys PW3 which finishes the proof.

We discuss the behaviour of relative CW-complexes under base field extension. Let K denote a real closed overfield of R. {The letter "S" is now needed for spheres.} Let (M,A) be a relative CW-complex over R. We choose characteristic maps $\psi_\sigma : E^n(R) \to M^n$ and corresponding attaching maps $\varphi_\sigma : S^{n-1}(R) \to M^{n-1}$ for all the cells σ of (M,A). {We write more precisely $E^n(R), S^{n-1}(R)$ instead of E^n, S^{n-1}.} Then (M(K),A(K)) is a relative CW-complex over K with the sets of cells

$$\Sigma(M(K),A(K)) = \{\sigma(K) \mid \sigma \in \Sigma(M,A)\}.$$

The n-skeleton $M(K)^n$ of (M(K),A(K)) coincides with the subspace $M^n(K)$ of M(K). The cell $\sigma(K)$ has the characteristic map $(\psi_\sigma)_K$ and the attaching map $(\varphi_\sigma)_K$. We have $\bar{\sigma}(K) = \overline{\sigma(K)}$ and $\partial(\sigma(K)) = (\partial\sigma)(K)$. Thus, if τ and σ are cells of (M,A), then $\tau \prec \sigma$ iff $\tau(K) \prec \sigma(K)$. It is now evident that the closed subcomplexes of $(M(K),A(K))$ are just the base extensions $L(K)$ of the closed subcomplexes L of (M,A).

We now prove two "main theorems" on CW-complexes which keep the same spirit as the two main theorems 5.2 on homotopy sets. In fact they are consequences of the latter.

Notations 7.9. Let (M,A) be a relative patch complex over K and (N,B) a relative patch complex over R. If (f,α) is a patch watching homotopy equivalence from (M,A) to $(N(K),B(K))$ then usually we regard α as a bijection from $\Sigma(M,A)$ to $\Sigma(N,B)$ rather than to $\Sigma(N(K),B(K))$. If α is any bijection from $\Sigma(M,A)$ to $\Sigma(N,B)$ and L is a subcomplex of $\Sigma(M,A)$ then ${}^{\alpha}L$ denotes the union of B and all cells ${}^{\alpha}\sigma$ with σ running through $\Sigma(L,A)$. It is a subcomplex of (N,B).

Theorem 7.10 (First main theorem on CW-complexes). Let (B,D) be a relative CW-complex over R. Let (M,C) be a relative CW-complex over K, and A a closed subcomplex of (M,C). Let (g,β) be a cell watching homotopy equivalence from (A,C) to $(B(K),D(K))$. Then there exists a relative CW-complex (N,D) over R such that B is a closed subcomplex of (N,D) (with the pregiven cell structure on (B,D)) and a cell watching homotopy equivalence (f,α) from (M,C) to $(N(K),D(K))$ such that f extends g and α extends β. In particular ($A = \emptyset$, $B = \emptyset$, $R = R_o$), every CW-complex M over K is homotopy equivalent to $N(K)$ for some CW-complex N over the field R_o of real algebraic numbers.

Proof. We denote the n-skeleton of (M,A) by $M^n (n \geq -1)$. We will build up N and (f,α) "skeleton by skeleton" proceeding in a similar way as in the proof of Theorem 7.8 above {but there the N_n were not the skeletons of (N,B)}. For every $\sigma \in \Sigma(M,A)$ we choose a characteristic map $\psi_\sigma : E^d(K) \to M^d$ $(d = \dim \sigma)$ and denote the corresponding attaching map by $\varphi_\sigma : S^{d-1}(K) \to M^{d-1}$.

We will construct a family of relative CW-complexes $((N^n,D) \mid n \geq -1)$ and a family $((f_n,\alpha_n) \mid n \geq -1)$ of patch watching homotopy equivalences (f_n,α_n) from (M^n,C) to $(N^n(K),D(K))$ inductively, starting with $N^{-1} = B$, $f_{-1} = g$, $\alpha_{-1} = \beta$, such that N^{n-1} is the (n-1)-skeleton of (N^n,B) and (f^n,α_n) extends (f_{n-1},α_{n-1}). Then we will be done by use of Lemma 7.7.

Assume that (N^{n-1},D) and (f_{n-1},α_{n-1}) are already constructed for some $n \geq 0$. We have

$$M^n = (\Sigma_n \times E^n(K)) \cup_{\varphi_n} M^{n-1}$$

using the notations 7.2. {In particular $\Sigma_n := \Sigma_n(M,A)$, considered here as a discrete space over K.} Let $\lambda := \alpha_{n-1}$. Let $u := f_{n-1} \circ (\varphi_n)_K$, a map from $\Sigma_n \times S^{n-1}(K)$ to $N^{n-1}(K)$. For any $\sigma \in \Sigma_n$ let $Q(\sigma)$ and $L(\sigma)$ denote the smallest closed subcomplexes of (M,C) containing σ and $\partial\sigma$ respectively (as in the proof of Th. 7.8). Let $u_\sigma : S^{n-1}(K) \to {}^{\lambda}L(\sigma)$ denote the map obtained by restricting u to $\{\sigma\} \times S^{n-1}(K)$. By the first main theorem 5.2.i on homotopy sets (or already III.3.1) there exists a map $v_\sigma : S^{n-1}(R) \to {}^{\lambda}L(\sigma)$ such that u_σ is homotopic to $(v_\sigma)_K$. We combine the v_σ into a map v from $\Sigma_n \times S^{n-1}(R)$ to N^{n-1}. Then u is homotopic to v_K. {Abusively we regard Σ_n also as a discrete space over R.} We define

$$N^n := (\Sigma_n \times E^n(R)) \cup_v N^{n-1} \quad .$$

This space contains N^{n-1} as a closed subspace, and we have an obvious patch decomposition $\Sigma(N^n,N^{n-1})$ of (N^n,N^{n-1}) together with a tautological bijection γ from $\Sigma_n = \Sigma(M^n,M^{n-1})$ to $\Sigma(N^n,N^{n-1})$. This patch decomposition and the patch decomposition of (N^{n-1},D) together equip (N^n,D) with the structure of a CW-complex such that N^{n-1} is the (n-1)-skeleton of (N^n,D). Moreover λ and γ combine into a bijection α_n from $\Sigma(M^n,C)$

to $\Sigma(N^n,D)$ which obeys the axiom PW1 (modified in the obvious way, since now the image of α_n consists of cells over R instead of K).

Using Corollary 4.9 we obtain a homotopy equivalence

$$f_\sigma : Q(\sigma) = E^n(K) \cup_{\varphi_\sigma} L(\sigma) \rightarrow [E^n(R) \cup_{v_\sigma} {}^\lambda L(\sigma)](K)$$

for every $\sigma \in \Sigma_n$ which extends the homotopy equivalence $f_{n-1}|L(\sigma)$ from $L(\sigma)$ to ${}^\lambda L(\sigma)(K)$. These maps f_σ and f_{n-1} fit together into a map $f_n : M^n \rightarrow N^n(K)$ which extends f_{n-1}. By construction the pair (f_n,α_n) fulfills PW2. It is now easily verified by use of Corollary 2.14 and Theorems 6.10, 2.13 that (f_n,α_n) also fulfills PW3, hence is a homotopy equivalence from (M^n,C) to $(N^n(K),D(K))$. q.e.d.

<u>Theorem 7.11</u> (Second main theorem on CW-complexes). Let (B,D) be a relative semialgebraic CW-complex over $\mathbb{R}$. Let (M,C) be a relative topological CW-complex and A a closed subcomplex of (M,C). Let (g,β) be a cell watching topological homotopy equivalence from (A,C) to (B_{top},D_{top}). Then there exists a relative semialgebraic CW-complex (N,D) over $\mathbb{R}$ such that B is a closed subcomplex of (N,D) (with the pregiven cell structure on (B,D)) and a cell watching topological homotopy equivalence (f,α) from (M,C) to (N_{top},D_{top}) such that f extends g and α extends β. In particular ($A = \emptyset$, $B = \emptyset$), every topological CW-complex M is homotopy equivalent to the underlying topological space N_{top} of a semialgebraic CW-complex N over $\mathbb{R}$.

The proof follows the same pattern as the preceding one. Instead of the first main theorem on homotopy sets one uses the second one. Further one uses some well known facts from the theory of topological CW-complexes (in particular the topological relative Whitehead theorem [W, p. 222]).

Theorem 7.10 is a useful tool to transfer results from the homotopy theory of topological CW-complexes to semialgebraic CW-complexes. Taking also into account Theorem 7.8, we sometimes can transfer the results further to arbitrary weak polytopes. In very favourable cases we can even go on to arbitrary spaces using partially complete cores (cf. IV, §9). We give an example.

Theorem 7.12 (Homotopy excision theorem). Let A and B be closed subspaces of a space M with $A \cup B = M$. Let x be a point in $A \cap B$. Assume that $(A, A \cap B)$ is n-connected and $(B, A \cap B)$ is m-connected for some numbers $n \geq 1$ and $m \geq 0$. Then the homomorphism

$$j_* : \pi_q(A, A \cap B, x) \to \pi_q(M, B, x)$$

induced by the inclusion $j : (A, A \cap B) \to (M, B)$ is bijective for $1 \leq q < n+m$ and surjective for $q = n+m$.

Proof. a) We first consider the case that M is a CW-complex and A, B are subcomplexes of M. The claim holds for $R = \mathbb{R}$ by the homotopy excision theorem for topological CW-complexes, cf. e.g. [Sw, 6.21]. By base extension from R_o to $\mathbb{R}$ we see that is also holds for $R = R_o$. For an arbitrary real closed base field R we now obtain the claim by use of Theorem 7.10. Indeed, by that theorem there exists a CW-complex M' over R_o, closed subcomplexes A' and B' of M', and a homotopy equivalence $f : M \to M'(R)$ such that $f(A) \subset A'(R)$, $f(B) \subset B'(R)$ and f yields homotopy equivalences from A to $A'(R)$, B to $B'(R)$, and $A \cap B$ to $(A' \cap B')(R)$ by restriction. The claim holds for (M, A, B, x) with x running through $A \cap B$ since it holds for (M', A', B', y) with y running through $A' \cap B'$.
b) Assume now that M is a weak polytope. Then we can verify the claim in a similar way by using Theorem 7.8 instead of 7.10.
c) Assume now that M is an arbitrary space but R is sequential. Then we obtain the claim for (M, A, B, x) from the claim for $(P(M), P(A), P(B), x)$.

{Notice that $P(A) \cap P(B) = P(A \cap B)$.}

d) Assume finally that R is not sequential. Then we choose a sequential real closed field extension K of R. The claim holds for $(M(K),A(K),B(K),x)$. We conclude by the first main theorem for homotopy sets 5.2.i (or already III.6.3) that the claim holds for (M,A,B,x). q.e.d.

Corollary 7.13. Let (M,A) be a closed pair of spaces with A m-connected and (M,A) n-connected for some numbers $m \geq 0$ and $n \geq 1$. Assume that A is partially complete near $M \setminus A$ (cf. IV, §8). Then the homomorphism

$$p_* : \pi_r(M,A,x) \to \pi_r(M/A,*)$$

induced by the natural projection $p : (M,A) \to (M/A,*)$ is, for any base point x in A, bijective if $2 \leq r \leq m+n$ and surjective if $r = m+n+1$.

Proof. If A is a weak polytope then the cone $CA = I \times A/1 \times A$ exists. In this case we obtain the claim by applying Theorem 7.12 to the triple of spaces $(M \cup CA, M, CA)$ and then using Proposition 2.15, cf. [Sw, p. 84]. In the general case that A is only partially complete near $M \setminus A$ we now obtain the claim by two steps similar to the steps c) and d) in the proof above. q.e.d.

We finally write down some consequences of our central results 7.8, 7.10, 7.11 which will be widely used in the next chapter. Here the full power of the theorems is not needed.

Definition 4. i) A system of CW-complexes over R, or CW-system for short, is a finite family $(P_o,P_1,\ldots,P_r)$ with P_o a CW-complex over R and every P_i a closed subcomplex of P_o. In the cases $r = 1$, $r = 2$, ... we call it a CW-pair, a CW-triple, etc. Notice that every component P_i of a CW-system $(P_o,\ldots,P_r)$ is a CW-complex. We call the CW-system $(P_o,\ldots,P_r)$ decreasing if it is a decreasing system of spaces, i.e.

P_{i+1} is a subcomplex of P_i for $1 \leq i \leq r-1$.

ii) A CW-approximation of a decreasing system of spaces $(M_o, M_1, \ldots, M_r)$ over R is a map

$$\varphi : (P_o, P_1, \ldots, P_r) \to (M_o, M_1, \ldots, M_r)$$

with $(P_o, \ldots, P_r)$ a CW-system over R and every component $\varphi_i : P_i \to M_i$ of φ a homotopy equivalence. Notice that every CW-approximation is a WP-approximation (§4, Def. 5).

Remark. In topology CW-approximations of systems of topological spaces are widely used, but there one only demands that the components of the map are weak homotopy equivalences, cf. [W,Chap. 5, §3], which, for topological spaces, means less than homotopy equivalence. We call them here topological CW-approximations.

Theorem 7.14. Every decreasing system of spaces $(M_o, \ldots, M_r)$ has a CW-approximation

$$\varphi : (P_o, \ldots, P_r) \to (M_o, \ldots, M_r) .$$

If M_r is contractible then P_r can be chosen as a one-point space. If M_i is complete, then P_i can be chosen as a finite CW-complex.

Proof. We know already that every decreasing system of spaces has a WP-approximation (Th. 4.10.i). Thus it suffices to consider the case that $(M_o, M_1, \ldots, M_r)$ is already a decreasing WP-system. Applying Theorem 7.8 we obtain a map

$$\psi : (M_o, M_1, \ldots, M_r) \to (P_o, P_1, \ldots, P_r)$$

with $(P_o, \ldots, P_r)$ a decreasing CW-system, at will with the additional properties stated above, and all components ψ_i of ψ homotopy equivalences. Since both systems of spaces are closed we know by 2.13 that ψ is a homotopy equivalence. A homotopy inverse φ of ψ is a CW-approximation of $(M_o,\dots,M_r)$. q.e.d.

<u>Theorem 7.15.</u> i) Let K be a real closed overfield or R. Let $(M_o,M_1,\dots,M_r)$ be a decreasing WP-system over K. Then there exist a decreasing CW-system $(P_o,P_1,\dots,P_r)$ over R together with a homotopy equivalence

$$\varphi : (P_o(K),P_1(K),\dots,P_r(K)) \to (M_o,M_1,\dots,M_r) .$$

If M_r is contractible then P_r can be chosen as a one-point space. If M_i is semialgebraic then P_i can be chosen as a finite CW-complex.
ii) If $(Q_o,\dots,Q_r)$ is a decreasing WP-system over R and f a map from $(Q_o(K),\dots,Q_r(K))$ to $(M_o,\dots,M_r)$ then there exists a map $g : (Q_o,\dots,Q_r) \to (P_o,\dots,P_r)$, unique up to homotopy, such that $\varphi \circ g_K \simeq f$. If f is a homotopy equivalence then also g is a homotopy equivalence.

<u>Proof.</u> In order to prove part i) we may assume, by the preceding theorem, that $(M_o,\dots,M_r)$ is already a CW-system. Using Theorem 7.10 we obtain a map

$$\psi : (M_o,\dots,M_r) \to (P_o(K),\dots,P_r(K))$$

such that $(P_o,\dots,P_r)$ is a decreasing CW-system over R, at will having the additional properties claimed above, and every component $\psi_i : M_i \to P_i(K)$ of ψ is a homotopy equivalence. This map of systems is a homotopy equivalence by 2.13. A homotopy inverse φ of ψ is a map as desired. The last claims in the theorem now follow from Theorem 6.14 and the first main theorem on homotopy sets 5.2.i. q.e.d.

Using instead of the first main theorems 7.10 and 5.2.i the corresponding second main theorems 7.11 and 5.2.ii we obtain

Theorem 7.16. i) Let $(M_o,\dots,M_r)$ be a topological CW-system. Then there exists a semialgebraic CW-system $(P_o,\dots,P_r)$ over $\mathbb{R}$ together with a topological homotopy equivalence

$$\varphi : (P_o,\dots,P_r) \to (M_o,\dots,M_r) .$$

If M_r is contractible then P_r can be chosen as a one-point space. If M_i is a finite CW-complex then also P_i can be chosen as a finite CW-complex.

ii) If $(Q_o,\dots,Q_r)$ is a WP-system over $\mathbb{R}$ and f is a continuous map from $(Q_o,\dots,Q_r)_{top}$ to $(M_o,\dots,M_r)$ then there exists a weakly semialgebraic map g from $(Q_o,\dots,Q_r)$ to $(P_o,\dots,P_r)$, unique up to homotopy, such that $\varphi\circ g \simeq f$ in the topological sense. If f is a topological weak homotopy equivalence then g is a homotopy equivalence.

Chapter VI - Homology and cohomology

We now have enough homotopy theory at our disposal to build up generalized homology and cohomology for (weakly semialgebraic) spaces over an arbitrary real closed field. This will be done in the present chapter.

In §7 we shall present the semialgebraic analogues of two variants of Brown's representation theorem for contravariant homotopy functors ([Bn], [Sw, Chap. 9]). This section may be regarded as an addendum to Chapter V. It is independent of the preceding sections in the present chapter, up to some easy results in §1 and some notations, and thus can be read right now. (We will give the necessary cross references.)

Brown's representation theorem leads to a description of homology and cohomology theories by spectra (§8). In the present chapter we could bring spectra into play at a much earlier stage. This would give alternative and sometimes easier proofs of some results in cohomology, but not so in homology. Already in classical algebraic topology it is a rather long way from Brown's representation theorem to a description of reduced homology theories by spectra [Sw, Chap. 14]. We can transfer this description to the semialgebraic setting, but, to the authors opinion and taste, this approach to generalized homology of weakly semialgebraic spaces would be too much a mixture of topological and semialgebraic arguments and would use too much machinery.

§1 - The basic categories, suspensions and cofibers

In this section we set the stage for a discussion of generalized homology and cohomology. We first compile the basic categorial notations. (Some of them have been used before.) Let R be any real closed field.

Notations 1.1. a) $\mathcal{P}(R)$ denotes the category of weak polytopes over R[*)], and $\mathcal{P}(2,R)$ denotes the category of pairs of weak polytopes over R. Further $\mathcal{P}^*(R)$ denotes the category of pointed weak polytopes over R. This is a full subcategory of $\mathcal{P}(2,R)$ since we may regard a pointed space (M,x) as the pair of spaces (M,{x}). Alternatively we may view $\mathcal{P}^*(R)$ as the category of weak polytopes under the one point space * (cf. V, §2, Def. 4). We identify every weak polytope M over R with the pair (M,∅). In this way also $\mathcal{P}(R)$ becomes a full subcategory of $\mathcal{P}(2,R)$.

b) Similarly we regard the category WSA(R) of spaces over R and the category WSA*(R) of pointed spaces over R as full subcategories of the category WSA(2,R) of pairs of spaces over R.

c) In WSA(R) we further have the full subcategories of paracompact locally semialgebraic spaces LSA(R) and of semialgebraic spaces SA(R), finally the category $LSA_c(R) = LSA(R) \cap \mathcal{P}(R)$ and the category $SA_c(R) = SA(R) \cap \mathcal{P}(R)$ of polytopes over R. These categories lead to categories of pointed spaces LSA*(R), SA*(R),... and of pairs of spaces LSA(2,R), SA(2,R), ... All categories mentioned so far are full subcategories of WSA(2,R).

d) For any of these categories $\mathfrak{A}$ we denote by $H\mathfrak{A}$ the corresponding homotopy category. $H\mathfrak{A}$ has the same objects as $\mathfrak{A}$ but the morphisms of $H\mathfrak{A}$ are the homotopy classes of maps between objects in $\mathfrak{A}$. All these

---

*) This is a substitute of the more systematic but clumsier notation $WSA_c(R)$.

categories $H\mathfrak{A}$ are full subcategories of HWSA(2,R).

e) We denote by $\mathcal{W}$ the category of topological CW-complexes. More precisely this means the full subcategory of the category TOP of Hausdorff topological spaces which has as objects the spaces which admit a (topological) CW-decomposition. We denote by $\mathcal{W}(2)$ the category of closed pairs (M,A) of topological spaces such that M admits a CW-decomposition which makes A a subcomplex of M. Similarly we denote by $\mathcal{W}^*$ the category of pointed CW-complexes. These are the pairs $(M,A) \in \mathcal{W}(2)$ with A a one point space. $\mathcal{W}$ and $\mathcal{W}^*$ are full subcategories of $\mathcal{W}(2)$. We denote the category of finite topological CW-complexes by $\mathcal{W}_F$ and the corresponding full subcategories of $\mathcal{W}^*$ and $\mathcal{W}(2)$ by $\mathcal{W}_F^*$ and $\mathcal{W}_F(2)$. If $\mathfrak{A}$ is one of these categories then again $H\mathfrak{A}$ means the corresponding (topological) homotopy category.

All categories mentioned so far admit finite direct products and finite direct sums. Some of them, in fact the most important ones, admit arbitrary direct sums. We spell this out in a special case.

Remark 1.2. i) Let $(M_\lambda | \lambda \in \Lambda)$ be a family of pointed spaces over R. The pointed space $M := V(M_\lambda | \lambda \in \Lambda)$ (cf. IV, 1.8) together with the natural inclusions $i_\lambda : M_\lambda \hookrightarrow M$ is the direct sum of the family $(M_\lambda | \lambda \in \Lambda)$ in the category WSA*(R). If we replace the i_λ by their (pointed) homotopy classes $[i_\lambda]$ we obtain the direct sum of this family in HWSA*(R).

ii) Let $(M_\lambda | \lambda \in \Lambda)$ be a finite family of pointed spaces over R with base points x_λ. We equip $M := \prod(M_\lambda | \lambda \in \Lambda)$ with the base point $x := (x_\lambda | \lambda \in \Lambda)$. Then M, together with the natural projections $p_\lambda : M \to M_\lambda$, is the direct product of the family $(M_\lambda | \lambda \in \Lambda)$ in WSA*(R). Replacing the p_λ by their homotopy classes $[p_\lambda]$ we obtain the direct product of this family in HWSA*(R).

If Λ is finite then the pointed space $V(M_\lambda | \lambda\in\Lambda)$ can and will be identified with the closed subspace of $\prod(M_\lambda | \lambda\in\Lambda)$ consisting of the points $(y_\lambda | \lambda\in\Lambda)$ with $y_\lambda = x_\lambda$ for all indices λ except at most one.

As a consequence of some central results in Chapters III and V we can write down various natural equivalences between the homotopy categories listed above.

Remarks 1.3. i) The inclusions $H\mathcal{P}(R) \hookrightarrow HWSA(R)$, $H\mathcal{P}^*(R) \hookrightarrow HWSA^*(R)$, $H\mathcal{P}(2,R) \hookrightarrow HWSA(2,R)$ are equivalences of categories (Th. V.4.10). The same holds for the inclusions $HSA_c(R) \hookrightarrow HSA(R)$, $HSA_c^*(R) \hookrightarrow HSA^*(R)$, $HSA_c(2,R) \hookrightarrow HSA(2,R)$, and the analogous inclusions with "SA" replaced by "LSA" (III.§1, V.2.13).

ii) If K is a real closed field extension of R then the functor "base extension" $(M,A) \mapsto (M(K),A(K))$ from $\mathcal{P}(2,R)$ to $\mathcal{P}(2,K)$ yields an equivalence of categories

$$\kappa : H\mathcal{P}(R,2) \xrightarrow{\sim} H\mathcal{P}(K,2).$$

The restrictions of κ to $H\mathcal{P}(R)$ and $H\mathcal{P}^*(R)$ are equivalences from $H\mathcal{P}(R)$ to $H\mathcal{P}(K)$ and from $H\mathcal{P}^*(R)$ to $H\mathcal{P}^*(K)$ respectively. This follows from Theorems V.5.2.i and V.7.15. If $\mathfrak{A}$ is one of the "meta-categories" LSA_c, SA_c, LSA_c^*, SA_c^*, $LSA_c(2,-)$, $SA_c(2,-)$ then it is already clear from Chapter III that κ yields by restriction an equivalence from $\mathfrak{A}(R)$ to $\mathfrak{A}(K)$.

iii) The functor from $\mathcal{P}(2,\mathbb{R})$ to TOP(2) which sends every pair of weak polytopes (M,A) over $\mathbb{R}$ to the pair (M_{top},A_{top}) of underlying topological spaces yields a full embedding

$$\lambda : H\mathcal{P}(2,\mathbb{R}) \hookrightarrow HTOP(2),$$

i.e. an isomorphism of $H\mathcal{P}(2,\mathbb{R})$ onto a full subcategory $\mathcal{C}(2)$ of HTOP(2). Moreover every object in $\mathcal{C}(2)$ is isomorphic to an object in $\mathcal{W}\mathcal{O}(2)$, and every object in the image categories $\mathcal{C}$, $\mathcal{C}^*$, $\mathcal{C}_F$, $\mathcal{C}_F^*$, $\mathcal{C}_F(2)$ of $H\mathcal{P}(\mathbb{R})$, $H\mathcal{P}^*(\mathbb{R})$, $HSA_c(\mathbb{R})$, $HSA_c^*(\mathbb{R})$, $HSA_c(2,\mathbb{R})$ is isomorphic to an object in $\mathcal{W}$, $\mathcal{W}^*$, $\mathcal{W}_F$, $\mathcal{W}_F^*$, $\mathcal{W}_F(2)$ respectively. All this follows from Theorems V.5.2.ii and V.7.16. As a consequence we can modify λ to an equivalence of categories

$$\lambda' : H\mathcal{P}(2,\mathbb{R}) \xrightarrow{\sim} \mathcal{W}(2)$$

which restricts to equivalences from $H\mathcal{P}(\mathbb{R})$, $H\mathcal{P}^*(\mathbb{R})$, $HSA_c(\mathbb{R})$, $HSA_c^*(\mathbb{R})$, $HSA_c(2,\mathbb{R})$ to $\mathcal{W}$, $\mathcal{W}^*$, $\mathcal{W}_F$, $\mathcal{W}_F^*$, $\mathcal{W}_F(2)$ respectively.

Notice that in remarks ii) and iii) we may replace the letter "$\mathcal{P}$" by "WSA", and similarly SA_c, LSA_c, by SA, LSA, as a consequence of i). But in the following partially complete spaces will play a dominant role.

We now focus attention to the category $\mathcal{P}^*(R)$ of pointed weak polytopes over a fixed real closed field R. Since we have equivalences $H\mathcal{P}^*(R_o) \xrightarrow{\sim} H\mathcal{P}^*(R)$, $H\mathcal{P}^*(R_o) \xrightarrow{\sim} H\mathcal{P}^*(\mathbb{R})$, $H\mathcal{P}^*(\mathbb{R}) \xrightarrow{\sim} H\mathcal{W}^*$, the reader may believe that the homotopy theory of pointed weak polytopes over R is somewhat identical to the homotopy theory of topological CW-complexes, and that this in particular holds for reduced homology and cohomology theories, which in topology are known to be a part of stable homotopy theory (cf. e.g. the book [Sw]). We shall verify this to a large extent in the present chapter.

Unfortunately the remarks 1.3 do not yet imply a meta theorem like this, since the equivalence between two categories is not such a strong statement as one might wish. We shall resort to the more explicit form of our main theorems in V,§7 and other homotopy results.

In the following we most often denote a pointed space (M,x_o) briefly by M and the set of homotopy classes of maps from M to a pointed space

N by [M,N] (instead of [M,N]* which would fit better with Def. 4 in V, §2).

Definition 1. a) The smash product $M \wedge N$ of two pointed weak polytopes M,N over R is the space $M \times N / M \vee N$ with base point $M \vee N / M \vee N$. Notice that if M and N carry the structure of pointed CW-complexes then also $M \wedge N$ is a pointed CW-complex in a natural way, since $M \vee N$ is a subcomplex of the product complex $M \times N$. Notice also that, in contrast to topology, we cannot define smash products of arbitrary pointed spaces.
b) We regard every sphere $S^n = S^n(R)$ as pointed by its north pole ∞. The suspension SM of a pointed weak polytope M is the smash product $S^1 \wedge M$. Using a fixed identification $S^1 = I/\{0,1\}$ we may write

$$SM = I \times M/(I \times x_o) \cup (0 \times M) \cup (1 \times M).$$

c) Any two maps $f : M \to M'$, $g : N \to N'$ between pointed weak polytopes induce a map $f \wedge g : M \wedge N \to M' \wedge N'$. In particular, f induces a map Sf from SM to SM'.
d) For every $k \in \mathbb{N}_o$ we denote the smash product $S^k \wedge M$ more briefly by $S^k M$. We have $S^o M = M$ and $S^k(S^1 M) = S^{k+1} M$. In particular, if $k > 0$, then $S^k M$ is the k-fold iteration of the suspension functor applied to M.

Every sphere S^k, $k \geq 1$, is equipped with a base point preserving map $\mu_k : S^k \to S^k \vee S^k$, unique up to homotopy, such that $(S^k, [\mu_k])$ is a cogroup object in $H\mathcal{P}^*(R)$, which for every $X \in \mathcal{P}^*(R)$ (and then also every $X \in WSA^*(R)$), gives the group structure on $\pi_k(X) = [S^k, X]$. This cogroup object is abelian for $k \geq 2$. Now it is evident from the distributivity of the smash product with respect to direct sums (= wedges) that also

$$\mu_k \wedge id_M : S^k M \to S^k M \vee S^k M$$

turns $S^k M$ into a cogroup in $H\mathcal{P}^*(R)$ for every pointed weak polytope M, which is abelian for $k \geq 2$. Thus, for every pointed space X and every $k \geq 1$, the set $[S^k M, X]$ is a group in a natural way, and this group is abelian if $k \geq 2$.

The comultiplication μ_1 can be defined as follows

$$\mu_1([t]) = \begin{cases} ([2t],[0]) & 0 \leq t \leq \frac{1}{2} \\ ([0],[2t-1]) & \frac{1}{2} \leq t \leq 1. \end{cases}$$

Here $[t]$ denotes the image of t in $S^1 = I/\{0,1\}$. If $k \geq 2$ then we can take as comultiplication μ_k the map $\mu_1 \wedge id_{S^{k-1}}$ from $S^k = S^1 \wedge S^{k-1}$ to $S^k \vee S^k = (S^1 \vee S^1) \wedge S^{k-1}$, cf. the formula in III, p. 265 for the multiplication in $\pi_k(X)$. By a purely formal argument (e.g. [Sw, 2.24]) one could equally well employ any other of the k factors in $S^k = S^1 \wedge \ldots \wedge S^1$. Thus we have the following fact.

<u>Lemma 1.4.</u> $S(\mu_k) \simeq \mu_{k+1}$ for every $k \geq 1$.

<u>Definition 2.</u> If M and N are pointed weak polytopes then the <u>suspension homomorphism</u>

$$S_{M,N} : [M,N] \to [SM,SN]$$

is defined by $S_{M,N}[f] := [Sf]$.

In general this is just a map from the pointed set $[M,N]$ to the group $[SM,SN]$. But if M is already a suspension SL of a weak polytope L then it follows from the preceding lemma that $S_{M,N}$ is a group homomorphism. More generally it can be shown in a purely formal way that $S_{M,N}$ is a group homomorphism if M is a cogroup object or N is a group object in $H\mathcal{P}^*(R)$. {Again it does not matter whether we use the factor S^1 in $SM = S^1 \wedge M$ or the cogroup M, resp. the group N to define multiplication on $[SM,SN]$.}

We will often denote the suspension homomorphism $S_{M,N}$ briefly by S, as long as no confusion is possible. (In much of the literature the letter Σ is used for the suspension homomorphism, but we need "Σ" to denote sets of patches.)

Theorem 1.5 (Freudenthal's suspension theorem). Let N be a pointed weak polytope which is n-connected for some $n \geq 0$. Then the suspension homomorphism

$$S_{S^q,N} : \pi_q(N) \to \pi_{q+1}(SN)$$

is bijective for $1 \leq q \leq 2n$ and surjective for $q = 2n+1$.

This can be deduced from Cor. V.7.13 as in topology, cf. [Sw, p. 85]. Of course we can obtain the theorem also by direct transfer from the topological Freudenthal suspension theorem. In §7 we will generalize the theorem to the suspension homomorphisms $S_{S^qM,N}$.

Definition 3. a) Let $f : M \to N$ be a map from a weak polytope M to a space N. The cofiber C(f) of f is the quotient $Z'(f)/O\times M$ of the "switched" mapping cylinder $Z'(f) = (I\times M) \cup_{1\times M,f} N$ by the partially complete subspace $O\times M$. {It will turn out to be slightly more convenient for us to use Z'(f) instead of Z(f).} We regard C(f) as a pointed space with natural base point $O\times M/O\times M$. The space N will be regarded as a (closed) subspace of C(f) by the embedding $N \hookrightarrow C(f)$ which is the composite of the natural embedding $N \hookrightarrow Z'(f)$ and the projection $Z'(f) \twoheadrightarrow C(f)$.
b) Let $f : M \to N$ be a map from a pointed weak polytope M to a pointed space N. The reduced cofiber $\tilde{C}(f)$ is the quotient $C(f)/I\times x_o$ with x_o the base point of M. This pointed space has the virtue, not shared by C(f), that the natural embedding $j : N \hookrightarrow \tilde{C}(f)$ preserves base points. We regard N as a pointed closed subspace of $\tilde{C}(f)$ via j. Notice that the natural projection $C(f) \twoheadrightarrow \tilde{C}(f)$ is a homotopy equivalence of pointed

spaces (cf. V.2.15).

If N is a weak polytope then also C(f) resp. $\tilde{C}(f)$ is a weak polytope. In contrast to topology we cannot define cofibers of maps between arbitrary spaces over R.

Examples 1.6. i) The cofiber of the identity map id_M of a weak polytope M is the cone $CM = I\times M/O\times M$ with its vertex as base point. If M has a base point x_o then the reduced cofiber of this map is the reduced cone

$$\tilde{C}M := I\times M/(O\times M) \cup (x_o\times I) .$$

This is the smash product $I\wedge M$, the unit interval I being equipped with the base point O. (We will always adopt this convention about I as a pointed space.)

ii) If A is a partially complete - hence closed - subspace of a space M then the cofiber C(i) of the inclusion $i : A \hookrightarrow M$ is the space $M \cup CA$ obtained by gluing the space CA to M along A by the map i. The notation $M \cup CA$ is justified since both M and CA are closed subspaces of C(i) and their union is C(i). The natural projection

$$p : M \cup CA \to M \cup CA/CA = M/A$$

is a homotopy equivalence between pointed spaces since CA is contractible.

iii) If A is a pointed partially complete subspace of a pointed space M then the reduced cofiber $\tilde{C}(i)$ of the inclusion $A \hookrightarrow M$ is the space $M \cup CA/I\times\{x_o\}$, which we justly denote by $M \cup \tilde{C}A$. Again the natural projection

$$\pi : M \cup \tilde{C}A \to M \cup \tilde{C}A/\tilde{C}A = M/A$$

is a homotopy equivalence between pointed spaces.

iv) Let $f : M \to N$ be a map from a pointed weak polytope M with base point x_o to a pointed space N. We may regard $I \times x_o$ as a closed subspace of the switched mapping cylinder $Z'(f)$. We call the quotient $\tilde{Z}(f) := Z'(f)/I \times x_o$ the <u>reduced switched mapping cylinder</u> of f. We have a natural closed embedding $i : M \hookrightarrow \tilde{Z}(f)$, $x \mapsto [(0,x)]$, preserving base points, which we regard as an inclusion. We have a commuting diagram of pointed spaces

$$\begin{array}{ccccc} M & \xrightarrow{\ i\ } & \tilde{Z}(f) & \longrightarrow & \tilde{C}(i) \\ \downarrow{\scriptstyle id_M} & & \downarrow{\scriptstyle \alpha} & & \downarrow{\scriptstyle \beta} \\ M & \xrightarrow[f]{} & N & \longrightarrow & \tilde{C}(f) \end{array}$$

with canonical projections α and

$$\beta : \tilde{C}(i) = \tilde{Z}(f) \cup \tilde{C}M \to \tilde{Z}(f)/M = \tilde{C}(f).$$

Both α and β are homotopy equivalences of pointed spaces. Thus up to a canonical homotopy equivalence, every reduced cofiber may be regarded as the reduced cofiber of an inclusion. {A similar remark holds for unreduced cofibers.}

<u>Lemma 1.7.</u> Let (M,A) be a closed pair of pointed spaces with A a weak polytope. For every pointed space X the natural sequence of pointed sets

$$[M \cup CA, X] \to [M,X] \xrightarrow{i^*} [A,X]$$

is exact.

This is obvious from the definitions (cf. [Pu, p. 305]). We could replace the first term in the exact sequence by $[M/A,X]$.

Let now $f : M \to N$ be a map between pointed weak polytopes. Clearly

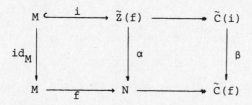

$\tilde{C}(f)/N = SM$. We denote the canonical projection $\tilde{C}(f) \twoheadrightarrow SM$ by q.

<u>Lemma 1.8.</u> We have an infinite commutative diagram

$$\begin{array}{ccccccccccccc}
M & \xrightarrow{f} & N & \xrightarrow{j} & \tilde{C}(f) & \xrightarrow{j'} & \tilde{C}(j) & \xrightarrow{j''} & \tilde{C}(j') & \xrightarrow{j'''} & \tilde{C}(j'') & \longrightarrow \\
\| & & \| & & \| & & \downarrow & & \downarrow & & \downarrow & \\
M & \xrightarrow[f]{} & N & \xrightarrow[j]{} & \tilde{C}(f) & \xrightarrow[q]{} & SM & \xrightarrow[Sf]{} & SN & \xrightarrow[Sj]{} & S\tilde{C}f & \longrightarrow
\end{array}$$

Here $j, j', j'', \ldots$ are the natural injections of the target spaces of the preceding maps into their cofibers. The vertical maps are the obvious natural projections. They all are homotopy equivalences of pointed weak polytopes. The lower long sequence moves three steps to the right if we apply the suspension functor.

All this can be checked as in topology (cf. [Pu, p. 308ff]).

<u>Definition 4.</u> We call the upper long sequence in the diagram the <u>long cofiber sequence</u> of f and the lower long sequence the <u>Puppe sequence</u> of f.

As a consequence of our discussion of cofibers above and the lemmas 1.7, 1.8 we obtain the following theorem of basic importance.

<u>Theorem 1.9</u> (Barratt [Ba], Puppe [Pu]). Let $f : M \to N$ be a map between pointed weak polytopes and let X be a further pointed weak polytope. Applying the functor $[-,X]$ to the Puppe sequence of f we obtain a long exact sequence of pointed sets and, starting from the fourth term, groups

$$[M,X] \leftarrow [N,X] \leftarrow [\tilde{C}(f),X] \leftarrow [SM,X] \leftarrow [SN,X] \leftarrow [S\tilde{C}(f),X] \leftarrow [S^2M,X] \leftarrow \ldots$$

Starting from the seventh term the groups are abelian.

In the special case of an inclusion map we obtain

Corollary 1.10. If (M,A) is a pair of pointed weak polytopes and X is another pointed weak polytope then we have a natural long exact sequence

$$[A,X] \leftarrow [M,X] \leftarrow [M/A,X] \leftarrow [SA,X] \leftarrow [SM,X] \leftarrow [S(M/A),X] \leftarrow [S^2A,X] \leftarrow \dots$$

Definition 5. We call the sequence of maps

$$A \to M \to M/A \to SA \to SM \to S(M/A) \to \dots$$

used in this corollary the suspension sequence of the closed pair (M,A).

If $f : M \to N$ is any map between pointed weak polytopes then the Puppe sequence of f can be identified in a natural way with the suspension sequence of the pair $(\tilde{Z}(f),M)$, cf. 1.6.iv.

§2 - Reduced cohomology of weak polytopes

The category $\mathcal{P}^*(R)$ of pointed weak polytopes over R and its homotopy category $H\mathcal{P}^*(R)$ both are equipped with a distinguished endomorphism, the suspension functor S. Using S we can give natural and well working definitions of reduced homology and cohomology theories on the category $\mathcal{P}^*(R)$, as is done in topology on the category $\mathcal{HO}^*$ of pointed CW-complexes [Sw, Chap. 7].

We do cohomology first. Let Ab denote the category of abelian groups. In the following we will consider many contravariant functors $F : H\mathcal{P}^*(R) \to Ab$. If $f : M \to N$ is any map between pointed weak polytopes we will often abusively denote all the group homomorphisms $F([f])$ by f^* as long as no confusion is to be feared.

Definition 1. A reduced (semialgebraic) cohomology theory k^* over R is a family $(k^n | n \in \mathbb{Z})$ of contravariant functors $k^n : H\mathcal{P}^*(R) \to Ab$ together with a family $(\sigma^n | n \in \mathbb{Z})$ of natural equivalences (= isomorphisms between functors) $\sigma^n : k^{n+1} \circ S \xrightarrow{\sim} k^n$ such that the following two axioms hold.

Exactness axiom. For every pair (M,A) of pointed weak polytopes over R and every $n \in \mathbb{Z}$ the sequence

$$k^n(M/A) \xrightarrow{p^*} k^n(M) \xrightarrow{i^*} k^n(A)$$

is exact. Here i denotes the inclusion $A \hookrightarrow M$ and p denotes the projection $M \twoheadrightarrow M/A$.

Wedge axiom. For every family $(M_\lambda | \lambda \in \Lambda)$ of pointed weak polytopes and every $n \in \mathbb{Z}$ the map

$$(i_\lambda^*) : k^n(\vee(M_\lambda | \lambda \in \Lambda)) \to \prod(k^n(M_\lambda) | \lambda \in \Lambda)$$

is an isomorphism. Here i_λ denotes the natural embedding of M_λ into M.

Actually it suffices to demand each of these axioms for $n \geq n_o$ with some bound $n_o \in \mathbb{Z}$. Then they follow for the other n by use of the natural equivalences σ^n since $S(M/A) = SM/SA$ and

$$S(V(X_\lambda | \lambda \in \Lambda)) = V(SX_\lambda | \lambda \in \Lambda).$$

Moreover, if the functors k^n and the equivalences σ^n are defined only for $n \geq n_o$ such that the axioms above hold for these n, then we can extend the family of these functors k^n to a reduced cohomology theory by defining

$$k^{n_o - r}(X) := k^{n_o}(S^r X)$$

for $r > 0$.

Notice also that the wedge axiom for Λ finite is a consequence of the exactness axiom. Indeed, if M_1 and M_2 are two pointed weak polytopes, then $M_1 \vee M_2/M_2 = M_1$ and $M_1 \vee M_2/M_1 = M_2$. By the exactness axiom the diagram

$$M_1 \underset{p_1}{\overset{i_1}{\rightleftarrows}} M_1 \vee M_2 \underset{i_2}{\overset{p_2}{\rightleftarrows}} M_2$$

of natural injections and projections becomes a direct sum diagram of abelian groups under k^n.

We call the equivalences $\sigma^n : k^{n+1} \circ S \xrightarrow{\sim} k^n$ the <u>suspension isomorphisms</u> of the cohomology theory. We usually denote them all by the same letter σ omitting the index n. In the following we also say more briefly "cohomology theory" instead of "reduced cohomology theory". (Starting from §4 we will need to be more careful since then also unreduced theories will be studied.)

We draw some consequences from the axioms of a given cohomology theory k^*.

If (M,A) is any pair of pointed weak polytopes then we define, for every $n \in \mathbb{Z}$, a homomorphism

$$\delta = \delta^n(M,A) : k^n(A) \to k^{n+1}(M/A)$$

by composing $\sigma^{-1} : k^n(A) \xrightarrow{\sim} k^{n+1}(SA)$ with $q^* : k^{n+1}(SA) \to k^{n+1}(M \cup \tilde{C}A)$ and $(\pi^*)^{-1} : k^{n+1}(M \cup \tilde{C}A) \to k^{n+1}(M/A)$. Here q denotes the natural projection from $M \cup \tilde{C}A$ to $M \cup \tilde{C}A/M = SA$, as in Lemma 1.8, and π denotes the natural projection from $M \cup \tilde{C}A$ to M/A which is a homotopy equivalence (Ex. 1.6.iii). As a consequence of Lemma 1.8 we deduce from the exactness axiom

<u>Proposition 2.1.</u> For every pointed WP-pair (M,A) the long sequence of abelian groups (going to infinity at both sides)

$$\xrightarrow{\delta} k^n(M/A) \xrightarrow{p^*} k^n(M) \xrightarrow{i^*} k^n(A) \xrightarrow{\delta} k^{n+1}(M/A) \to$$

is exact.

<u>Corollary 2.2.</u> For every map $f : M \to N$ between pointed weak polytopes we have a natural exact sequence of abelian groups (going to infinity at both sides)

$$\longrightarrow k^n(\tilde{C}(f)) \xrightarrow{j^*} k^n(N) \xrightarrow{f^*} k^n(M) \to k^{n+1}(\tilde{C}(f)) \to \qquad .$$

This follows from 2.1 since, up to canonical homotopy equivalences, the Puppe sequence of f and the suspension sequence of $(\tilde{Z}(f),M)$ are the same (cf. end of §1).

Let (M,A,B) be triple of pointed weak polytopes with $M = A \cup B$. For every $n \in \mathbb{Z}$ we define a homomorphism

$$\Delta = \Delta^n(M,A,B) : k^n(A \cap B) \to k^{n+1}(M),$$

as the composite

$$k^n(A \cap B) \xrightarrow[\delta]{} k^{n+1}(A/A \cap B) \xrightarrow[\gamma]{\sim} k^{n+1}(M/B) \xrightarrow[p^*]{} k^{n+1}(M)$$

with $\delta = \delta^n(A, A \cap B)$, γ the inverse of the group isomorphism induced by the natural space isomorphism $A/A \cap B \xrightarrow{\sim} M/B$, and $p : M \to M/B$ the natural projection.

<u>Proposition 2.3.</u> For every WP-triple (M,A,B) with $M = A \cup B$ the infinite "Mayer-Vietoris sequence"

$$\xrightarrow{\Delta} k^n(M) \xrightarrow[\alpha]{} k^n(A) \oplus k^n(B) \xrightarrow[\beta]{} k^n(A \cap B) \xrightarrow{\Delta} k^{n+1}(M) \to$$

is exact. Here α and β are defined by

$$\alpha(Z) = (j_1^* Z, j_2^* Z), \quad \beta(u,v) = i_1^* u - i_2^* v$$

with inclusion maps $j_1 : A \hookrightarrow M$, $j_2 : B \hookrightarrow M$, $i_1 : A \cap B \hookrightarrow A$, $i_2 : A \cap B \hookrightarrow B$.

This follows from Proposition 2.1 in a well known way, cf. [ES, p. 39ff], [Sw, p. 105]. Notice that

$$\Delta^n(M,B,A) = -\Delta^n(M,A,B),$$

cf. [ES, p. 38], [Sw, p. 106].

<u>Proposition 2.4.</u> Let $(M_n \mid n \in \mathbb{N})$ be an admissible filtration (cf. V, §2, Def. 3) of a pointed weak polytope M. Then there exists, for every $q \in \mathbb{Z}$, a natural exact sequence

$$0 \to \varprojlim_n{}^1 k^{q-1}(M_n) \xrightarrow{\alpha} k^q(M) \xrightarrow{\beta} \varprojlim_n k^q(M_n) \to 0 \quad .$$

Here the homomorphism β is induced, of course, by the inclusions $M_n \hookrightarrow M$. The homomorphism α is explained in [Mi], [Sw, p. 128].

This follows by applying the preceding proposition to the <u>reduced</u>

telescope $\tilde{\mathrm{Tel}}(\mathfrak{M}) := \mathrm{Tel}(\mathfrak{M})/x_o \times I_\infty$ of the family $\mathfrak{M} := (M_n \mid n \in \mathbb{N})$ and suitable closed subspaces A,B of $\tilde{\mathrm{Tel}}(\mathfrak{M})$ with $A \cup B = \tilde{\mathrm{Tel}}(\mathfrak{M})$ (cf. [Mi], [Sw, p. 128], see also step b) in the proof of Thm. 6.6 below).

NB. $\mathrm{Tel}(\mathfrak{M})$ had been defined in V, §4 and had been identified with a closed subspace of $M \times I_\infty$. Of course, x_o denotes the base point of M. It is contained in every M_n.

Our main goal in the following is to prove that, in vague terms, the cohomology theories over any real closed field R correspond uniquely with the topological cohomology theories (cf. [Sw, Chap. 7]) on the category $\mathcal{W}^*$ of pointed topological CW-complexes. In order to express this in precise terms we need more terminology.

Definition 2. Let k* and l* be two cohomology theories over R with families of suspension isomorphisms $(\sigma^n \mid n \in \mathbb{Z})$ and $(\tau^n \mid n \in \mathbb{Z})$.

a) A *natural transformation* $T : k^* \to l^*$ from k* to l* is a family of natural transformations between functors $(T^n : k^n \to l^n \mid n \in \mathbb{Z})$ such that, for every $X \in \mathcal{P}^*(R)$ and $n \in \mathbb{Z}$, the square

$$\begin{array}{ccc} k^{n+1}(SX) & \xrightarrow{T^{n+1}(SX)} & l^{n+1}(SX) \\ \sigma^n(X) \downarrow \cong & & \cong \downarrow \tau^n(X) \\ k^n(X) & \xrightarrow[T^n(X)]{} & l^n(X) \end{array}$$

commutes.

b) We call T a *natural equivalence*, or an *isomorphism*, from k* to l* if, in addition $T^n(X)$ is an isomorphism for every $X \in \mathcal{P}^*(R)$ and every $n \in \mathbb{Z}$.

Proposition 2.5. Let $T, U : k^* \rightrightarrows l^*$ be two natural transformations between cohomology theories over R.

a) Assume that $T^n(S^o) = U^n(S^o)$ for every $n \in \mathbb{Z}$. Then $T = U$.

b) Assume that $T^n(S^o)$ is bijective for every $n \in \mathbb{Z}$. Then T is a natural equivalence.

The proof runs similar to the proof in [Sw, p. 123f] using the wedge axiom and Propositions 2.1, 2.4.

Proposition 2.5 tells us that a reduced cohomology theory k^* is uniquely determined, in a restricted sense made precise there, by the sequence of abelian groups $(k^n(S^o) \mid n \in \mathbb{Z})$. These groups are called the <u>coefficient groups</u> of cohomology theory k^*.

We want to compare cohomology theories over different base fields and later cohomology theories over $\mathbb{R}$ with topological cohomology theories.

Let K be a real closed overfield of R. Every cohomology theory k^* over K "restricts" to a cohomology theory $(k^*)^R$ over R in the following obvious way. We define, for $n \in \mathbb{Z}$ and $X \in \mathcal{P}^*(R)$,

(2.6)
$$(k^n)^R(X) := k^n(X(K)) ,$$
$$(\sigma^n)^R(X) := \sigma^n(X(K)) .$$

{Notice that $(SX)(K) = S(X(K))$.} We further define, for any map $f : X \to Y$

$$(k^n)^R([f]) := k^n([f_K]) .$$

Every natural transformation $T : k^* \to l^*$ from k^* to another cohomology theory l^* over K restricts to a natural transformation T^R from $(k^*)^R$ to $(l^*)^R$, defined by

$$(T^n)^R(X) = T^n(X(K)) .$$

Let $\mathrm{Hom}(k^*,l^*)$ denote the set of natural transformations from k^* to l^*.

Proposition 2.7. The restriction map $T \mapsto T^R$ from $\mathrm{Hom}(k^*, l^*)$ to $\mathrm{Hom}((k^*)^R, (l^*)^R)$ is bijective. A natural transformation T over K is a natural equivalence iff T^R is a natural equivalence.

The proof is an easy exercise in category theory. One uses the fact that the base extension functor $H\mathcal{P}^*(R) \to H\mathcal{P}^*(K)$ is an equivalence of categories which commutes with suspensions.

Let $\widetilde{\mathrm{Coho}}(R)$ denote the category whose objects are the cohomology theories over R and whose morphisms are the natural transformations between them. {The tilda in this notation is a reminder that we are dealing with reduced cohomology theories.} For every real closed field extension K of R (more precisely, every embedding $R \hookrightarrow K$ into a real closed field K) we have a restriction functor

$$\mathrm{res}_R^K : \widetilde{\mathrm{Coho}}(K) \to \widetilde{\mathrm{Coho}}(R)$$

which maps an object k^* of $\widetilde{\mathrm{Coho}}(K)$ to $(k^*)^R$ and a morphism T of $\widetilde{\mathrm{Coho}}(K)$ to T^R. We call $(k^*)^R$ and T^R the restrictions of k^* and T to R.

Theorem 2.8 (First main theorem for cohomology theories). res_R^K is an equivalence from the category $\widetilde{\mathrm{Coho}}(K)$ to $\widetilde{\mathrm{Coho}}(R)$. In particular, up to isomorphism the cohomology theories over K correspond uniquely to the cohomology theories over R.

Proof. Proposition 2.7 means that res_R^K is fully faithful. It remains to construct, for a given cohomology theory l^* over R a cohomology theory k^* over K together with a natural equivalence $T : l^* \xrightarrow{\sim} (k^*)^R$.

Let $M \in \mathcal{P}^*(K)$ be given. We denote by $I(M,R)$ the set of all pairs $(X,[\varphi])$ consisting of a pointed weak polytope X over R and the homotopy class $[\varphi]$ of a homotopy equivalence $\varphi : X(K) \to M$. Such pairs exist by

Theorem V.7.15.*) If $(X,[\varphi])$ and $(Y,[\psi])$ are two elements of $I(M,R)$ then, again by V.7.15, there exists, up to homotopy, a unique map $\chi : X \to Y$ such that $\psi \circ \chi_K \simeq \varphi$, and χ is a homotopy equivalence.

We have a generalized inverse system of abelian groups $(l^n(X) \mid (X,[\varphi]) \in I(M,R))$ with transition maps

$$\chi^* : l^n(Y) \xrightarrow{\sim} l^n(X) ,$$

χ as above, for <u>any</u> two pairs $(X,[\varphi])$, $(Y,[\psi])$ in $I(M,R)$. (Since the transition maps are isomorphisms this system could also be regarded as a generalized direct system.) We define

$$k^n(M) := \varprojlim (l^n(X) \mid (X,[\varphi]) \in I(M,R)) .$$

Roughly speaking, we put $k^n(M) = l^n(X)$ for any $(X,[\varphi]) \in I(M,R)$ without making a definite choice about the "approximating pair" $(X,[\varphi])$.

For any $(X,[\varphi]) \in I(M,R)$ we denote the natural projection from $k^n(M)$ to $l^n(X)$ by φ^*, abusively neither specifying the space X nor the number n.

If $f : M \to N$ is a map from M to another pointed weak polytope N over K and if $(X,[\varphi]) \in I(M,R)$, $(Y,[\psi]) \in I(N,R)$ are given, then there exists, up to homotopy, a unique map $g : X \to Y$ over R such that the diagram

$$\begin{array}{ccc} X(K) & \xrightarrow{\varphi} & M \\ {\scriptstyle g_K}\downarrow & & \downarrow{\scriptstyle f} \\ Y(K) & \xrightarrow{\psi} & N \end{array}$$

commutes up to homotopy. We define $k^n([f]) : k^n(N) \to k^n(M)$ as the dotted arrow which makes the square

---

*) We ignore the problem whether $I(M,R)$ is a set. It can be settled by a suitable interpretation of "all".

$$\begin{array}{ccc} k^n(M) & \xrightarrow[\sim]{\varphi^*} & l^n(X) \\ {\scriptstyle k^n[f]}\ \uparrow & & \uparrow\ {\scriptstyle l^n[g]} \\ k^n(N) & \xrightarrow[\psi^*]{\sim} & l^n(Y) \end{array}$$

commutative. The homomorphism $k^n[f]$ does not depend on the choice of $(X,[\varphi])$ and $(Y,[\psi])$. In this way we obtain a family $(k^n \mid n\in\mathbb{Z})$ of contravariant functors from $H\mathcal{P}^*(K)$ to Ab.

We now look for a suspension isomorphism

$$\sigma^n(M) \ : \ k^{n+1}(SM) \xrightarrow{\sim} k^n(M) \ .$$

We choose some $(X,[\varphi]) \in I(M,R)$. Then $(SX,[S\varphi]) \in I(SM,R)$. We define $\sigma^n(M)$ as the dotted arrow such that the square

$$\begin{array}{ccc} k^{n+1}(SM) & \xrightarrow[\sim]{(S\varphi)^*} & l^{n+1}(SX) \\ {\scriptstyle \sigma^n(M)}\ \downarrow & & \downarrow\ {\scriptstyle \sigma^n(X)} \\ k^n(M) & \xrightarrow[\varphi^*]{\sim} & l^n(X) \end{array}$$

commutes. Clearly $\sigma^n(M)$ is an isomorphism of abelian groups which is independent of the choice of $(X,[\varphi])$. It is natural in M.

We check the wedge axiom for the functors k^n. Let $(M_\lambda \mid \lambda\in\Lambda)$ be a family in $\mathcal{P}^*(K)$. For every $\lambda \in \Lambda$ we choose a pair $(X_\lambda,[\varphi_\lambda]) \in I(M_\lambda,R)$. Let $M := V(M_\lambda \mid \lambda\in\Lambda)$, $X := V(X_\lambda \mid \lambda\in\Lambda)$ and $\varphi = V\varphi_\lambda \ : X(K) \to M$. Then $(X,[\varphi])$ is an element of $I(M,R)$. Let $j_\lambda \ : X_\lambda \hookrightarrow X$ and $i_\lambda \ : M_\lambda \hookrightarrow M$ denote the natural embeddings. We have a commuting square

$$\begin{array}{ccc} k^n(M) & \xrightarrow[\sim]{\varphi^*} & l^n(X) \\ {\scriptstyle (i_\lambda^*)}\ \downarrow & & \downarrow\ {\scriptstyle (j_\lambda^*)} \\ \prod_{\lambda\in\Lambda} k^n(M_\lambda) & \xrightarrow[\prod\varphi_\lambda^*]{\sim} & \prod l^n(X_\lambda) \end{array} \ .$$

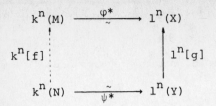

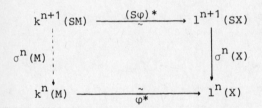

By the wedge axiom for l^n the right vertical arrow is bijective. Thus also the left vertical arrow is bijective.

We now check the exactness axiom for k^n. Let (M,A) be a pair of pointed weak polytopes over K. We choose a pair (X,B) of pointed weak polytopes over R together with a homotopy equivalence $(\varphi,\psi) : (X(K),B(K)) \to (M,A)$. This is possible by Theorem V.7.15. {(X,B) can even be chosen as a pointed CW-pair.} φ induces a homotopy equivalence $\chi : X(K)/B(K) \xrightarrow{\sim} M/A$. Moreover $X(K)/B(K) = (X/B)(K)$. Using the pairs $(X,[\varphi]) \in I(M,R)$, $(B,[\psi]) \in I(A,R)$ and $(X/B,\chi) \in I(M/A,R)$ it is now easy to deduce the exactness of the sequence

$$k^n(A) \to k^n(M) \to k^n(M/A)$$

from the exactness of

$$l^n(B) \to l^n(X) \to l^n(X/B) .$$

Thus $(k^n \mid n \in \mathbb{Z})$ is indeed a cohomology theory over K.

For any $X \in \mathcal{P}^*(R)$ the pair $(X,[id_{X(K)}])$ is an element of $I(X(K),R)$. We denote the natural projection $k^n(X(K)) \xrightarrow{\sim} l^n(X)$ associated to this pair by $T^n(X)$. It depends functorially on X in $H\mathcal{P}^*(R)$. Thus we obtain a natural equivalence $T^n : (k^n)^R \to l^n$ of functors. The T^n fit together to a natural equivalence $T : (k^*)^R \to l^*$ of cohomology theories. Theorem 2.8 is proved.

We call the cohomology theory k^* over K constructed here the <u>base extension of</u> l^* <u>to</u> K, and write $k^* = l^*_K$. We have established a <u>canonical</u> isomorphism $T : (l^*_K)^R \to l^*$ and thus feel justified to say that $(l^*_K)^R = l^*$. More explicitely, for every $X \in \mathcal{P}^*(R)$ and every $n \in \mathbb{Z}$,

$$(2.9) \qquad l^n_K(X(K)) = l^n(X) .$$

Conversely let k^* be a cohomology theory over K. Let $M \in \mathcal{P}^*(K)$, $n \in \mathbb{Z}$. Then

$$((k^n)^R)_K(M) = \varprojlim ((k^n)^R(X) \mid (X,[\varphi]) \in I(M,R))$$
$$\underset{(2.6)}{=} \varprojlim (k^n(X(K)) \mid (X,[\varphi]) \in I(M,R)) .$$

We have a canonical group isomorphism $U^n(M) : k^n(M) \xrightarrow{\sim} ((k^n)^R)_K(M)$ whose component at any $(X,[\varphi])$ is $k^n([\varphi])$. This gives us a canonical natural equivalence

$$U : k^* \xrightarrow{\sim} ((k^*)^R)_K .$$

By use of U we identify

(2.10) $$((k^*)^R)_K = k^* .$$

We now look for a theorem similar to Theorem 2.8 which relates the semialgebraic cohomology theories over $\mathbb{R}$ to the topological cohomology theories. A <u>topological</u> (reduced) <u>cohomology theory</u> k^* is a sequence $(k^n \mid n \in \mathbb{Z})$ of contravariant functors $k^n : H\mathcal{W}^* \to Ab$ on the homotopy category of pointed topological CW-complexes (cf. notations 1.1) together with suspension isomorphisms $\sigma^n : k^{n+1} \circ S \xrightarrow{\sim} k^n$ which fulfills the exactness axiom and the wedge axiom (cf. [Sw, Chap. 7], here S means of course the topological suspension functor).

Let $\tilde{\mathcal{L}}$ denote the category of topological cohomology theories. {The morphisms are the natural transformations, defined as above.} We want to define a restriction functor $\rho : \tilde{\mathcal{L}} \to \widetilde{Coho}(\mathbb{R})$. This will be slightly more complicated than the definition of the functors res_R^K above, since the underlying topological space M_{top} of a weak polytope over $\mathbb{R}$ is not necessarily a topological CW-complex. We will indicate the steps of the definition of ρ leaving the easy details to the reader.

Let k^* be a topological cohomology theory. For any $M \in \mathcal{P}^*(\mathbb{R})$ let $J(M)$ denote the set of all pairs $(X,[\varphi])$ with X a pointed CW-complex and $\varphi : M \to X$ a homotopy equivalence of pointed spaces. Such pairs exist by Theorem V.7.8. For any $n \in \mathbb{Z}$ the family $(k^n(X_{top}) \mid (X,[\varphi]) \in J(M))$ can be regarded as a direct system of abelian groups with an obvious and canonical transition isomorphism between any two members of the family. We define

$$(k^n)^{sa}(M) := \underrightarrow{\lim} \ (k^n(X_{top}) \mid (X,[\varphi]) \in J(M)).$$

One extends this definition to a definition of a contravariant functor $(k^n)^{sa} : H\mathcal{P}^*(\mathbb{R}) \to Ab$ in the obvious way. One then defines suspension isomorphisms

$$\sigma^n : (k^{n+1})^{sa} \circ S \xrightarrow{\sim} (k^n)^{sa}$$

in a natural straightforward way, and verifies for the functors $(k^n)^{sa}$ the exactness axiom and the wedge axiom. For exactness one needs CW-approximations of pairs of pointed weak polytopes, i.e. CW-triples with a one point space as last component. They exist by Theorem V.7.14 (or already V.7.8).

In this way a cohomology theory $(k^*)^{sa}$ over $\mathbb{R}$ is established which we call the <u>semialgebraic restriction</u> of k^*. If M is a pointed semialgebraic CW-complex then we may identify

$$(2.11) \qquad (k^n)^{sa}(M) = k^n(M_{top})$$

via the injection at $(M,[id_M]) \in J(M)$.

Every natural transformation $T : k^* \to l^*$ between topological cohomology theories leads to a natural transformation $T^{sa} : (k^*)^{sa} \to (l^*)^{sa}$ in a straightforward way such that, for every $M \in \mathcal{P}^*(\mathbb{R})$, $(X,[\varphi]) \in J(M)$, and $n \in \mathbb{Z}$, the square

$$\begin{array}{ccc} k^n(X_{top}) & \xrightarrow[\sim]{\varphi^*} & (k^n)^{sa}(M) \\ {\scriptstyle T^n(X_{top})}\Big\downarrow & & \Big\downarrow{\scriptstyle (T^n)^{sa}(M)} \\ l^n(X_{top}) & \xrightarrow[\varphi^*]{\sim} & (l^n)^{sa}(M) \end{array}$$

commutes. Here the horizontal arrows denote the natural injections at the "coordinate" $(X,[\varphi])$.

We now have established the desired restriction functor $\rho : k^* \mapsto (k^*)^{sa}$, $T \mapsto T^{sa}$, from $\tilde{\mathcal{L}}$ to $\widetilde{\mathrm{Coho}}(\mathbb{R})$.

<u>Theorem 2.12</u> (Second main theorem for cohomology theories). The restriction functor $\rho : \tilde{\mathcal{L}} \to \widetilde{\mathrm{Coho}}(\mathbb{R})$ is a natural equivalence. In particular, the topological cohomology theories correspond with the semialgebraic cohomology theories over $\mathbb{R}$ uniquely up to isomorphism.

The proof runs similar to the proof of the first main theorem 2.8. At crucial points Theorem V.7.16 is needed. We leave the details to the reader but mention that, given a semialgebraic cohomology theory l^* over $\mathbb{R}$, we can construct a topological cohomology theory l^*_{top} together with a natural equivalence T from $(l^*_{top})^{sa}$ to l^* in a canonical way (namely such that we obtain a functor $l^* \mapsto l^*_{top}$ right adjoint to $k^* \mapsto (k^*)^{sa}$, and T is one of the adjunction morphisms).

This is done as follows. For any $M \in \mathcal{W}^*$ let $I(M)$ denote the set of pairs $(X,[\varphi])$ with X a pointed semialgebraic CW-complex (over $\mathbb{R}$) and $[\varphi]$ the homotopy class of a topological homotopy equivalence $\varphi : X \xrightarrow{\sim} M$. Such pairs exist by V.7.16.i. We define the group $l^n_{top}(M)$ as the projective limit of the inverse system of abelian groups $(l^n(X) \mid (X,[\varphi]) \in I(M))$ with transition isomorphisms between any two members of the system emanating from Theorem V.7.16.ii.

If X is a pointed semialgebraic CW-complex then $I(X_{top})$ contains the pair $(X,[id_X])$. The projection $l^n_{top}(X_{top}) \to l^n(X)$ at the "coordinate" $(X,[id_X])$ is an isomorphism which is natural in X. Thus we may identify

$$(2.13)\qquad l^n_{top}(X_{top}) = l^n(X) \ .$$

If $M \in \mathcal{P}^*(\mathbb{R})$ then

$$\begin{aligned}(l^n_{top})^{sa}(M) &= \varinjlim (l^n_{top}(X_{top}) \mid (X,[\varphi]) \in J(M)) \\ &= \varinjlim (l^n(X) \mid (X,[\varphi]) \in J(M)) \ .\end{aligned}$$

We define $T^n(M) : (l^n_{top})^{sa}(M) \xrightarrow{\sim} l^n(M)$ as the isomorphism whose component at the coordinate $(X,[\varphi])$ is the isomorphism $l^n[\varphi]$.

We call l^*_{top} the <u>topological extension</u> of the semialgebraic cohomology theory l^*. Since we have a canonical natural equivalence T from $(l^*_{top})^{sa}$ to l^* we feel justified to identify, using T,

$$(2.14)\qquad (l^*_{top})^{sa} = l^* \ .$$

Given a topological cohomology theory k^* we have, for any $M \in \mathcal{W}^*$ and $n \in \mathbb{Z}$,

$$\begin{aligned}((k^n)^{sa})_{top}(M) &= \varprojlim ((k^n)^{sa}(X) \mid (X,[\varphi]) \in I(M)) \\ &\underset{(2.11)}{=} \varprojlim (k^n(X_{top}) \mid (X,[\varphi]) \in I(M)) \ .\end{aligned}$$

Thus we have a natural isomorphism

$$U^n(M) : k^n(M) \xrightarrow{\sim} ((k^n)^{sa})_{top}(M)$$

whose component at the coordinate $(X,[\varphi])$ is $k^n[\varphi]$. These isomorphisms fit together to a natural transformation U from k^* to $((k^*)^{sa})_{top}$. Again we identify, using U,

$$(2.15)\qquad ((k^*)^{sa})_{top} = k^* \ .$$

As a consequence of the main theorems 2.8 and 2.13 and the subsequent discussions we learn that every topological cohomology theory k^* induces, for every real closed field R, a semialgebraic cohomology theory $_Rk^*$ over R as follows:

$$(2.16) \qquad {}_Rk^* := (((k^*)^{sa})^{R_o})_R \ ,$$

with R_o the field of real algebraic numbers. Moreover, in this way the topological cohomology theories correspond with the cohomology theories over R in a unique way. We have

$$_{\mathbb{R}}k^* = (k^*)^{sa} \ , \ ({}_{\mathbb{R}}k^*)_{top} = k^* \ ,$$

and if $R \hookrightarrow K$ is an embedding of a real closed field into a real closed K then, with respect to this embedding,

$$({}_Kk^*)^R = {}_Rk^*, \ ({}_Rk^*)_K = {}_Kk^* \ .$$

By these wonderful correspondences and equalities we feel justified to use the following simplified notations.

<u>Notations 2.17.</u> If k^* is any topological cohomology theory and if M is a pointed weak polytope over any real closed field R then we denote the cohomology group $({}_Rk^n)(M)$ simply by $k^n(M)$. Similarly for any map $f : M \to N$ between pointed weak polytopes over R we denote the group homomorphism $({}_Rk^n)[f]$ simply by $k^n[f]$ or, more briefly, by f^*.

In particular we now can evaluate all the well known reduced topological cohomology theories (singular cohomology; orthogonal, unitary, and symplectic K-theory, cobordism theories ...) on pointed weak polytopes and maps between them over any real closed field.

§3 - Cellular homology

Having settled (to the same extent as in classical algebraic topology) the question which reduced cohomology theories exist over a given real closed field R we now may be brief about reduced homology theories. We use an obvious "dual" notation to the one in §2, Definition 1.

Definition 1. A reduced (semialgebraic) homology theory k_* over R is a family $(k_n | n \in \mathbb{Z})$ of covariant functors $k_n : H\mathcal{P}^*(R) \to Ab$ together with a family $(\sigma_n | n \in \mathbb{Z})$ of natural equivalences $\sigma_n : k_n \xrightarrow{\sim} k_{n+1} \circ S$ such that the following two axioms hold.

Exactness axiom. For every pair (M,A) of pointed weak polytopes over R and every $n \in \mathbb{Z}$ the sequence

$$k_n(A) \xrightarrow{i_*} k_n(M) \xrightarrow{p_*} k_n(M/A)$$

is exact.

Wedge axiom. For every family $(M_\lambda | \lambda \in \Lambda)$ of pointed weak polytopes and every $n \in \mathbb{Z}$ the map

$$(i_{\lambda *}) : \bigoplus_{\lambda \in \Lambda} k_n(M_\lambda) \to k_n(\bigvee_{\lambda \in \Lambda} M_\lambda)$$

is an isomorphism.

Dually to Definition 2 in §2 we define natural transformations between (reduced) homology theories over R and establish the category $\tilde{Ho}(R)$ of homology theories over R. Analogously we define the category $\tilde{\mathfrak{h}}$ of topological homology theories which live on $H\mathcal{W}^*$ instead of $H\mathcal{P}^*(R)$, cf. [Sw. Chap. 7].

With the only exception of Proposition 2.4 all the properties, theorems, and the corollary in §2 have obvious "duals" which can be proved in an analogous way. The homology counterpart of Proposition 2.4 is simpler then that proposition and reads as follows (cf. [Mi], [Sw, p. 121f] for a proof).

Proposition 3.1. Let $(M_n | n \in \mathbb{N})$ be an admissible filtration of a pointed weak polytope M over R. Then, for any reduced homology theory k_* over R and every $q \in \mathbb{Z}$, the natural map

$$\varinjlim_n k_q(M_n) \to k_q(M)$$

is an isomorphism.

The topological reduced homology theories correspond with the reduced homology theories over any real closed field R uniquely up to isomorphism, and we are justified to use a notation analogous to 2.17. Thus we can evaluate any reduced topological homology theory on pointed weak polytopes and maps between them over any real closed field R.

Definition 2. A reduced homology theory k_* over R - and similarly a reduced topological homology theory - is called ordinary, if $k_q(S^0) = 0$ for $q \neq 0$.

Remark 3.2. It will become obvious from Theorem 3.3 below (and is well known in the topological theory, cf. [Sw, Chap. 10]) that, for a given abelian group G, there exists, up to isomorphism, a unique ordinary reduced homology theory k over R with $k_0(S^0) \cong G$. This is the theory coming from reduced singular homology $H_*(-,G)$ on the category $\mathcal{WO}^*$.

Let k_* be a reduced ordinary homology theory over R and $G := k_0(S^0)$. Going to more concrete terms we want to explicate an almost combinatorial "cellular" description of the groups $k_n(M)$ of a given CW-complex M over R, which is completely analogous to the well known topological case, cf. [Sw, Chap. 10].

Recall that the <u>degree</u> deg(f) of a map $f : S^n \to S^n$ is the integer m with $[f] = m[id_{S^n}]$ in $\pi_n(S^n)$, provided $n > 0$. {N.B. This makes sense also if f does not preserve base points, since $\pi_1(S^n)$ acts trivially on $\pi_n(S^n)$.} If $n = 0$ we put $deg(f) = +1$ if $f = id_{S^0}$, $deg(f) = -1$ if f interchanges the two points of S^0, and $deg(f) = 0$ in the remaining cases.

For every cell σ of M we fix a characteristic map $\psi_\sigma : E^n \to M^n$ with $n := \dim \sigma$ (cf. V, §7). Let $\varphi_\sigma : S^{n-1} \to M^{n-1}$ denote the associated attaching map obtained from ψ_σ by restriction. For every cell τ of dimension n-1 we have a well defined base-point preserving map $\varphi_\sigma^\tau : S^{n-1} \to S^{n-1}$ such that the following diagram commutes.

$$\begin{array}{ccccc} S^{n-1} & \xrightarrow{\varphi_\sigma} & M^{n-1} & \longrightarrow & M^{n-1}/M^{n-1} \setminus \tau \\ \Big\downarrow \varphi_\sigma^\tau & & & & \Big\uparrow \cong \ \overline{\psi}_\tau \\ S^{n-1} & \longleftarrow & & & E^{n-1}/S^{n-2} \end{array} .$$

Here the unadorned arrows are canonical projections, and $\overline{\psi}_\tau$ is an isomorphism induced by ψ_τ. We put

$$[\sigma:\tau] = deg(\varphi_\sigma^\tau).$$

Notice that $[\sigma:\tau] = 0$ if τ is not an immediate face of σ.

We define the <u>cellular chain complex</u> $C.(M)$ as follows. If $n < 0$ then $C_n(M) = 0$. If $n \geq 0$ then $C_n(M)$ is the free abelian group generated by the set $\Sigma_n(M)$ of cells of dimension n. Here the base point x_o of M is <u>not</u> admitted as an element of $\Sigma_o(M)$, since M has to be regarded as a relative CW-complex $(M,\{x_o\})$. The boundary map $\partial : C_n(M) \to C_{n-1}(M)$ is given by the formula $(n \geq 0)$

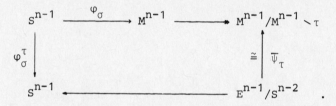

$$\partial(\sigma) := \sum_{\tau} [\sigma:\tau]\tau$$

with τ running through $\Sigma_{n-1}(M)$. Every cellular map $f : M \to N$ to a second pointed CW-complex N induces a chain map $C.(f) : C.(M) \to C.(N)$ in the obvious way. It will soon become obvious that indeed $\partial\partial = 0$.

Theorem 3.3. Let k_* be an ordinary homology theory and $G := k_o(S^o)$. There exist canonical isomorphisms

$$\alpha_M : H_n(C.(M) \otimes_{\mathbf{Z}} G) \to k_n(M)$$

such that, for every cellular map $f : M \to N$, the square

$$\begin{array}{ccc} H_n(C.(M) \otimes G) & \xrightarrow[\alpha_M]{\sim} & k_n(M) \\ {\scriptstyle H_n(C.(f) \otimes G)}\downarrow & & \downarrow{\scriptstyle f_*} \\ H_n(C.(N) \otimes G) & \xrightarrow[\alpha_N]{\sim} & k_n(N) \end{array}$$

commutes.

This can be derived directly from the axioms as in the topological theory, cf. [Sw, Chap. 10] where also the explicit description of α_M is given. The starting point is the observation that M^n/M^{n-1} is the wedge of copies of the sphere S^n, one copy for each $\sigma \in \Sigma_n(M)$. This implies $k_q(M^n/M^{n-1}) = 0$ for $q \neq n$ and $k_n(M^n/M^{n-1}) = C_n(M) \otimes_{\mathbf{Z}} G$. The case $G = \mathbf{Z}$ also gives an easy proof that for the boundary maps in $C.(M)$ indeed $\partial\partial = 0$ [loc.cit.].

In the special case that M is a geometric simplicial complex, hence $M = |K|_R$ with K the abstraction of M (III, §4), we can order [ES, p. 67]*) the abstract simplicial complex K in one of many ways and then have canonical characteristic maps at our disposal. Now $C.(M)$ coincides with the chain complex $C.(K)$ of the abstract pointed and ordered

---

*) cf. also p. 314 below.

simplicial complex K, and we obtain the following truly combinatorial description of $k_n(M)$.

Corollary 3.4. For every $n \in \mathbb{Z}$ and every abstract pointed ordered simplicial complex K there exists a canonical isomorphism

$$\alpha_K : H_n(K,G) \xrightarrow{\sim} k_n(|K|_R)$$

such that, for any simplicial map $f : K \to L$, the following square commutes:

$$\begin{array}{ccc} H_n(K,G) & \xrightarrow[\alpha_K]{\sim} & k_n(|K|_R) \\ {\scriptstyle H_n(f)}\downarrow & & \downarrow{\scriptstyle (|f|_R)_*} \\ H_n(L,G) & \xrightarrow[\alpha_L]{\sim} & k_n(|L|_R) \end{array} .$$

If the reduced homology theory k_* is not ordinary then again a cellular description of the groups $k_*(M)$ for M a pointed CW-complex is possible, although this can be enormously complicated. As in the topological theory [Sw, Chap. 15] one proves

Theorem 3.5 (Atiyah-Hirzebruch, G.W. Whitehead). There exists a natural homological spectral sequence $(E^r_{p,q}, d^r)$ with $E^2_{p,q} = H_p(C.(M) \otimes k_q(S^o))$ converging to $k_*(M)$.

In cohomology duals of these theorems 3.3 and 3.5 can be proved in an analogous way.

§4 - Homology of pairs of weak polytopes

We need some formalities about the relations between spaces with and without base points.

Notations 4.1. a) If M is a weak polytope over R then M^+ denotes the direct sum $M \sqcup \{+\}$ with a one point space $\{+\}$, regarded as a pointed weak polytope with base point +. {We hesitate to use this notation for arbitrary spaces since this would conflict with Def. 3 in IV, §6.}
b) If N is a pointed space over R then N^o denotes the space N forgetting the base point.

We have a natural bijection

(4.2) $$[M,N^o] = [M^+,N]$$

where, of course, the left hand side denotes the set of (free) homotopy classes of maps from M to N^o while the right hand side denotes the set of base point preserving homotopy classes of maps from M^+ to N. If X is a second weak polytope then

(4.3) $$M^+ \wedge X^+ = (M \times X)^+.$$

If $(M_\lambda \mid \lambda \in \Lambda)$ is a family of weak polytopes then

(4.4) $$\bigvee_{\lambda\in\Lambda} M_\lambda^+ = (\bigsqcup_{\lambda\in\Lambda} M_\lambda)^+ .$$

Our objective in this section and the next one is to extend an arbitrary reduced homology or cohomology theory over R in the correct way from the category $H\mathcal{P}*(R)$ to the homotopy category HWSA(2,R) of all pairs of spaces over R and to understand the formal properties of the extended theory.

In this section we will focus attention at WP-pairs (= pairs of weak polytopes). The category $\mathcal{P}(2,R)$ of these pairs and the homotopy category $H\mathcal{P}(2,R)$ both have a natural endomorphism E which maps a pair (M,A) to the pair $(A,\emptyset)$.

If F is a covariant or contravariant functor from $H\mathcal{P}(2,R)$ to Ab then usually we write F(M) instead of $F(M,\emptyset)$. {Recall that we regard $H\mathcal{P}(R)$ as a full subcategory of $H\mathcal{P}(2,R)$, cf. 1.1.} If $f : (M,A) \to (N,B)$ is a map between WP-pairs then most often we denote the homomorphism F[f] by f_* in the covariant case and by f^* in the contravariant case. Analogous conventions will be obeyed for other space categories.

In the last sections we have been most of the time more explicit on cohomology than on homology. For justice we now give preference to homology.

Definition 1. A homology theory (more precisely, an unreduced semialgebraic homology theory) h_* over R is a family $(h_n | n \in \mathbb{Z})$ of covariant functors $h_n : H\mathcal{P}(2,R) \to Ab$ together with a family $(\partial_n | n \in \mathbb{Z})$ of natural transformations $\partial_n : h_n \to h_{n-1} \circ E$ which satisfy the following axioms.

Exactness. For every WP-pair (M,A) the long sequence

$$\longrightarrow h_n(A) \xrightarrow[i_*]{} h_n(M) \xrightarrow[j_*]{} h_n(M,A) \xrightarrow[\partial_n(M,A)]{} h_{n-1}(A) \longrightarrow \ldots$$

is exact. Here i and j denote the inclusions from $(A,\emptyset)$ to $(M,\emptyset)$ and from $(M,\emptyset)$ to (M,A). The sequence is called the homology sequence of the WP-pair (M,A).

Excision. If A and B are closed subspaces of a weak polytope M with $M = A \cup B$ then the inclusion $i : (A, A\cap B) \to (M,B)$ induces an isomorphism

$$i_* : h_n(A, A\cap B) \xrightarrow{\sim} h_n(M,B)$$

for every $n \in \mathbb{Z}$.

Additivity. If M is the direct sum (IV,1.10) of a family $(M_\lambda | \lambda \in \Lambda)$ of weak polytopes then the inclusions $i_\lambda : M_\lambda \to M$ induce an isomorphism

$$(i_{\lambda *}) : \bigoplus_{\lambda \in \Lambda} h_n(M_\lambda) \xrightarrow{\sim} h_n(M)$$

for every $n \in \mathbb{Z}$.

These axioms are the precise analogues of the axioms of a (strongly additive) homology theory on the category $\mathcal{W}\mathcal{O}(2)$ of pairs of topological CW-complexes [Sw, Chap. 7]. We call such a homology here a topological homology theory, more precisely an unreduced topological homology theory.

We draw some consequences from the axioms by arguments very well known from topology and mostly not repeated here, cf. [ES, Chap. 1], [Sw, Chap. 7]. Let h_* be a homology theory over R.

For any point x of a weak polytope M over R there is a unique retraction from M to $\{x\}$. Thus the homology exact sequence of the pair $(M,\{x\})$ splits and we have a canonical isomorphism

(4.5) $$h_n(M) \xrightarrow{\sim} h_n(*) \oplus h_n(M,\{x\}) .$$

If M is contractible then $\{x\} \hookrightarrow M$ is a homotopy equivalence and we conclude that $h_n(M,\{x\}) = 0$. (More generally, whenever (M,A) is a pair with A a strong deformation retract of M then $h_n(M,A) = 0$ for all n.)

Proposition 4.6 (Homology sequence of a triple). Let (M,A,C) be a decreasing WP-triple, and let $i : (A,C) \to (M,C)$, $j : (M,C) \to (M,A)$ denote the associated inclusion maps of WP-pairs. Let i_1 denote the inclusion $(A,\emptyset) \to (A,C)$. Finally, for any $n \in \mathbb{Z}$, we introduce the

the composite homomorphism

$$\partial_n(M,A,C) : h_n(M,A) \xrightarrow[\partial_n(M,A)]{} h_{n-1}(A) \xrightarrow[(i_1)_*]{} h_{n-1}(A,C) \quad .$$

The long sequence

$$\xrightarrow[\partial_{n+1}(M,A,C)]{} h_n(A,C) \xrightarrow[i_*]{} h_n(M,C) \xrightarrow[j_*]{} h_n(M,A) \xrightarrow[\partial_n(M,A,C)]{} h_{n-1}(A,C) \rightarrow$$

is exact.

<u>Proposition 4.7</u> (Mayer-Vietoris sequence). Let A,B,C be closed subspaces of a weak polytope M with $A \cup B = M$ and $A \cap B \supset C$. Let $\Delta_n = \Delta_n(M,A,B,C)$ denote the composite

$$h_n(M,C) \xrightarrow[I_{1*}]{} h_n(M,B) \xleftarrow[j_{1*}]{\sim} h_n(A,A\cap B) \xrightarrow[\partial_n(A,A\cap B,C)]{} h_{n-1}(A\cap B,C)$$

with inclusion maps j_1, I_1. The following long sequence is exact.

$$\xrightarrow{\Delta_{n+1}} h_n(A\cap B,C) \xrightarrow{\alpha} h_n(A,C) \oplus h_n(B,C) \xrightarrow{\beta} h_n(M,C) \xrightarrow{\Delta_n} h_{n-1}(A\cap B,C) \rightarrow$$

where $\alpha(x) = (i_{1*}(x), i_{2*}(x))$ and $\beta(u,v) = i_{3*}(u) - i_{4*}(v)$ with inclusion maps i_1, i_2, i_3, i_4. Moreover,

$$\Delta_n(M,B,A,C) = -\Delta_n(M,A,B,C) \quad .$$

We now study the restrictions $\tilde{h}_n$ of the functors h_n to the full subcategory $H\mathcal{P}^*(R)$ of $H\mathcal{P}(2,R)$. If M is a pointed weak polytope with base point x then, by definition,

$$\tilde{h}_n(M) = h_n(M^o,\{x\}) \quad .$$

For every WP-pair (M,A) with A not empty we regard M/A as a pointed space with base point A/A. If $A = \emptyset$ then we put $M/A = M^+$.

<u>Proposition 4.8.</u> For every WP-pair (M,A) and every $n \in \mathbb{Z}$ the natural projection $p : (M,A) \rightarrow (M/A,A/A)$ induces an isomorphism

$$p_* : h_n(M,A) \xrightarrow{\sim} \tilde{h}_n(M/A) \ .$$

<u>Proof.</u> This is evident if $A=\emptyset$. Assume now that A is not empty. Let $M \cup CA$ denote the cofiber of the inclusion $A \to M$ (cf. 1.6.ii). We have a commuting triangle of WP-pairs

$$\begin{array}{ccc} (M,A) & \xhookrightarrow{\ i\ } & (M\cup CA, CA) \\ & {\scriptstyle p}\searrow \quad \swarrow{\scriptstyle \pi} & \\ & (M/A, A/A) & \end{array}$$

with π a homotopy equivalence. {Illegally here $M \cup CA$, CA, M/A, A/A mean spaces without base point. We omitted the upper index $^{\circ}$.} By the excision axiom $h_n[i]$ is an isomorphism. Thus also $h_n[p]$ is an isomorphism. q.e.d.

Given a pointed weak polytope $M = (N,x)$ we obtain an isomorphism $\sigma_n(M)$ from $\tilde{h}_n(M)$ to $\tilde{h}_{n+1}(SM)$ for every $n \in \mathbb{Z}$ as follows. Recall that $SM = (\tilde{C}N/N,\bar{x})$ with $\tilde{C}N$ the reduced cone of N and $\bar{x}$ its natural base point, which is the image of the point $x \in N \subset \tilde{C}N$ under the natural projection $p : \tilde{C}N \to \tilde{C}N/N$. {Again illegally, we regard $\tilde{C}N$ and $\tilde{C}N/N$ as spaces without base point omitting the index $^{\circ}$.} Since $\tilde{C}N$ is contractible $h_n(\tilde{C}N,\{x\}) = 0$. The homology sequence of the triple $(\tilde{C}N,N,\{x\})$ (cf. Prop. 4.6) shows that the boundary maps $\partial_n := \partial_n(\tilde{C}N,N,\{x\})$ are isomorphisms. We define $\sigma_n(M)$ as the composite

$$h_n(N,\{x\}) \xrightarrow[\partial^{-1}_{n+1}]{\sim} h_{n+1}(\tilde{C}N,N) \xrightarrow[p_*]{} h_{n+1}(\tilde{C}N/N,\{\bar{x}\}) \quad .$$

The second arrow is an isomorphism by Proposition 4.8 above. Thus $\sigma_n(M)$ is an isomorphism. It is natural in M.

<u>Proposition 4.9.</u> $(\tilde{h}_n | n \in \mathbb{Z})$ together with the family $(\sigma_n | n \in \mathbb{Z})$ of natural equivalences $\sigma_n : \tilde{h}_n \to \tilde{h}_{n+1} \circ S$ is a reduced homology theory over R.

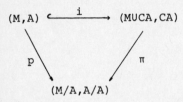

Proof. We obtain the exactness axiom (§3, Def. 1) for a given pointed WP-pair (M,A) from 4.8 and the homology sequence 4.6 of the triple $(M^o,A^o,\{x\})$, with x the base point of M and A. It remains to verify the wedge axiom for a given family $(M_\lambda \mid \lambda \in \Lambda)$ of pointed weak polytopes. We consider the direct sum $N := \coprod_\lambda M_\lambda^o$ and its closed subspace $B := \coprod_\lambda \{x_\lambda\}$, with x_λ the base point of M_λ. We have $N/B = \bigvee_\lambda M_\lambda$. We denote this pointed weak polytope by M. Since B is a retract of N the homology sequence of the pair (N,B) (cf. Def. 1) splits into short exact sequences. Using (4.8) and the additivity axiom for h_* we obtain short exact sequences

$$0 \to \bigoplus_\lambda h_n(\{x_\lambda\}) \to \bigoplus_\lambda h_n(M_\lambda^o) \to \tilde{h}_n(M) \to 0 \quad .$$

The cokernel of the first map in the sequence is $\bigoplus_\lambda \tilde{h}_n(M_\lambda)$ (cf. 4.5). Thus we obtain an isomorphism from $\bigoplus_\lambda \tilde{h}_n(M_\lambda)$ to $\tilde{h}_n(M)$. It is easily checked that this isomorphism is induced by the inclusions $i_\lambda : M_\lambda \hookrightarrow M$.

q.e.d.

We now start with a reduced homology theory k_* over R and want to "extend" it to an unreduced homology theory $\hat{k}_*$. For any pair (M,A) of weak polytopes over R we define

$$\hat{k}_n(M,A) := k_n(M/A) \quad ,$$

and thus indeed obtain extensions $\hat{k}_n : H\mathcal{P}(2,R) \to Ab$ of the functors $k_n : H\mathcal{P}^*(R) \to Ab$. Notice that for any weak polytope M,

$$\hat{k}_n(M) := \hat{k}_n(M,\emptyset) = k_n(M^+) \quad .$$

Let (M,A) be a pair of weak polytopes over R. Then (M^+,A^+) is a pair of pointed weak polytopes. We identify $M^+/A^+ = M/A$. Let p denote the natural projection from M^+ to M/A. The homomorphism $k_n[p]$ from $k_n(M^+)$ to $k_n(M/A)$ coincides with the homomorphism $\hat{k}_n[j]$ induced by the inclusion map $j : (M,\emptyset) \to (M,A)$. Applying the dual of Proposition 2.1 to the pair (M^+,A^+) we obtain a long exact sequence

$$\xrightarrow[\partial_{n+1}]{} \hat{k}_n(A) \xrightarrow[i_*]{} \hat{k}_n(M) \xrightarrow[j_*]{} \hat{k}_n(M,A) \xrightarrow[\partial_n]{} \hat{k}_{n-1}(A) \rightarrow$$

with homomorphisms $\partial_n = \partial_n(M,A)$ which naturally depend on the pair (M^+,A^+) and hence on (M,A).

Proposition 4.10. $(\hat{k}_n | n\in\mathbb{Z})$ together with this family $(\partial_n | n\in\mathbb{Z})$ of natural transformations $\partial_n : \hat{k}_n \rightarrow \hat{k}_{n-1} \circ E$ is a homology theory over R.

Proof. The exactness axiom has just been established. Excision is evident by the definition of $\hat{k}_n$. Additivity follows from (4.4). q.e.d.

As for reduced homology theories there is an obvious notion of a natural transformation $T : h_* \rightarrow g_*$ between homology theories (cf. [Sw, Chap. 7]), and we can introduce the category Ho(R) of homology theories over R with these natural transformations as morphisms.

The assignment $h_* \mapsto \tilde{h}_*$ extends naturally to a functor Red : Ho(R) $\rightarrow$ $\tilde{\text{Ho}}$(R). Similarly the assignment $k_* \mapsto \hat{k}_*$ extends naturally to a functor Unred : $\tilde{\text{Ho}}$(R) $\rightarrow$ Ho(R).

Theorem 4.11. These functors are quasi-inverse natural equivalences between Ho(R) and $\tilde{\text{Ho}}$(R), i.e. there exist natural equivalences $id_{Ho(R)} \xrightarrow{\sim}$ Unred $\circ$ Red and $id_{\tilde{Ho}(R)} \xrightarrow{\sim}$ Red $\circ$ Unred. In particular, up to isomorphism, the homology theories over R correspond uniquely with the reduced homology theories over R.

Since we have been very explicit on similar matters in §2 we now can safely leave the proof of this theorem to the reader.

Analogous natural equivalences $h_* \mapsto \tilde{h}_*$, $k_* \mapsto \hat{k}_*$ exist between the categories $\mathfrak{h}$ and $\tilde{\mathfrak{h}}$ of unreduced and reduced topological homology theories, as is well known and proved in the same way.

Henceforth we usually will identify $\hat{\hat{k}}_* = k_*$, $\hat{h}_* = h_*$, both in the semialgebraic and the topological setting.

If $K \supset R$ is a real closed base field extension then it is clear from §2 that for every homology theory h_* over K we have a homology theory h_*^R over R, called the <u>restriction of</u> h_* <u>to</u> R, such that $(h_*^R)^\sim = (\tilde{h}_*)^R$.

Conversely, every homology theory g_* over R yields a homology theory $(g_*)_K$ over K, called the <u>extension of</u> g_* <u>to</u> K, such that $((g_*)_K)^\sim = (\tilde{g}_*)_K$. In this way the homology theories over R correspond with the homology theories over K uniquely up to natural equivalence.

For any WP-pair (M,A) over R we have formulas, in analogy to (2.6) and (2.9),

$$
\begin{aligned}
(4.11)\qquad & h_n^R(M,A) = h_n(M(K),A(K)) \\
& (g_n)_K(M(K),A(K)) = g_n(M,A)
\end{aligned}
$$

and similar formulas for the boundary maps ∂_n .

Every topological homology theory h_* "restricts" to a semialgebraic homology theory h_*^{sa} over $\mathbb{R}$ such that $(h_*^{sa})^\sim = (\tilde{h}_*)^{sa}$. Conversely, every semialgebraic homology theory g_* over $\mathbb{R}$ "extends" to a topological homology theory $(g_*)_{top}$ such that $((g_*)_{top})^\sim = (\tilde{g}_*)_{top}$. In this way the semialgebraic homology theories over $\mathbb{R}$ correspond uniquely with the topological ones, up to natural equivalence. For every pair (M,A) of semialgebraic CW-complexes over $\mathbb{R}$,

$$
\begin{aligned}
(4.12)\qquad & h_n^{sa}(M,A) = h_n(M_{top},A_{top}) \\
& (g_n)_{top}(M_{top},A_{top}) = g_n(M,A) \ .
\end{aligned}
$$

Consequently, given a topological homology theory h_*, there exists over each real closed field R a homology theory

$$ {}_Rh_* = ((h_*^{sa})^{R_o})_R $$

which corresponds canonically to h_*.

The pendant ${}_RH_*(-,G)$ over R of singular homology $H_*(-,G)$ with coefficients in some abelian group G can be evaluated on a CW-pair (M,A) over R by a cellular procedure analogous to Theorem 3.3. For an extraordinary topological homology theory h_* we can describe ${}_Rh_n(M,A)$ in a cellular way analogous to Theorem 3.5.

We now define unreduced cohomology theories.

Definition 2. A (semialgebraic unreduced) cohomology theory h^* over R is a family $(h^n \mid n\in\mathbb{Z})$ of contravariant functors $h^n : H\mathcal{P}(2,R) \to Ab$ together with a family $(\delta^n \mid n\in\mathbb{Z})$ of natural transformations $\delta^n : h^n \circ E \to h^{n+1}$ such that the following three axioms hold (notations as in Def. 1).

Exactness. For every pair (M,A) of weak polytopes the long sequence

$$ \to h^n(M,A) \xrightarrow{j^*} h^n(M) \xrightarrow{i^*} h^n(A) \xrightarrow[\delta^n(M,A)]{} h^{n+1}(M,A) \to $$

is exact.

Excision. If A and B are closed subspaces of a weak polytope M with $A \cup B = M$ then for every $n \in \mathbb{Z}$, the natural map

$$ i^* : h^n(M,B) \to h^n(A, A\cap B) $$

is an isomorphism.

Additivity. If M is the direct sum of a family $(M_\lambda \mid \lambda\in\Lambda)$ of weak polytopes then, for every $n \in \mathbb{Z}$, the natural map

$$(i_\lambda^*) : h^n(M) \to \prod_{\lambda \in \Lambda} h^n(M_\lambda)$$

is an isomorphism

All our considerations on homology theories can be repeated word by word for cohomology theories. They lead to notations and results analogous to the ones above, which we will use without further explanation.

§5 - Homology of pairs of spaces

We want to extend a given homology theory h_* over R from the category $H\mathcal{P}(2,R)$ to the homotopy category HWSA(2,R) of all pairs of spaces over R. This will be done by use of WP-approximations (cf. V, §4) in a way similar to the procedure in §2.

We first deal with the simpler problem how to extend a single covariant functor $F : H\mathcal{P}(2,R) \to Ab$ to HWSA(2,R) in a canonical way. Given a pair of spaces (M,A) over R let I(M,A) denote the set of triples $(X,B,[\varphi])$ with (X,B) a WP-pair over R and $[\varphi]$ the homotopy class of a WP-approximation $\varphi : (X,B) \to (M,A)$. {It turns out to be convenient to write a triple $(X,B,[\varphi])$ instead of the class $[\varphi]$, although this is redundant notation. As elsewhere we ignore set theoretic difficulties.} Given any two triples $(X,B,[\varphi])$, $(Y,C,[\psi])$ in I(M,A) there exists, up to homotopy, a unique map $\chi : (X,B) \to (Y,C)$ with $\psi \circ \chi \simeq \varphi$, and χ is a homotopy equivalence (Th. V.4.10.ii). We thus have a canonical isomorphism $\chi_* : F(X,B) \xrightarrow{\sim} F(Y,C)$.

We define an abelian group $\check{F}(M,A)$ as the direct limit of the family of abelian groups $(F(X,B) \mid (X,B,[\varphi]) \in I(M,A))$ with these isomorphisms χ_* as transition maps between any two members of the family. For every $(X,B,[\varphi]) \in I(M,A)$ let

$$\varphi_* : F(X,B) \to \check{F}(M,A)$$

denote the canonical map emanating from the definition of $\varinjlim$. It is an isomorphism.

Let now [f] be a homotopy class of maps from (M,A) to a second pair of spaces (M',A'). Then we obtain a natural homomorphism $\check{F}[f]$ from $\check{F}(M,A)$ to $\check{F}(M',A')$ as follows. We choose WP-approximations

$\varphi : (X,B) \to (M,A)$ and $\varphi' : (X',B') \to (M',A')$. By Cor. V.6.15 there exists up to homotopy a unique map $g : (X,B) \to (X',B')$ such that $\varphi' \circ g \simeq f \circ \varphi$. We define $\check{F}[f]$ as the dotted arrow which makes the square

$$\begin{array}{ccc} F(X,B) & \xrightarrow{F[g]} & F(X',B') \\ \varphi_* \downarrow \cong & & \cong \downarrow \varphi'_* \\ F(M,A) & \xrightarrow{\check{F}[f]} & F(M',A') \end{array}$$

commutative. It is easily seen that $\check{F}[f]$ does not depend on the choice of φ and φ'. In this way we obtain a covariant functor

$$\check{F} : HWSA(2,R) \to Ab.$$

It is clear from the last sentence in V.6.15 that $\check{F}$ turns every weak homotopy equivalence $f : (M,A) \to (M',A')$ into an isomorphism $\check{F}[f]$.

Given a WP-pair (M,A) over R there exists a distinguished element $(M,A,[id_{(M,A)}])$ in $I(M,A)$. Thus we have a canonical isomorphism

$$(id_{(M,A)})_* : F(M,A) \xrightarrow{\sim} \check{F}(M,A) .$$

These isomorphisms fit together to a natural equivalence from the functor F to the restriction $\check{F}|H\mathcal{P}(2,R)$ of $\check{F}$. We feel justified to identify

$$\check{F}|H\mathcal{P}(2,R) = F,$$

and thus we write $\check{F}(M,A) = F(M,A)$ and $\check{F}[f] = F[f]$ for any WP-pair (M,A) over R and any map f between such pairs.

If $\varphi : (X,B) \to (M,A)$ is a WP-approximation of a pair of spaces over R then the homomorphism φ_* from above coincides with $\check{F}[\varphi]$. We now feel free to abbreviate $\check{F}[f]$ by f_* for any map between pairs of spaces over R.

We call $\check{F}$ the <u>extension of the functor</u> F <u>to</u> HWSA(2,R) <u>by WP-approximation</u>. It is easily seen that every natural transformation $T : F \to G$ from F to a second functor $G : H\mathcal{P}(2,R) \to Ab$ extends in a unique way to a natural transformation $\check{T} : \check{F} \to \check{G}$, called the <u>extension of</u> T <u>by WP-approximation</u>.

By an analogous procedure, using classical CW-approximation [W, p. 224] we can extend any covariant functor $F : H\mathcal{W}(2) \to Ab$ on the homotopy category of pairs of topological CW-complexes canonically to a functor $\check{F} : HTOP(2) \to Ab$ which turns topological weak homotopy equivalences into isomorphisms.

More generally all this works for functors on $H\mathcal{P}(2,R)$ or $H\mathcal{W}(2)$ with values in the category Gr of groups or the category Set of sets instead of Ab and, of course, also for contravariant functors. The whole point is that for diagrams in Ab or Gr or Set there is a distinguished way to produce direct and inverse limits.

If h_* is a homology theory over R then we obtain for every h_n a functor $\check{h}_n$ on HWSA(2,R). We want to understand the properties of the family $(\check{h}_n | n \in \mathbb{Z})$ of these functors. Here the difficult question is what sort of excision theorem holds for the $\check{h}_n$. It turns out to be convenient to ignore this question for some time first dealing with easier matters.

<u>Definition 1.</u> Let $\mathfrak{A}$ be a space category. {We are a bit vague here. Important cases are $\mathfrak{A} = \mathcal{P}(R)$, WSA(R), $\mathcal{W}$, TOP with the usual notion of subspace.} Let $H\mathfrak{A}(2)$ denote the homotopy category of pairs of spaces in $\mathfrak{A}$, and let E denote the endomorphism $(M,A) \mapsto (A,\emptyset)$ of $H\mathfrak{A}(2)$. {We use the letter E uniformly for all $\mathfrak{A}$.}

a) A <u>prehomology theory</u> h_* on $\mathfrak{A}$ is a sequence $(h_n | n \in \mathbb{Z})$ of covariant functors $h_n : H\mathfrak{A}(2) \to Ab$ together with a sequence $(\partial_n | n \in \mathbb{Z})$ of natural transformations $\partial_n : h_n \to h_{n-1} \circ E$ such that, for every pair $(M,A) \in \mathfrak{A}$, the "long homology sequence"

$$\xrightarrow[\partial_{n+1}(M,A)]{} h_n(A) \xrightarrow[i_*]{} h_n(M) \xrightarrow[j_*]{} h_n(M,A) \xrightarrow[\partial_n(M,A)]{} h_{n-1}(A) \to$$

is exact. Here i and j denote the inclusions from $(A,\emptyset)$ to $(M,\emptyset)$ and from $(M,\emptyset)$ to (M,A).

b) A natural transformation $T : h_* \to h'_*$ between prehomology theories on $\mathfrak{A}$ is a sequence $(T_n | n \in \mathbb{Z})$ of natural transformations $T_n : h_n \to h'_n$ such that for every $(M,A) \in \mathfrak{A}(2)$ the following square commutes.

$$\begin{array}{ccc} h_n(M,A) & \xrightarrow{T_n(M,A)} & h'_n(M,A) \\ \downarrow{\scriptstyle \partial_n(M,A)} & & \downarrow{\scriptstyle \partial'_n(M,A)} \\ h_{n-1}(A) & \xrightarrow[T_{n-1}(A,\emptyset)]{} & h'_{n-1}(A) \end{array}$$

{As in previous sections we write $h_q(A)$ instead of $h_q(A,\emptyset)$.} We call T a natural equivalence (or isomorphism) from h_* to h'_* if every T_n is a natural equivalence (i.e. every $T_n(M,A)$ is an isomorphism).

Let h_* be a prehomology theory on $\mathcal{P}(R)$. Every natural transformation $\partial_n : h_n \to h_{n-1} \circ E$ extends to a natural transformation $\check{\partial}_n : \check{h}_n \to (h_{n-1} \circ E)\check{}$, and it is pretty obvious that $(h_{n-1} \circ E)\check{} = \check{h}_{n-1} \circ E$.

Proposition 5.1. The sequence of functors $(\check{h}_n | n \in \mathbb{Z})$ together with the sequence of natural transformations $(\check{\partial}_n | n \in \mathbb{Z})$ is a prehomology theory $\check{h}_*$ on WSA(R).

Proof. Given a pair (M,A) of spaces over R we choose a WP-approximation $\varphi : (X,B) \to (M,A)$. We then can compare the h_*-homology sequence of (X,B) with the $\check{h}_*$-homology sequence of (M,A) using the isomorphisms φ_*, ψ_*, χ_* induced by φ and its restrictions $\psi : (X,\emptyset) \to (M,\emptyset)$,

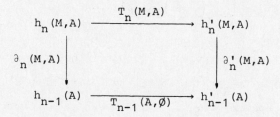

$\chi : (B,\emptyset) \to (A,\emptyset)$ which are again WP-approximations. Since the h_*-homology sequence of (X,B) is exact the same holds for the $\check{h}_*$-homology sequence of (M,A). q.e.d.

Proposition 5.2. For every natural transformation $T : h_* \to g_*$ between prehomology theories on $\mathcal{P}(R)$ the sequence $(\check{T}_n | n \in \mathbb{Z})$ is a natural transformation $\check{T} : \check{h}_* \to \check{g}_*$ between prehomology theories on WSA(R). If T is an isomorphism then $\check{T}$ is an isomorphism.

We leave the easy proof to the reader. We call $\check{h}_*$ (resp. $\check{T}_*$) the extension of h_* (resp. T_*) to HWSA(R) by WP-approximations.

Similarly, by classical CW-approximation every prehomology theory h_* on $\mathcal{W}$ extends to a prehomology theory $\check{h}_*$ on TOP and every natural transformation (resp. isomorphism) $T : h_* \to g_*$ between such theories extends to a natural transformation (resp. isomorphism) $\check{T} : \check{h}_* \to \check{g}_*$.

Given any homology theory g_* over R we want to understand the formal properties of the prehomology theory $\check{g}_*$. For all questions we have in mind it does not matter whether we look at $\check{g}_*$ or at a prehomology theory isomorphic to $\check{g}_*$. We know from §4 that there exists a topological homotopy theory h_*, unique up to isomorphism, such that g_* is isomorphic to ${}_R h_*$. By Proposition 5.2 the prehomology theory $\check{g}_*$ is isomorphic to $({}_R h_*)^\vee$. Thus it suffices to study the theory $({}_R h_*)^\vee$. This allows us to simplify notations drastically.

Notations 5.3. For the rest of this section, and in §6 as well, h_* is a fixed topological homology theory (living on $H\mathcal{W}(2)$).

i) If (M,A) is a pair of topological spaces then we write $h_n(M,A)$ and $\partial_n(M,A)$ instead of $\check{h}_n(M,A)$ and $\check{\partial}_n(M,A)$.

ii) If (M,A) is a pair of spaces over R then we write $h_n(M,A)$ and

$\partial_n(M,A)$ instead of $({}_Rh_n)^\vee(M,A)$ and $({}_R\partial_n)^\vee(M,A)$, and rather often instead of $\partial_n(M,A)$ more briefly, but abusively, ∂_n or even ∂.

iii) If $M = (N,x)$ is a pointed space over R or a pointed topological space we define

$$\tilde{h}_n(M) := h_n(N,\{x\}) .$$

iv) If M is a pointed weak polytope over R then we denote the suspension isomorphism $\tilde{h}_n(M) \to \tilde{h}_{n+1}(SM)$ of the reduced homology theory ${}_R\tilde{h}_*$ by $\sigma_n(M)$. Similarly, if M is a pointed topological space, we denote its topological suspension [Sw, Chap. 2] by SM and the natural isomorphism $\tilde{h}_n(M) \xrightarrow{\sim} \tilde{h}_n(SM)$ (cf. [Sw, Chap. 7]) again by $\sigma_n(M)$. In both cases we often write more briefly σ_n, or even σ, instead of $\sigma_n(M)$.

Theorem 5.4 (First and second main theorem for homology). i) Let K be a real closed field extension of a real closed field R. For every pair of spaces (M,A) over R and every $n \in \mathbb{Z}$ there exists a natural isomorphism $\kappa_n(M,A) : h_n(M,A) \xrightarrow{\sim} h_n(M(K),A(K))$ such that all the squares

$$\begin{array}{ccc} h_n(M,A) & \xrightarrow{\partial_n(M,A)} & h_{n-1}(A) \\ \kappa_n(M,A) \downarrow \cong & & \cong \downarrow \kappa_{n-1}(A,\emptyset) \\ h_n(M(K),A(K)) & \xrightarrow[\partial_n(M(K),A(K))]{} & h_{n-1}(A(K)) \end{array}$$

commute.

ii) For every pair of weakly semialgebraic spaces (M,A) over $\mathbb{R}$ there exists a natural isomorphism

$$\lambda_n(M,A) : h_n(M,A) \xrightarrow{\sim} h_n(M_{top},A_{top})$$

such that all the squares

$$\begin{array}{ccc} h_n(M,A) & \xrightarrow{\partial_n(M,A)} & h_{n-1}(A) \\ \lambda_n(M,A) \downarrow \cong & & \cong \downarrow \lambda_{n-1}(A,\emptyset) \\ h_n(M_{top},A_{top}) & \xrightarrow[\partial_n(M_{top},A_{top})]{} & h_{n-1}(A_{top}) \end{array}$$

commute.

Indication of proof. In §4 we have already established such isomorphisms $\kappa_n(M,A)$, $\lambda_n(M,A)$ for pairs of weak polytopes (M,A), and then have identified $h_n(M,A)$ with $h_n(M(K),A(K))$ and $h_n(M_{top},A_{top})$ respectively by these isomorphisms. We now define these isomorphisms for an arbitrary pair of spaces (M,A) over R (resp. $\mathbb{R}$) leaving the verification of the various commutativities to the reader.

i) Let (M,A) be a pair of spaces over R. If $\varphi : (X,B) \to (M,A)$ is a WP-approximation of (M,A) then φ_K is a WP-approximation of $(M(K),A(K))$. Thus the direct system $(h_n(X,B) \mid (X,B,[\varphi]) \in I(M,A))$ from above can be regarded as a subsystem of $(h_n(Y,C) \mid (Y,C,[\psi]) \in I(M(K),A(K))$ in a canonical way. We define $\kappa_n(M,A)$ as the natural map from the direct limit of the first system to the direct limit of the second one. It is an isomorphism.

ii) Let (M,A) be a pair of weakly semialgebraic spaces over $\mathbb{R}$. Let $I'(M,A)$ denote the subset of $I(M,A)$ consisting of all triples $(X,B,[\varphi])$ with $\varphi : (X,B) \to (M,A)$ a semialgebraic CW-approximation. Further let $J(M_{top},A_{top})$ denote the set of triples $(Y,C,[\psi])$ with $\psi : (Y,C) \to (M_{top},A_{top})$ a topological CW-approximation.We have an evident map $\alpha_n(M,A) : \varinjlim (h_n(X,B) \mid (X,B,[\varphi]) \in I'(M,A)) \to h_n(M,A)$. The system $(h_n(X,B) \mid (X,B,[\varphi])) \in I'(M,A)$ can be regarded canonically as a subsystem of $(h_n(Y,C) \mid (Y,C,[\psi]) \in J(M_{top},A_{top}))$. We define $\beta_n(M,A)$ as the induced map from the direct limit of the first system to the direct limit of the second system, which is $h_n(M_{top},A_{top})$. Both $\alpha_n(M,A)$ and $\beta_n(M,A)$ are isomorphisms. We define

$$\lambda_n(M,A) = \beta_n(M,A) \circ \alpha_n(M,A)^{-1} .$$

(End of our explanations concerning Th. 5.4).

We now are justified to identify $h_n(M,A)$ with $h_n(M(K),A(K))$ (resp. $h_n(M_{top},A_{top})$) by the isomorphism $\kappa_n(M,A)$ (resp. $\lambda_n(M,A)$). This means that most often we write

$$h_n(M,A) = h_n(M(K),A(K)), \text{ resp. } h_n(M,A) = h_n(M_{top},A_{top}) ,$$

and, of course, also

$$\partial_n(M,A) = \partial_n(M(K),A(K)), \text{ resp. } \partial_n(M,A) = \partial_n(M_{top},A_{top}) .$$

Example 5.5. Let h_* be singular homology $H_*(-,G)$ with coefficients in some abelian group G. For (M,A) a pair of locally semialgebraic spaces over R the groups $H_q(M,A,G)$ defined now coincide, up to natural isomorphisms, with the groups $H_q(M,A,G)$ inventend by Delfs ([D], $[D_1]$, $[DK_3]$) and described in III, §7.

This is evident for M and A partially complete from the simplicial description 3.4 of these groups. It is then also clear in general since there exist "enough" pairs (K,L) of partially complete subspaces $K \subset M$ and $L \subset A$ (with $L \subset K$, of course) which are strong deformation retracts of M and A respectively (cf. formula 3.6 in $[DK_3]$).

In the following we shall not make use of this fact. On the contrary, the considerations of the present chapter will give the results, say, in III, §7 anew and in a somewhat easier way than Delfs' theory does. But one should keep in mind that Delfs' theory has a deeper content than the present one by its inherent connection with sheaf cohomology which is absent here. For example, by his theory it is clear the the groups $H^q(M,G)$ of a locally semialgebraic space M {defined as $({}_Rh^q)^\vee(M)$

for $h^* = H^*(-,G)$ = singular cohomology with coefficients in G} coincide with the cohomology groups $H^q(M,G_M)$ of the constant sheaf G_M on M with stalks G.

Open Question B. In which generality is it true for a space M over R that $H^q(M,G) = H^q(M,G_M)$? How about a sheaf theoretic interpretation of the relative cohomology groups $H^q(M,A,G)$?

We return to an arbitrary topological homology theory h_*.

Proposition 5.6. The long homology sequence of any triple (M,A,C) of spaces with $A \supset C$, defined as in 4.6, is exact.

This is a formal consequence of the exactness of the long homology sequence of a pair of spaces, cf. [ES, p. 25ff.].

Proposition 5.7. Let $((M_\lambda,A_\lambda)\mid\lambda\in\Lambda)$ be a family of spaces over R. Let M and A denote the direct sums of the M_λ and the A_λ respectively. The inclusion maps $i_\lambda : (M_\lambda,A_\lambda) \hookrightarrow (M,A)$ induce an isomorphism

$$(i_{\lambda *}) : \bigoplus_{\lambda\in\Lambda} h_n(M_\lambda,A_\lambda) \xrightarrow{\sim} h_n(M,A)$$

for every $n \in \mathbf{Z}$.

Proof. It suffices to prove this in the special case that all A_λ are empty. Then we obtain the result in general by use of the long homology sequences of (M,A) and the (M_λ,A_λ) and the five-lemma. For any $\lambda \in \Lambda$ we choose a WP-approximation $\varphi_\lambda : X_\lambda \to M_\lambda$. Let $X := \bigsqcup(X_\lambda \mid \lambda\in\Lambda)$. The map $\varphi := \bigsqcup\varphi_\lambda$ from X to M is a WP-approximation of M. We know that the claim holds for the family $(X_\lambda\mid\lambda\in\lambda)$. Using the isomorphisms $h_n[\varphi_\lambda]$ and $h_n[\varphi]$ we see that the claim holds for $(M_\lambda\mid\lambda\in\Lambda)$. q.e.d.

§6 - Excision and limits

The conventions from the preceding section remain in force. In particular, h_* is an arbitrary topological homology theory which will be evaluated on pairs of spaces over R (cf. 5.3).

Definition 1. A triple of spaces (M,A,B) is called a triad if $M = A \cup B$. We call a triad (M,A,B) excisive (for the homology theory h_*) if the inclusion map $i : (A, A\cap B) \to (M,B)$ induces an isomorphism

$$i_* : h_n(A, A\cap B) \xrightarrow{\sim} h_n(M,B)$$

for every $n \in \mathbb{Z}$.

Proposition 6.1. If (M,A,B) is an excisive triad and C is a subspace of $A \cap B$ then the Mayer-Vietoris of the quadruple (M,A,B,C), defined as in 4.7, is exact.

This is again a formal consequence of the exactness of the long homology sequences of pairs, cf. [Sw, p. 105f], [ES, p. 37ff].

We come to the crucial question: Which triads are excisive? The following proposition witnesses the "tameness" of weakly semialgebraic spaces in contrast to arbitrary topological spaces.

Proposition 6.2. Every triad (M,A,B) of spaces over R with A and B closed in M is excisive.

Proof. We know from §4 that triads of weak polytopes are excisive. If the field R is sequential then, for every space X over R, the partially complete core $p_X : P(X) \to X$ exists and is a WP-approximation of X (cf. V.4.7). Now (P(M),P(A),P(B)) is a triad of weak polytopes and $P(A) \cap P(B) = P(A \cap B)$. We have a commuting square

$$\begin{array}{ccc}
(P(A),P(A)\cap P(B)) & \xrightarrow{P(i)} & (P(M),P(B)) \\
\Big\downarrow (p_A,p_{A\cap B}) & & \Big\downarrow (p_M,p_B) \\
(A,A\cap B) & \xrightarrow[i]{} & (M,B)
\end{array}$$

and $P(i)$ is just the inclusion map between the pairs above. The verticals are WP-approximations. Since $(P(M),P(A),P(B))$ is excisive we conclude that (M,A,B) is excisive.

Now consider the case that R is not sequential. Then we choose a real closed overfield K of R which is sequential. $(M(K),A(K),B(K))$ is again a triad with $A(K)$ and $B(K)$ closed in $M(K)$. Thus $(M(K),A(K),B(K))$ is excisive. Since $A(K)\cap B(K) = (A\cap B)(K)$ we conclude by the first main theorem for homology 5.4.i that (M,A,B) is excisive. q.e.d.

<u>Remark 6.3.</u> If (M,A,B) is an excisive triad then also (M,B,A) is excisive.

This can be seen in much the same way as in topology by applying Proposition 6.2 to the triads (M',A',B') and (M',B',A'), where M' is the subspace $(A\times 0)\cup(C\times I)\cup(B\times 1)$ of $M\times I$, with $C := A\cap B$, and A',B' are the closed subspaces $(A\times 0)\cup(C\times[0,\frac{1}{2}]),(C\times[\frac{1}{2},1])\cup(B\times 1)$ of M', cf. [Sw, p. 103f

<u>Theorem 6.4.</u> Let (M,A) be a closed pair of spaces over R and $f : A\to N$ be a map over R which is partially proper near $M\setminus A$. Then the "push out map" $g : (M,A)\to(M\cup_f N,N)$ (cf. IV.8.4) induces an isomorphism

$$g_* : h_n(M,A) \xrightarrow{\sim} h_n(M\cup_f N,N)$$

for every $n\in\mathbb{Z}$.

<u>Proof.</u> i) We first consider the case that M and N are weak polytopes. Now we have a commuting diagram

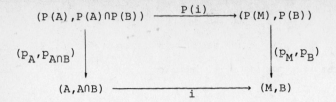

$$\begin{array}{ccc} (M,A) & \xrightarrow{\ g\ } & (M \cup_f N, N) \\ {\scriptstyle p}\downarrow & & \downarrow{\scriptstyle q} \\ (M/A,*) & \xrightarrow[\ \bar{g}\ ]{} & (M \cup_f N/N,*) \end{array}$$

with $\bar{g}$ an isomorphism and canonical projections p and q. We know from §4 that $h_n[p]$ and $h_n[q]$ are isomorphisms. Of course also $h_n[\bar{g}]$ is an isomorphism. Thus $h_n[g]$ is an isomorphism.

ii) We now prove the theorem in the case that the field R is sequential. Then we have a commuting square

$$\begin{array}{ccc} (P(M),P(A)) & \xrightarrow{\ P(g)\ } & (P(M \cup_f N),P(N)) \\ {\scriptstyle p}\downarrow & & \downarrow{\scriptstyle q} \\ (M,A) & \xrightarrow[\ g\ ]{} & (M \cup_f N, N) \end{array}$$

with WP-approximations p,q. Moreover $P(M \cup_f N) = P(M) \cup_{P(f)} P(N)$ (IV.9.14), and P(g) is the pushout map between the pairs above. We know from step (i) that $h_n[P(g)]$ is bijective and conclude that $h_n[g]$ is bijective.

iii) If finally R is not sequential then we choose a sequential real closed field $K \supset R$. We know from step (ii) that $h_n[g_K]$ is bijective and conclude by the main theorem 5.4.i that $h_n[g]$ is bijective. q.e.d.

In the special case that N is the one-point space the theorem reads as follows.

<u>Corollary 6.5.</u> For every closed pair of spaces (M,A) with A partially complete near $M \setminus A$ and every $n \in \mathbb{Z}$ we have a natural isomorphism

$$h_n(M,A) \xrightarrow{\sim} \tilde{h}_n(M/A) .$$

We now start out to prove a very general theorem about the homology of direct limits of spaces.

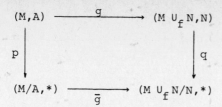

Theorem 6.6. Let (M,A) be a pair of spaces and $(X_\lambda \mid \lambda \in \Lambda)$ a directed system of subspaces of M which is an admissible covering of M (cf. Def. 7 in IV, §3). For every $q \in \mathbb{Z}$ the natural map

$$\varinjlim_{\lambda\in\Lambda} h_q(X_\lambda, A \cap X_\lambda) \to h_q(M,A)$$

is an isomorphism.

Proof. a) It suffices to prove this in the case that A is empty. Then the claim follows in general by using long homology sequences and the five-lemma. Thus henceforth we consider a single space M instead of a pair (M,A).

b) We first prove the claim for an admissible filtration $(X_n \mid n \in \mathbb{N})$ of M (cf. V, §2, Def. 3). Then the theorem can be verified by use of Milnor's trick [Mi]. We consider the telescope $T \subset M \times I_\infty$ of the family $(X_n \mid n \in \mathbb{N})$, cf. V, §4. The natural projection $p : T \to M$ is a homotopy equivalence (V.4.5). In T we have the closed subspaces

$$A := \bigsqcup (X_n \times [n-1,n] \mid n \text{ even})$$

$$B := \bigsqcup (X_n \times [n-1,n] \mid n \text{ odd})$$

with $A \cup B = T$. By Proposition 6.2 the triple (T,A,B) is excisive. A close inspection of the Mayer-Vietoris sequence of the quadruple $(T,A,B,A\cap B)$ gives the claim, cf. [Mi].

c) We now prove the theorem in the case that the directed system $(X_\lambda \mid \lambda\in\Lambda)$ is an exhaustion of M. Omitting superfluous indices we may assume that the exhaustion is faithful (Prop. IV.1.14). We enlarge, without loss of generality, our directed family to a faithful lattice exhaustion (Prop. IV.1.15). The space M carries a patch decomposition such that $(X_\lambda \mid \lambda\in\Lambda)$ is cofinal in the family of all finite closed subcomplexes with respect to this decomposition, cf. V, §1.

Henceforth we assume that M is a patch complex and $(X_\lambda \mid \lambda \in \Lambda)$ is the family of finite closed subcomplexes of M. We look at the chunks M_n and belts $M(n)$ of M and also at the chunks $X_{\lambda,n}$ and belts $X_\lambda(n)$ of the subcomplex X_λ. Since X_λ is closed in M, we have $X_{\lambda,n} = M_n \cap X_\lambda$. Moreover $X_\lambda(n)$ is just the direct sum of all closed patches $\bar{\sigma}$ of M of height n with $\bar{\sigma} \subset X_\lambda$. In particular, $X_\lambda(n)$ is empty for n large, keeping λ fixed. We have $\partial X_\lambda(n) = X_\lambda(n) \cap \partial M(n)$.

By Theorem 6.4 the pushout map $g : (M(n), \partial M(n)) \to (M_n, M_{n-1})$ yields isomorphisms

$$g_* : h_q(M(n), \partial M(n)) \xrightarrow{\sim} h_q(M_n, M_{n-1}) \ .$$

For every $\lambda \in \Lambda$ the map g restricts to the pushout map g_λ from $(X_\lambda(n), \partial X_\lambda(n))$ to $(X_{\lambda,n}, X_{\lambda,n-1})$ which yields isomorphisms

$$g_{\lambda *} : h_q(X_\lambda(n), \partial X_\lambda(n)) \xrightarrow{\sim} h_q(X_{\lambda,n}, X_{\lambda,n-1}) \ .$$

We obtain a commuting square with bijective vertical arrows

$$\begin{array}{ccc}
\varinjlim_\lambda h_q(X_\lambda(n), \partial X_\lambda(n)) & \longrightarrow & h_q(M(n), \partial M(n)) \\
\downarrow \cong & & \downarrow \cong \\
\varinjlim h_q(X_\lambda \cap M_n, X_\lambda \cap M_{n-1}) & \xrightarrow{\alpha} & h_q(M_n, M_{n-1})
\end{array} \ .$$

$(M(n), \partial M(n))$ is the direct sum of the pairs $(\bar{\sigma}, \partial\sigma)$ with σ running through all patches of height n and every $(X_\lambda(n), \partial X_\lambda(n))$ is the direct sum of some of these pairs $(\bar{\sigma}, \partial\sigma)$. Moreover, every $\bar{\sigma}$ is contained in some space $X_\lambda(n)$. Thus, by the additivity of h_q, the upper horizontal arrow is an isomorphism. This implies that α is an isomorphism. Using long homology sequences and the five-lemma we see by induction on n that the natural map

$$\varinjlim_\lambda h_q(X_\lambda \cap M_n) \to h_q(M_n)$$

is an isomorphism for every n. The directed system $(X_\lambda \cap M_n \mid (\lambda,n) \in \Lambda \times \mathbb{N})$ of subspaces of M gives us a natural commuting square of homomorphisms

$$\begin{array}{ccc} \varinjlim_{(\lambda,n)} h_q(X_\lambda \cap M_n) & \longrightarrow & \varinjlim_{n} h_q(M_n) \\ \downarrow & & \downarrow \\ \varinjlim_{\lambda} h_q(X_\lambda) & \longrightarrow & h_q(M) \end{array} .$$

The upper horizontal arrow is bijective by what has just been proved. The vertical arrows are bijective by step b) of the proof. Thus the lower horizontal arrow is bijective, as claimed.

d) We obtain the theorem in general by somewhat repeating the last argument. We choose an exhaustion $(M_\alpha \mid \alpha \in I)$ of M. For every $\lambda \in \Lambda$ the family $(X_\lambda \cap M_\alpha \mid \alpha \in I)$ is an exhaustion of X_λ. The directed system $(X_\lambda \cap M_\alpha \mid (\lambda,\alpha) \in \Lambda \times I)$ of subspaces of M gives us a natural commuting square

$$\begin{array}{ccc} \varinjlim_{(\lambda,\alpha)} h_q(X_\lambda \cap M_\alpha) & \longrightarrow & \varinjlim_{\alpha} h_q(M_\alpha) \\ \downarrow & & \downarrow \\ \varinjlim_{\lambda} h_q(X_\lambda) & \longrightarrow & h_q(M) \end{array} .$$

The vertical arrows are bijective by step c) of the proof. The upper horizontal arrow is trivially bijective since every M_α is contained in some X_λ. Thus the lower horizontal arrow is bijective, as claimed. q.e.d.

We now can prove that the elements in our homology groups "live" in polytopes in the given space. For any pair of spaces (M,A) let $\gamma_c(M,A)$ denote the set of pairs (K,L) with K and L complete semialgebraic subsets of M and A respectively and, of course, $L \subset K$. This set $\gamma_c(M,A)$ is a directed system of pairs of polytopes under the partial ordering given by inclusion.

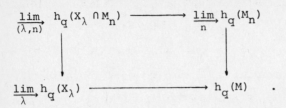

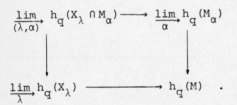

Theorem 6.7. For every $q \in \mathbb{Z}$ the natural map

$$\varinjlim (h_q(K,L) \mid (K,L) \in \gamma_c(M,A)) \to h_q(M,A)$$

is an isomorphism.

Proof. i) We choose an exhaustion $(M_\alpha \mid \alpha \in I)$ of M. We first prove the surjectivity of the map. Let $\xi \in h_q(M,A)$ be given. If X and Y are subspaces of M and A with $X \supset Y$ then we call an element η of $h_q(X,Y)$ a preimage of ξ if η maps onto ξ under $i_* : h_q(X,Y) \to h_q(M,A)$ with i the inclusion $(X,Y) \hookrightarrow (M,A)$.

By the preceding Theorem 6.6 there exists some $\alpha \in I$ and some $\eta \in h_q(M_\alpha, A \cap M_\alpha)$ which is a preimage of ξ. We choose a good triangulation of $(M_\alpha, A \cap M_\alpha)$, cf. III, §2. Let K and L denote the cores of M_α and $A \cap M_\alpha$ in this triangulation. Then $(K,L) \in \gamma_c(M,A)$. Moreover K and L are strong deformation retracts of M and A respectively. Using long homology sequences and the five-lemma we conclude that the inclusion map from (K,L) to $(M_\alpha, A \cap M_\alpha)$ gives an isomorphism in homology. Thus η has a preimage in $h_q(K,L)$ and this is also a preimage of ξ.

ii) We now prove injectivity. We are given elements $\xi_1 \in h_q(K_1,L_1)$ and $\xi_2 \in h_q(K_2,L_2)$ for some pairs (K_1,L_1) and (K_2,L_2) in $\gamma_c(M,A)$, such that ξ_1 and ξ_2 have the same image in $h_1(M,A)$. We have to find a pair $(K,L) \in \gamma_c(M,A)$ with $K_1 \cup K_2 \subset K$ and $L_1 \cup L_2 \subset L$ such that ξ_1 and ξ_2 have the same image in $h_q(K,L)$.

By Theorem 6.6 there exists some $\alpha \in I$ with $K_1 \cup K_2 \subset M_\alpha$ such that ξ_1 and ξ_2 have the same image in $h_q(M_\alpha, A \cap M_\alpha)$. We choose a good triangulation of the system of spaces $(M_\alpha, A \cap M_\alpha, K_1, K_2, L_1, L_2)$. Let K and L denote the cores of M_α and $A \cap M_\alpha$. Then $K_1 \cup K_2 \subset K$ and $L_1 \cup L_2 \subset L$. The inclusion map from (K,L) to $(M_\alpha, A \cap M_\alpha)$ gives an isomorphism in homology. Thus ξ_1 and ξ_2 have the same image in $h_q(K,L)$. q.e.d.

We return to excision problems.

Lemma 6.8. Let $(X_\lambda \mid \lambda \in \Lambda)$ be a directed system of subspaces of a space M which is an admissible covering of M. Let (M,A,B) be a triad such that, for every $\lambda \in \Lambda$, the triad $(X_\lambda, A \cap X_\lambda, B \cap X_\lambda)$ is excisive. Then (M,A,B) is excisive.

This is an easy consequence of Theorem 6.6. We are now next door to a rather general excision result. In order to state it we need two definitions.

Definition 2. A subset X of M is called **basic** if, for every $L \in \bar{\gamma}(M)$ {i.e. L closed and semialgebraic in M}, the set $X \cap L$ is locally closed in the semialgebraic space L. This means that $X \cap L$ is open in its closure $\overline{X \cap L}$ or, in other terms, $X \cap L = U \cap A$ with $U \in \mathring{\gamma}(L)$ and $A \in \bar{\gamma}(L)$.

N.B. It suffices to check this property with L running through the sets of a given exhaustion of M. By the way, then $X \cap L$ is locally closed in L for every $L \in \gamma(M)$.

If the space M is locally **semialgebraic** then every basic subset X of M is an intersection $U \cap A$ with $U \in \mathring{\mathcal{T}}(M)$ and $A \in \bar{\mathcal{T}}(M)$, and we can choose $A = \bar{X}$. We also can write, in this case,

$$X = \{x \in M \mid f(x) \geq 0, \; g(x) > 0\}$$

with two locally semialgebraic functions f and g on M (cf. I.4.15). The word "basic" alludes to the possibility in general to give such a "basic description" at least for the sets $X \cap L$ in L.

Definition 3. A triad (M,X,Y) is called **basic** if X and Y are basic in M and, for every $L \in \bar{\gamma}(M)$, there exist closed semialgebraic subsets A

and B of L with $A \subset X \cap L$, $B \subset Y \cap L$, and $A \cup B = L$. Again it suffices to check this last property with L running through the sets of a given exhaustion of M.

Examples 6.9. i) Every triad (M,A,B) with A and B closed in M ("closed triad") is basic.

ii) Every triad (M,U,V) with U and V open in M ("open triad") is basic. Indeed, in this case there even exist sets $A \in \overline{\mathcal{T}}(M)$, $B \in \overline{\mathcal{T}}(M)$ with $A \subset U$, $B \subset V$ and $A \cup B = M$ by the shrinking lemma V.3.4.

iii) More generally, let X and Y be basic subsets of M such that $M = \mathring{X} \cup \mathring{Y}$, with $\mathring{X}$ and $\mathring{Y}$ denoting the interiors of X and Y in the strong topology of M. Then (M,X,Y) is a basic triad. Indeed, let $L \in \overline{\gamma}(M)$. Then the interiors U and V of $L \cap X$ and $L \cap Y$ in L contain $L \cap \mathring{X}$ and $L \cap \mathring{Y}$ respectively. Thus $U \cup V = L$. By the shrinking lemma (in its semialgebraic version [DK_5, 1.4]) there exist closed semialgebraic subsets A and B of L with $A \subset U$, $B \subset V$ and $A \cup B = L$.

Theorem 6.10. Every basic triad (M,X,Y) is excisive.

Proof. Applying the preceding Lemma 6.8 to an exhaustion of M we see that it suffices to consider the case that M is semialgebraic. Now there exist closed semialgebraic subsets A and B of M with $A \subset X$, $B \subset Y$, and $A \cup B = M$. We choose a good triangulation of the system $(M,X,Y,X\cap Y,A,B)$ and then assume without loss of generality that this system is a tame system of (finite) simplicial complexes (cf. III, §2). We now may replace our system $(M,X,Y,X\cap Y,A,B)$ by its core $(\text{co}\, M, X\cap \text{co}\, M, Y\cap \text{co}\, M, \ldots)$ since this core is a strong deformation retract of the given system (Prop. III.2.1). Thus we assume henceforth, without loss of generality, that the finite simplicial complex M is closed. Then the subcomplexes A and B are also closed.

Let C and D denote the cores of X and Y. Then $C \cap D$ is the core of $X \cap Y$. We know from III,§1 that $C, D, C\cap D$ are strong deformation retracts of $X, Y, X\cap Y$ respectively. Moreover, $A \subset C$, $B \subset D$, hence $C \cup D = M$. We look at the commuting square

$$\begin{array}{ccc} h_q(C, C\cap D) & \longrightarrow & h_q(M,D) \\ \downarrow & & \downarrow \\ h_q(X, X\cap Y) & \longrightarrow & h_q(M,Y) \end{array}$$

with all arrows induced by inclusions. The upper horizontal map is an isomorphism by Prop. 6.2. Using long exact sequences and the five-lemma we see that the vertical maps are isomorphisms. Thus the lower horizontal map is an isomorphism, which means that, indeed, (M,X,Y) is excisive.

{N.B. The whole argument works more generally if M is locally semi-algebraic.} q.e.d.

Let now h^* be a topological cohomology theory. We define excisiveness of triads with respect to h^* as in Definition 1. The analogues of our results 6.2 - 6.5 remain true but the theory of limits is more difficult since the higher derivatives $\varprojlim^{(n)}$ of the inverse limit functor may come into play. Without any extra effort we can state the following theorem, which is evident by Milnor's telescope trick [Mi].

Theorem 6.11. Let (M,A) be a pair of spaces over R and $(X_n \mid n\in \mathbb{N})$ an admissible filtration of M. For every $q \in \mathbb{Z}$ there exists a natural exact sequence of abelian groups

$$0 \to \varprojlim{}^{(1)} h^{q-1}(X_n, A\cap X_n) \to h^q(M,A) \to \varprojlim h^q(X_n, A\cap X_n) \to 0 \quad .$$

If M is a space of countable type (IV, §4) then M has an exhaustion $(M_n \mid n\in \mathbb{N})$. In this case we obtain the following excision result by the above methods.

Theorem 6.12. If M is of countable type then every basic triad (M,X,Y) is excisive for h*.

This also holds for M locally semialgebraic since then every connected component of M is of countable type. One can give a direct proof in this special case without using a limit argument (cf. Proof of Theorem 6.10).

§7 - Representation theorems, pseudo-mapping spaces

Some notations. We study contravariant functors F from $H\mathcal{P}^*(R)$ to the category Set* of pointed sets. If $f : M \to N$ is a map between pointed weak polytopes then, as before, we shall often denote the induced map $F([f])$ by f^*. If A is a pointed closed subspace of a pointed weak polytope M and x is an element of F(M) then we shall often denote the element $i^*(x)$, with $i : A \hookrightarrow M$ the inclusion, by $x|A$. We call $x|A$ the restriction of x to A.

An isomorphism in $H\mathcal{P}^*(R)$, i.e. a homotopy equivalence, will be denoted by our usual sign " $\xrightarrow{\sim}$ " for isomorphisms. If there exists a homotopy equivalence $M \xrightarrow{\sim} N$, which we do not care to make explicit, then we write $M \simeq N$.

If F is the functor $[-,L]$ on $H\mathcal{P}^*(R)$ for some pointed space L over R then it follows from Remark 1.2.i and an easy homotopy extension argument (cf. [Sw, p. 153]) that F fulfills the following two axioms.

(W) Wedge axiom. If $(M_\lambda \mid \lambda \in \Lambda)$ is any family in $\mathcal{P}^*(R)$ then the natural map

$$F(\vee(M_\lambda \mid \lambda\in\Lambda)) \to \prod(F(M_\lambda) \mid \lambda\in\Lambda),$$
$$x \mapsto ((x|M_\lambda) \mid \lambda\in\Lambda),$$

is bijective.

(MV) Mayer-Vietoris axiom. If X is a pointed weak polytope and A_1, A_2 are pointed closed subspaces of X with $X = A_1 \cup A_2$ and if elements $x_1 \in F(A_1)$, $x_2 \in F(A_2)$ are given with $x_1|A_1 \cap A_2 = x_2|A_1 \cap A_2$, then there exists some $y \in F(X)$ with $y|A_1 = x_1$ and $y|A_2 = x_2$.

We are about to state a strong converse of this fact, a refined version of Brown's representation theorem [Bn].

Definition 1. Let $F : H\mathcal{P}^*(R) \to Set^*$ be a contravariant functor. Let L be a pointed weak polytope and let φ denote the map from the one point space $*$ to L. Let u be an element of F(L) such that $\varphi^*(u)$ is the distinguished element of $F(*)$. Then u yields a natural transformation

$$T_u : [-,L] \to F$$

sending an element $[f] \in [X,L]$ to $f^*(u) \in F(X)$. In this way, as follows from Yoneda's lemma [Mt, p. 97], the natural transformations from $[-,L]$ to F correspond uniquely to these elements of F(L). The functor F is called representable, if there exists a space L and an element $u \in F(L)$ such that T_u is an isomorphism (= natural equivalence) of functors. In this case L is called a classifying space and u is called a universal element for F.

Theorem 7.1. Every contravariant functor $F : H\mathcal{P}^*(R) \to Set^*$ which fulfills the axioms (W) and (MV) is representable. More precisely the following holds. Given a weak polytope C and an element $v \in F(C)$ there exists a relative CW-complex (L,C) and an element $u \in F(L)$ with $u|C = v$ such that L is a classifying space and u is a universal element for F. If C is already a pointed CW-complex L can be chosen as a pointed CW-complex containing C as a subcomplex.

We refer the reader to [Sw, p. 155ff] where a detailed proof of the analogous topological theorem is given. This - very beautiful - proof can be adapted word by word to the semialgebraic setting.

Remark 7.2. The proof of Theorem 7.1 gives us for free the following refinements of that theorem.

i) Assume that, for some $n \in \mathbb{N}_0$, the map $T_v(X) : [X,C] \to F(X)$ is bijective for $X = S^q$ with $0 \leq q < n$ and surjective for $X = S^n$. Then the relative CW-complex (L,C) can be chosen without cells of dimension $\leq n$.

ii) If $T_v(X)$ is bijective for every sphere $X = S^q$ then C is already a classifying space of F with universal element v.

Remark 7.3. If F is a contravariant functor from $H\mathcal{P}^*(R)$ to the category of groups then it follows from Theorem 7.1 in a purely formal way that there exists a (base point preserving) map $\mu : L \times L \to L$, unique up to homotopy, such that $(L,[\mu])$ is a group in $H\mathcal{P}^*(R)$ and $T_u(X)$ is an isomorphism of groups for every $X \in \mathcal{P}^*(R)$. If the values of F are abelian groups then $(L,[\mu])$ is an abelian group in $H\mathcal{P}^*(R)$.

Theorem 7.1 applies to the functors k^n of a reduced cohomology theory over R. Indeed the axiom (W) there holds by definition and (MV) holds by Prop. 2.3. We will exploit this application in §8.

Here we present some other applications of the representation theorem 7.1. First we consider again the functor $F = [-,N]$ on $H\mathcal{P}^*(R)$ with N any pointed space over R. By Theorem 7.1 there exists a pointed CW-complex L over R together with a map $u : L \to N$ which represents F. Taking into account that N has a WP-approximation (V.4.1) this means that u is a CW-approximation of N, as defined in V, §7. Thus we have proved anew the existence of CW-approximations for pointed spaces. More generally we obtain a CW-approximation of a pair (N,B) of pointed spaces by first choosing a CW-approximation $C \to B$ and then applying Theorem 7.1 and a homotopy extension argument to the functor $F = [-,N]$ and the homotopy class $[v] \in F(C)$ of the map $v : C \to B \hookrightarrow N$. Starting from this one obtains a second proof of Theorem V.7.14 on CW-approximations of arbitrary decreasing systems of spaces.

Let now N be a pointed space and M a pointed weak polytope over R. The functor $X \mapsto [X \wedge M, N]$ on $H\mathcal{P}^*(R)$ clearly fulfills the wedge and the Mayer-Vietoris-axiom. By Theorem 7.1 there exists a pointed weak polytope, which we denote by Map(M,N), and a map

$$\varepsilon_{M,N} : \mathrm{Map}(M,N) \wedge M \to N$$

such that, for every $X \in \mathcal{P}^*(R)$, the map

$$[X,\mathrm{Map}(M,N)] \to [X \wedge M,N],$$

$$[f] \mapsto [\varepsilon_{M,N} \circ (f \wedge id_M)]$$

is bijective. We call Map(M,N) a <u>pseudo-mapping space</u> and $\varepsilon_{M,N}$ an <u>evaluation map</u>. Composing $\varepsilon_{M,N}$ with the switching isomorphism from $M \wedge \mathrm{Map}(M,N)$ to $\mathrm{Map}(M,N) \wedge M$ we obtain a map

$$\varepsilon'_{M,N} : M \wedge \mathrm{Map}(M,N) \to N$$

such that, for every $X \in \mathcal{P}^*(R)$, the map

$$[X,\mathrm{Map}(M,N)] \to [M \wedge X,N],$$

$$[f] \mapsto [\varepsilon'_{M,N} \circ (id_M \wedge f)]$$

is bijective.

The pseudo-mapping space is determined by M and N only up to homotopy equivalence. To obtain straight notations let us agree that we choose, once and for all, for every pair M,N as above a pseudo-mapping space and an evaluation map and denote <u>these</u> by Map(M,N) and $\varepsilon_{M,N}$. Then we obtain a bifunctor

$$\mathrm{Map}(-,-) : H\mathcal{P}^*(R) \times HWSA^*(R) \to H\mathcal{P}^*(R)$$

which is contravariant in the first and covariant in the second argument. Without spelling out all the details, we remark that there exist natural homotopy equivalences

(7.4) $\quad \mathrm{Map}(M,N_1 \times N_2) \xrightarrow{\sim} \mathrm{Map}(M,N_1) \times \mathrm{Map}(M,N_2)$

(7.5) $\quad \mathrm{Map}(M_1 \wedge M_2,N) \xrightarrow{\sim} \mathrm{Map}(M_1,\mathrm{Map}(M_2,N))$

for $M, M_1, M_2 \in \mathcal{P}^*(R)$, $N_1, N_2, N \in WSA^*(R)$. These are consequences of the natural bijections

$$[X \wedge M, N_1 \times N_2] \cong [X \wedge M, N_1] \times [X \wedge M, N_2],$$
$$[X \wedge (M_1 \wedge M_2), N] = [(X \wedge M_1) \wedge M_2, N].$$

The string of equations

$$\pi_o(\mathrm{Map}(M,N)) = [S^o, \mathrm{Map}(M,N)] = [M \wedge S^o, N] = [M,N]$$

tells us that the connected components of Map(M,N) can be interpreted as the homotopy classes of maps from M to N. Of course, it would be much better to have a space whose points correspond uniquely with the maps from M to N in a natural way, as one has in topology. But, as we will see, already a pseudo-mapping space instead of a true mapping space can be useful.

We turn to a special type of pseudo-mapping spaces, the pseudo-loop spaces.

Definitions 2. i) For any pointed space N we denote the pseudo-mapping space $\mathrm{Map}(S^1, N)$ by ΩN and call it a pseudo-loop space of N. We denote the switched evaluation map $\varepsilon'_{S^1,N}$ by η_N. Thus η_N is a map from $S\Omega N$ to N which induces, for every $X \in \mathcal{P}^*(R)$, a bijection $[f] \mapsto [\eta_N \circ (Sf)]$ from $[X, \Omega N]$ to $[SX, N]$.
ii) Since S^1 is a cogroup in $H\mathcal{P}^*(R)$ (cf. §1) we have, by Remark 7.3, a multiplication $\mu : \Omega N \times \Omega N \to \Omega N$, unique up to homotopy, such that $(\Omega N, [\mu])$ is a group in $H\mathcal{P}^*(R)$ and the above bijections $[X, \Omega N] \xrightarrow{\sim} [SX, N]$ are group isomorphisms. Once and for all we choose, for every $N \in WSA^*(R)$, such a multiplication μ and denote it by μ_N. Both maps η_N and μ_N are regarded as ingredients of the pseudo-loop space ΩN.
iii) For every map $f : N \to N'$ between pointed spaces we choose, once and for all, a map $g : \Omega N \to \Omega N'$ such that, for every $X \in \mathcal{P}^*(R)$, the

diagram

$$\begin{array}{ccc} [SX,N] & \xrightarrow{f_*} & [SX,N'] \\ \uparrow \cong & & \uparrow \cong \\ [X,\Omega N] & \xrightarrow[g_*]{} & [X,\Omega N'] \end{array}$$

commutes. The vertical arrows here mean, of course, the canonical bijections. We denote this map g by Ωf. Since the maps f_* are group homomorphisms we have a diagram

$$\begin{array}{ccc} \Omega N \times \Omega N & \xrightarrow{\mu_N} & \Omega N \\ \downarrow \Omega f \times \Omega f & & \downarrow \Omega f \\ \Omega N' \times \Omega N' & \xrightarrow[\mu_{N'}]{} & \Omega N' \end{array}$$

which commutes up to homotopy. All this means that we have obtained a functor $N \mapsto \Omega N$, $[f] \mapsto [\Omega f]$ from HWSA*(R) to the category of group objects in $H\mathcal{P}$*(R).

iv) The pseudo-loop functor Ω : HWSA*(R) $\to$ H *(R) is right adjoint (= adjoint in [Mt]) to the composite $j \circ S$ of the suspension functor $S : H\mathcal{P}^*(R) \to H\mathcal{P}^*(R)$ with the inclusion $j : H\mathcal{P}^*(R) \to HWSA^*(R)$. The maps $\eta_N : S\Omega N \to N$ are the associated left adjunction maps [Mt, p. 118]. We also have, for every $M \in \mathcal{P}^*(R)$, a right adjunction map $\zeta_M : M \to \Omega SM$, choosen once and for all in the homotopy class in $[M,\Omega SM]$ which is mapped to $[id_{SM}]$ under the natural bijection $[M,\Omega SM] \xrightarrow{\sim} [SM,SM]$. Thus, by definition,

$$id_{SM} \simeq \eta_{SM} \circ (S\zeta_M) ,$$

which characterizes ζ_M up to homotopy.

v) For any $r \in \mathbb{N}$ we denote the r-fold iteration of the pseudo-loop functor by Ω^r. By (7.5) we have, for every $N \in WSA^*(R)$, a homotopy equivalence

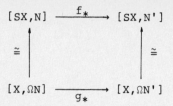

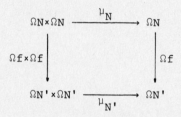

$\Omega^r N \simeq \mathrm{Map}(S^r,N)$.

From the definitions it is obvious that, for every $q \geq 0$ and $r \geq 0$,

(7.6) $\quad \pi_q(\Omega^r N) = \pi_{q+r}(N)$.

We illustrate the usefulness of the pseudo-loop functor by an example.

Theorem 7.7 (General suspension theorem). Let M and N be pointed weak polytopes. Assume that N is n-connected for some $n \in \mathbb{N}_o$. The suspension homomorphism (§1, Def. 2)

$$S_{M,N} : [M,N] \to [SM,SN]$$

is bijective if $\dim M \leq 2n$ and surjective if $\dim M = 2n+1$.

Proof. Freudenthal's suspension theorem 1.5 means that the adjunction map $\zeta_N : N \to \Omega SN$ is a $(2n+1)$-equivalence (cf. V, §6, Def. 5 and Def. 7). The claim now follows from Theorem V.6.13. q.e.d.

We digress for short from our general theme in order to indicate how "free" pseudo-mapping spaces can be obtained for spaces without base points.

Theorem 7.8. Let M be a weak polytope and L a space over R. The contravariant functor $X \mapsto [X\times M,L]$ from $H\mathcal{P}(R)$ to the category of sets is representable. Thus there exists a weak polytope $\mathrm{Map}(M,L)$ and a map

$$\varepsilon_{M,L} : \mathrm{Map}(M,L)\times M \to L$$

such that, for every $X \in \mathcal{P}(R)$, the map

$$[X,\mathrm{Map}(M,L)] \to [X\times M,L]$$
$$[f] \mapsto [\varepsilon_{M,L} \circ (f\times id_M)]$$

is bijective.

Proof. We use the notations 4.1. We choose some $y \in L$ and denote the pointed space (L,y) by N. Then $N^{o} = L$. Let $T := \mathrm{Map}(M^{+},N)^{o}$. Then for any $X \in \mathcal{P}(R)$, in slightly sloppy notation,

$$[X\times M,L] = [(X\times M)^{+},N] = [X^{+}\wedge M^{+},N] = [X^{+},\mathrm{Map}(M^{+},N)] = [X,T]$$

(cf. 4.2, 4.3) q.e.d.

Corollary 7.9. If $M \in \mathcal{P}(R)$ and $N \in WSA^{*}(R)$ then

$$\mathrm{Map}(M,N^{o}) \simeq \mathrm{Map}(M^{+},N)^{o} .$$

We return to pointed spaces and state another representation theorem. It is an analogue of a famous representation theorem due to Brown and Adams, cf. [Ad_1], [Sw, Th. 9.21]. It can be proved here word by word in the same way as there. In particular, no transfer principle is needed for the proof.

Theorem 7.10. Let F be a contravariant functor from the homotopy category $HSA_c^{*}(R)$ of pointed polytopes over R to the category of groups. Assume that F fulfills the Mayer-Vietoris axiom (MV) and the wedge axiom (W) for finite families $(M_\lambda \mid \lambda\in\Lambda)$ {cf. the beginning of the section. Of course now all the spaces involved have to be polytopes.}. Then there exists a group object L in $H\mathcal{P}^{*}(R)$ such that the functor $[-,L]$ on $HSA_c^{*}(R)$ is naturally equivalent (= isomorphic) to F.

Again we call L a classifying space of the functor F. Theorem 7.10 will not be needed in the sequel, but it gives us a hold that we are on the right track. It indicates that weak polytopes (with a homotopy group law) are natural and very useful objects even if one is only interested in complete semialgebraic spaces.

§8 - Ω-Spectra

We now have the prerequisites for drawing the connection between reduced cohomology theories and spectra. We will not delve deeply into the theory of spectra but will be content with the view point that suitable spectra serve to represent cohomology theories on the level of weak polytopes. This will allow us to work with a naive notion of maps - here called "homotopy maps" - between spectra, which in the topological setting already appears in G. Whitehead's fundamental paper [W_2] (there called "maps").

Definitions 1. a) A (*semialgebraic*) *spectrum* E *over* R is a family of pointed weak polytopes $(E_n | n \in \mathbb{Z})$ together with a family $(\varepsilon_n^E | n \in \mathbb{Z})$ of (base point preserving, as always) maps $\varepsilon_n^E : SE_n \to E_{n+1}$.

b) A spectrum E over R is called an *Ω-spectrum* if the maps $\eta_n^E : E_n \to \Omega E_{n+1}$, which are adjoint to the ε_n^E above, are homotopy equivalences (cf. §7 for the definition of the pseudo-loop functor Ω). In this case every E_n is an abelian group object in $HP^*(R)$ via the homotopy equivalence $\Omega(\eta_{n+1}) \circ \eta_n$ from E_n to $\Omega^2 E_{n+2}$ (cf. §7).

c) A *homotopy map* $f : E \to F$ from a spectrum E to a spectrum F is a family $(f_n | n \in \mathbb{Z})$ of maps $f_n : E_n \to F_n$ between spaces such that the diagrams

$$\begin{array}{ccc} SE_n & \xrightarrow{Sf_n} & SF_n \\ {\scriptstyle \varepsilon_n^E}\downarrow & & \downarrow{\scriptstyle \varepsilon_n^F} \\ E_{n+1} & \xrightarrow[f_{n+1}]{} & F_{n+1} \end{array}$$

commute up to homotopy. {For a more useful notion of maps between spectra see [Ad] or [Sw, Chap. 8].}

d) A *homotopy equivalence* between spectra E,F is a homotopy map $f : E \to F$ such that every map $f_n : E_n \to F_n$ is a homotopy equivalence.

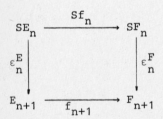

e) Analogously we define <u>topological spectra</u>, using the category $\mathcal{W}\mathcal{O}^*$ instead of $\mathcal{P}^*(R)$, and topological Ω-spectra, using the genuine (topological) loop functor Ω, further homotopy maps and homotopy equivalences between topological spectra.

<u>N.B.</u> The maps ε_n^E and η_n^E here have nothing to do with the ε's and η's in §7.

<u>Definition 2.</u> Let $f,g : E \rightrightarrows E'$ be two homotopy maps between semialgebraic or topological spectra. A <u>homotopy</u> F <u>from</u> f <u>to</u> g is a family $(F_n \mid n \in \mathbb{Z})$ of (base point preserving) homotopies $F_n : E_n \times I \to E'_n$ with $F_n(-,0) = f_n$ and $F_n(-1) = g_n$. For every $t \in I$ this family F gives a map $F(-,t) : E \to E'$, and we have $F(-,0) = f$, $F(-,1) = g$. This notion of homotopy fits well with the definition of homotopy equivalence above.

Given a spectrum E over R we define an abelian group $\tilde{H}^n(X,E)$ for every pointed weak polytope X and $n \in \mathbb{Z}$ as follows.

$$(8.1) \qquad \tilde{H}^n(X,E) := \varinjlim_k [S^k X, E_{n+k}].$$

Here the limit is taken with respect to the transition maps

$$[S^k X, E_{n+k}] \xrightarrow{S} [S^{k+1}X, SE_{n+k}] \xrightarrow{(\varepsilon^E_{n+k})_*} [S^{k+1}X, E_{n+k+1}] .$$

We know from §1 that these limits are indeed abelian groups in a natural way. For every $X \in \mathcal{P}^*(R)$ we have

$$\tilde{H}^{n+1}(SX,E) = \varinjlim_k [S^{k+1}X, E_{n+1+k}] = \varinjlim_k [S^k X, E_{n+k}] = \tilde{H}^n(X,E)$$

<u>Remark 8.2.</u> The family of contravariant functors $(\tilde{H}^n(-,E) \mid n \in \mathbb{Z})$ from $H\mathcal{P}^*(R)$ to Ab together with the identity maps $\tilde{H}^{n+1}(-,E) \circ S \to \tilde{H}^n(-,E)$ is a "weak" reduced cohomology theory $\tilde{H}^*(-,E)$ over R, i.e. it fulfills Definition 1 in §2 with the exception of the wedge axiom (which is only granted for finite families).

This is obvious (recall 1.2.i and 1.7). In the following we briefly write $E^n(X)$ instead of $\tilde{H}^n(X,E)$ and E^* instead of $\tilde{H}^*(-,E)$.

Definition 3. We have an obvious notion of natural transformations between weak reduced cohomology theories over R, cf. §2, Def. 2. Every homotopy map $f : E \to F$ between spectra over R induces a natural transformation from E^* to F^* in the evident way. We denote this natural transformation by U_f.

Proposition 8.3. If E is an Ω-spectrum then, for every $X \in \mathcal{P}^*(R)$ and $n \in \mathbb{Z}$, the evident map

$$[X,E_n] \to E^n(X)$$

is an isomorphism of abelian groups. E^* is a reduced cohomology theory.

Proof. We have commutative diagrams

$$\begin{array}{ccccc} & & [S^{k+1}X, SE_{n+k}] & & \\ & \overset{S}{\nearrow} & & \searrow^{(\varepsilon^E_{n+k})_*} & \\ [S^kX, E_{n+k}] & & & & [S^{k+1}X, E_{n+k+1}] \\ & \underset{(\eta^E_{n+k})_*}{\searrow} \cong & & \nearrow \cong & \\ & & [S^kX, \Omega E_{n+k+1}] & & \end{array}$$

which tell us that the transition maps for the inductive limit (8.1) are isomorphisms. (The unadorned arrow is the adjunction isomorphism made explicit in §7). Thus the functor E^n is isomorphic to $[-,E_n]$, and this implies that E^n obeys the full wedge axiom (cf. 1.2.i). q.e.d.

Remark 8.4. Moreover it is clear that the homotopy classes [f] of homotopy maps $f : E \to F$ between Ω-spectra correspond uniquely to the natural transformations $T : E^* \to F^*$ via $T = U_f$, and T is a natural equivalence iff f is a homotopy equivalence.

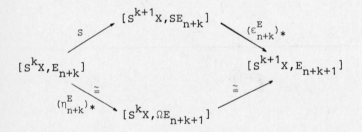

<u>Theorem 8.5.</u> For every reduced cohomology theory k^* over R there exists an Ω-spectrum E over R together with a natural equivalence $T : E^* \xrightarrow{\sim} k^*$.

<u>Proof.</u> We know from the wedge axiom (§2, Def. 1) and Proposition 2.3 that, for every $n \in \mathbb{Z}$, the functor $k^n : H\mathcal{P}^*(R) \to Ab$ fulfills the axioms (W) and (MV) required in the representation theorem 7.1. Thus there exists a pointed weak polytope E_n over R together with a natural equivalence $T^n : [-,E_n] \xrightarrow{\sim} k^n$. For every $X \in \mathcal{P}^*(R)$ we have a bijection

$$\alpha(X) : [X,E_n] \xrightarrow{\sim} [X,\Omega E_{n+1}]$$

such that the diagram

$$\begin{array}{ccccc} [X,E_n] & \xrightarrow[T^n(X)]{\sim} & k^n(X) & & \\ {\scriptstyle \alpha(X)}\downarrow & & & \overset{\sigma^n(X)}{\underset{\cong}{\searrow}} & \\ [X,\Omega E_{n+1}] & \xrightarrow{\sim} & [SX,E_{n+1}] & \xrightarrow[T^{n+1}(SX)]{\sim} & k^{n+1}(SX) \end{array}$$

commutes. The $\alpha(X)$ together form a natural equivalence α from $[-,E_n]$ to $[-,\Omega E_{n+1}]$. By the Yoneda lemma there exists a homotopy equivalence $\eta_n : E_n \to \Omega E_{n+1}$, unique up to homotopy, such that $\alpha = (\eta_n)_*$. The families $(E_n | n \in \mathbb{Z})$ and $(\eta_n | n \in \mathbb{Z})$ together define an Ω-spectrum, and the family $(T^n | n \in \mathbb{Z})$ is a natural equivalence from E^* to k^*. q.e.d.

<u>Remarks 8.6.</u> If F is a second Ω-spectrum and $V : F^* \xrightarrow{\sim} k^*$ is again a natural equivalence then it is evident from 8.4 that there exists a homotopy map $f : E \to F$, unique up to homotopy, such that $V \circ U_f = T$, and f is a homotopy equivalence. In short, the Ω-spectrum E in the theorem is determined by k^* uniquely up to homotopy equivalence.

We now consider spectra over different real closed fields and also topological spectra.

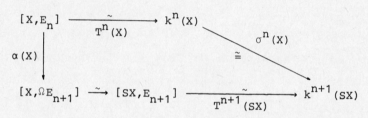

<u>Proposition 8.7.</u> Every spectrum over R is homotopy equivalent to a spectrum E over R with all spaces E_n being pointed CW-complexes and all maps ε_n^E being cellular.

This follows from V.7.14 and V.7.4.

<u>Definitions 4.</u> a) Let K be a real closed field extension of R, and let E be a spectrum over R. Replacing every space E_n by $E_n(K)$ and every map ε_n^E by $(\varepsilon_n^E)_K$ we obtain a spectrum E(K) over K which we call the <u>spectrum obtained from</u> E <u>by base field extension from</u> R <u>to</u> K. Every homotopy map $f : E \to F$ between spectra over R yields a homotopy map $f_K : E(K) \to F(K)$ in the obvious way.

b) Let E be a spectrum over $\mathbb{R}$ with all spaces E_n pointed CW-complexes. (This is no essential restriction of generality by the proposition above). Replacing the E_n by their underlying topological spaces $(E_n)_{top}$ we obtain a topological spectrum E_{top}, called the <u>topological spectrum underlying</u> E. Every homotopy map $f : E \to F$ between spectra over $\mathbb{R}$ can be read as a homotopy map between topological spectra $f_{top} : E_{top} \to F_{top}$.

<u>Theorem 8.8</u> (First main theorem for spectra). Let K be a real closed overfield of R.

i) If E and F are spectra over R then the homotopy classes of homotopy maps $f : E \to F$ correspond uniquely to the homotopy classes of homotopy maps $g : E(K) \to F(K)$ by the relation $[g] = [f_K]$. A homotopy map $f : E \to F$ is a homotopy equivalence iff f_K is a homotopy equivalence.

ii) If F is a spectrum over K then there exists a spectrum E over R together with a homotopy equivalence $\varphi : E(K) \to F$.

iii) A spectrum E over R is an Ω-spectrum iff E(K) is an Ω-spectrum.

All this follows from the first main theorem V.5.2.i and Theorem V.7.15.i, Analogously, using the theorems V.5.2.ii and V.7.16.i, we obtain

Theorem 8.9 (Second main theorem for spectra). i) If E and F are semialgebraic spectra over $\mathbb{R}$ then the (semialgebraic) homotopy classes of homotopy maps $f : E \to F$ correspond uniquely with the (topological) homotopy classes of homotopy maps $g : E_{top} \to F_{top}$ by the relation $[f_{top}] = [g]$. A homotopy map $f : E \to F$ is a (semialgebraic) homotopy equivalence iff f_{top} is a (topological) homotopy equivalence.
ii) Given a topological spectrum F there exists a semialgebraic spectrum E over $\mathbb{R}$, with every space E_n a semialgebraic CW-complex, and a (topological) homotopy equivalence $\varphi : E_{top} \to F$.
iii) A semialgebraic spectrum E over $\mathbb{R}$ is an Ω-spectrum iff E_{top} is a topological Ω-spectrum.

For k^* a reduced cohomology theory over R and K a real closed field extension of R we have constructed in §2 a reduced cohomology theory k^*_K over K. Similarly for l^* a reduced semialgebraic cohomology theory over $\mathbb{R}$ we have constructed a reduced topological cohomology theory l^*_{top}. As an immediate consequence of the definitions (cf. §2) one obtains

Proposition 8.10. i) For every Ω-spectrum E over R there exists an evident and canonical isomorphism

$$E(K)^* \xrightarrow{\sim} (E^*)_K$$

of reduced cohomology theories over K.
ii) If E is a semialgebraic Ω-spectrum over $\mathbb{R}$ with every E_n a semialgebraic CW-comples then there exists an evident and canonical isomorphism

$$(E_{top})^* \xrightarrow{\sim} (E^*)_{top}$$

of topological reduced cohomology theories.

By the results 8.5 - 8.10 the contents of the second half of §2 on the relations between semialgebraic cohomology theories over different real closed base fields and topological cohomology theories becomes rather obvious. Notice that these results have been obtained out of Chapter V and the representation theorem 7.1 without serious labour. Thus we have gained a rather comfortable new access to §2.

A particularly pleasant feature of the relation between cohomology theories and Ω-spectra is that, if K is a real closed field extension of R, then the restriction process for reduced cohomology theories $1^* \rightsquigarrow 1^{*R}$ is more natural than the extension process $1^* \rightsquigarrow 1^*_K$ while for spectra the process $E \rightsquigarrow E(K)$ is the natural one. An analogous remark pertains to "sa" and "top".

We will be rather brief about the connection between spectra and homology theories. Let E be a spectrum over R (not necessarily an Ω-spectrum). For any weak polytope X and any $n \in \mathbb{Z}$ we have a direct system of abelian groups $(\pi_{n+k}(E_k \wedge X) \mid k \gg 0)$ with obvious transition maps

$$\pi_{n+k}(E_k \wedge X) \xrightarrow{S} \pi_{n+k+1}(SE_k \wedge X) \xrightarrow[(\varepsilon_k^E \wedge id_X)_*]{} \pi_{n+k+1}(E_{k+1} \wedge X) \quad .$$

We define

$$\tilde{H}_n(X,E) := \varinjlim_k \pi_{n+k}(E_k \wedge X) \ .$$

In this way we obtain covariant functors $\tilde{H}_n(-,E)$ on $HP^*(R)$ with values in abelian groups. We define suspension isomorphisms in the (perhaps up to a sign) obvious way: $\sigma_n(X)$ is induced by the family of homomorphisms

$$\pi_{n+k}(E_k \wedge X) \xrightarrow{S} \pi_{n+k+1}(SE_k \wedge X) \xrightarrow[\varphi_*]{} \pi_{n+1+k}(E_k \wedge SX) \ ,$$

with φ the switching isomorphism from $S^1 \wedge E_k \wedge X$ to $E_k \wedge S^1 \wedge X$.

Theorem 8.11. i) The families $(\tilde{H}_n(-,E)\,|\,n\in\mathbb{Z})$ and $(\sigma_n\,|\,n\in\mathbb{Z})$ together form a reduced homology theory $\tilde{H}_*(-,E)$ over R.
ii) Given a reduced homology theory k_* over R there exists a spectrum E over R together with a natural equivalence $\tilde{H}_*(-,E) \xrightarrow{\sim} k_*$.

Proof. The theorem is well known to be true in the topological setting [W_2], [Sw, Chap. 8 and Chap. 14]. By our main theorems on homotopy sets (or groups), homology theories and spectra we know without further labour that the theorem holds in the semialgebraic setting.

Of course, it would be more satisfactory to prove the theorem directly in the semialgebraic setting. This is rather easy for the first statement (cf. [W_2, p. 249f.]; in contrast to cohomology the wedge axiom does not cause any troubles). In order to prove the second statement directly it seems advisable to use the modern sophisticated language of CW-spectra and maps between them as designed by Boardman and Adams (cf. [Ad, Part III] and some chapters in [Sw]), which anyway is indispensable for understanding the deeper aspects of stable homotopy theory. It is perfectly possible to transfer this language and theory to the semialgebraic setting, and the reader is invited to do so. He will have the choice to work only with CW-complexes or also with weak polytopes, the latter perhaps being more natural.

Chapter VII - Simplicial spaces

As already said in the introduction this chapter is written with an eye to applications in the theory of fibrations in the third volume [SFC]. From the viewpoint of the preceding chapters it would be sufficient to deal with simplicial sets instead of simplicial spaces, i.e. we could assume that the occuring simplicial spaces are discrete (cf. 1.2.ix and 2.5 below). This would trivialize the major part of §1-§5.

§1. The basic definitions

We have to recall some of the standard terminology on simplicial objects, cf. [La], [May], [Cu].

For every non negative integer $n \in \mathbb{N}_o$ we denote by [n] the set $\{0,1,2,\ldots,n\}$ equipped with its natural total ordering. Let Ord denote the category whose objects are the sets [n] and whose morphisms are the monotonic maps between these totally ordered sets. $\{\alpha : [n] \to [m]$ is called monotonic if $i \leq j$ implies $\alpha(i) \leq \alpha(j)$.$\}$

N.B. There exists a unique natural equivalence from the category $\widetilde{\text{Ord}}$ of all finite nonempty totally ordered sets to the small category Ord. In all our study below we could replace Ord by $\widetilde{\text{Ord}}$, but more often than not it seems to be more comfortable to work with Ord than with $\widetilde{\text{Ord}}$.

Definition 1. Let C be any category.

a) A simplicial object in C is a contravariant functor from Ord to C, i.e. a functor $X : \text{Ord}^o \to C$. The value $X([n])$ will usually be denoted by X_n. If $\alpha : [n] \to [m]$ is a monotonic map then the morphism

$X(\alpha) : X_m \to X_n$ will usually be denoted briefly by α^*. Notice that, if $\beta : [m] \to [p]$ is a second monotonic map, $(\beta\alpha)^* = \alpha^*\beta^*$. We call these morphisms α^* the <u>transition morphisms</u> of X.

b) Let X and Y be simplicial objects in C. A <u>simplicial morphism</u> $f : X \to Y$ in C is a natural transformation from the functor X to the functor Y.

c) The category of simplicial objects and simplicial morphisms in C is denoted by sC.

d) If X is a simplicial object in C then we define the <u>face morphisms</u>

$$d_i = d_i^m : X_n \to X_{n-1} \quad (0 \leq i \leq n)$$

and the <u>degeneracy morphisms</u>

$$s_i = s_i^n : X_n \to X_{n+1} \quad (0 \leq i \leq n)$$

by

$$d_i = (\delta^i)^*, \quad s_i = (\sigma^i)^*,$$

with $\delta^i : [n-1] \to [n]$ the monotonic injection which omits the value i and $\sigma^i : [n+1] \to [n]$ the monotonic surjection which takes the value i twice.

Notice that every monotonic map $\alpha : [q] \to [n]$ has a unique decomposition

$$\alpha = \delta^{i_1} \dots \delta^{i_s} \sigma^{j_1} \dots \sigma^{j_t}$$

with $n \geq i_1 > \dots > i_s \geq 0$ and $0 \leq j_1 < \dots < j_t < q$ and, of course, $q+s = n+t$ [La, p. 2f.]. In particular α has a unique factorization $\alpha = \gamma\beta$ with β a monotonic surjection and γ a monotonic injection. This implies

$$\alpha^* = s_{j_t} \dots s_{j_1} d_{i_s} \dots d_{i_1} = \beta^*\gamma^* .$$

It is now evident that we may think of a simplicial object X in C as a sequence $(X_n \mid n \in \mathbb{N}_0)$ of objects in C together with morphisms $d_i : X_n \to X_{n-1}$, $s_i : X_n \to X_{n+1}$ $(0 \leq i \leq n)$ which fulfill the appropriate identities. These are [La, p. 5]:

$$
\begin{array}{lll}
\text{a)} & d_i d_j = d_{j-1} d_i , & i < j. \\
\text{b)} & s_i s_j = s_{j+1} s_i , & i \leq j. \\
\text{c)} & d_i s_j = s_{j-1} d_i , & i < j. \\
\text{d)} & d_i s_i = d_{i+1} s_i = \mathrm{id}. & \\
\text{e)} & d_i s_j = s_j d_{i-1} , & i > j+1.
\end{array}
\qquad (1.1)
$$

Further we may think of a simplicial morphism $f : X \to Y$ as a sequence $(f_n \mid n \in \mathbb{N}_0)$ of morphisms $f_n : X_n \to Y_n$ in C which commute with the face and the degeneracy morphisms,

$$s_i f_n = f_n s_i, \; d_i f_n = f_n d_i \quad (0 \leq i \leq n) .$$

Definition 2. A simplicial space over R is a simplicial object X in the category WSA(R) of weakly semialgebraic spaces over R. A simplicial map $f : X \to Y$ between simplicial spaces over R is a simplicial morphism in WSA(R). If no confusion is possible about the base field R under consideration we call these X and f more briefly "simplicial spaces" and "simplicial maps".

This terminology prolongates our use of the words "space" and "map" in previous chapters.

Let us also agree upon the following: If P is one of the properties which we have previously defined for spaces (say "semialgebraic", "complete", "partially complete", "discrete" ...) then we say that a simplicial space X has property P iff every X_n has property P. If Q is a property which we have defined for maps between spaces then we say that a simplicial map $f : X \to Y$ between simplicial spaces has property Q iff every map $f_n : X_n \to Y_n$ has property Q.

We shall use all the vocabulary which naturally emanates from this agreement without much further explanation. For example, a simplicial weak polytope X means, of course, a simplicial space X over R such that every X_n is a weak polytope.

Examples 1.2. i) Let X be a simplicial algebraic variety over R, i.e. a simplicial object in the category C of algebraic schemes over R. Such objects play an important role for example in the Hodge theory of singular algebraic varieties [De] and in etale homotopy theory ([AM], [Frd]). The composite of $X : \mathrm{Ord}^o \to C$ and the functor $Z \mapsto Z(R)$ from C to WSA(R) is a semialgebraic simplicial space over R which we denote by X(R). In more concrete terms, X(R) is the sequence $(X_n(R) \mid n \in \mathbb{N}_o)$ together with the boundary maps $(d_i)_R : X_n(R) \to X_{n-1}(R)$ and the degeneracy maps $(s_i)_R : X_n(R) \to X_{n+1}(R)$.

ii) Similarly a simplicial algebraic variety X over $C := R(\sqrt{-1})$ gives us a semialgebraic simplicial space X(C) over R.

iii) Every space M over R gives us a constant simplicial space over R, namely the constant functor from Ord^o to WSA(R), which maps all objects [n] to M and all monotonic maps α to identity. We denote this simplicial space by $\underline{M}$.

iv) Let $f : M \to S$ be a map between spaces over R. Starting from f we obtain a simplicial space X over R as follows. X_n is the fibre product $(M/S)^{n+1} = M \times_S \dots \times_S M$ of n+1 copies of M over S. If $\alpha : [n] \to [m]$ is monotonic then $\alpha^* : X_m \to X_n$ is the map $(x_o,\dots,x_m) \mapsto (x_{\alpha(o)},\dots,x_{\alpha(n)})$. We have an obvious simplicial map from X to $\underline{S}$, which sends $(x_o,\dots,x_m)$ to $f(x_o) = \dots = f(x_m)$.

v) Let G be a weakly semialgebraic monoid, i.e. a weakly semialgebraic space with a monoid structure (associative, with unit element e) such that the multiplication map $G \times G \to G$, $(x,y) \to xy$, is weakly semialgebraic. Then we can define a simplicial space NG as follows: $(NG)_n$ is the n-fold product $G^n = G \times \ldots \times G$. If $\alpha : [m] \to [n]$ is monotonic then $\alpha^*(g_1,\ldots,g_n) = (h_1,\ldots,h_m)$ with h_i the ordered product of all g_k with $\alpha(i-1) < k \leq \alpha(i)$ {empty product = e}. Thus we have

$$d_i(g_1,\ldots,g_n) = (g_1,\ldots,g_{i-1},g_i g_{i+1},\ldots,g_n)$$

if $1 \leq i \leq n-1$,

$$d_o(g_1,\ldots,g_n) = (g_2,\ldots,g_n),$$
$$d_n(g_1,\ldots,g_n) = (g_1,\ldots,g_{n-1}),$$

and, for every $i \in [n]$,

$$s_i(g_1,\ldots,g_{n-1}) = (g_1,\ldots,g_i,e,g_{i+1},\ldots,g_{n-1}) .$$

We call NG the <u>nerve</u> of the monoid G. It will play a role only in the next volume [SFC] (cf. introduction of this book).

vi) If X and Y are simplicial spaces over R then we obtain a simplicial space $X \times Y$ over R by combining the functors X and Y into a functor $[n] \mapsto X_n \times Y_n$ from Ord^o to WSA(R). Clearly $X \times Y$ together with the obvious simplicial projection maps $pr_1 : X \times Y \to X$, $pr_2 : X \times Y \to Y$ is the <u>direct product</u> of X and Y in the category sWSA(R).

vii) For any family $(X_\lambda | \lambda \in \Lambda)$ of simplicial spaces we may form the direct sum $X := \sqcup(X_\lambda | \lambda \in \Lambda)$ in the obvious way, $X_n := \sqcup(X_{\lambda n} | \lambda \in \Lambda)$ (cf. IV.1.10).

viii) If $\tilde{R}$ is a real closed field extension of R then every simplicial space X over R yields a simplicial space $X(\tilde{R})$ over $\tilde{R}$ by composing the functor $X : \mathrm{Ord}^o \to \mathrm{WSA}(R)$ with the base field extension functor $\mathrm{WSA}(R) \to \mathrm{WSA}(\tilde{R})$. Every simplicial map $f : X \to Y$ over R yields a simplicial map $f_{\tilde{R}} : X(\tilde{R}) \to Y(\tilde{R})$.

ix) Every <u>simplicial set</u> K, i.e. simplicial object in the category Set of sets, gives us a simplicial space K_R by regarding every set K_n as a discrete space over R. These simplicial spaces K_R will be of primary importance for us, cf. §6-§8.

x) Conversely if X is a simplicial space, then we obtain a simplicial space X^δ by regarding the underlying simplicial set of X as a discrete simplicial space. The identity map $X^\delta \to X$ is a simplicial map from the space X^δ to X. We call X^δ the discretization of X.

In the following X is a simplicial space over R.

Definition 3. The points of X_n are called the n-simplices of X. An n-simplex x is called degenerate, if there exists a monotonic surjection $\alpha : [n] \twoheadrightarrow [q]$ with $n > q$ and some $y \in X_q$ such that $x = \alpha^*(y)$. Otherwise x is called nondegenerate. The set of degenerate n-simplices of X is denoted by DX_n and the set of nondegenerate n-simplices by NX_n {D = "degenerate", N = "new"}.

Proposition 1.3. i) If $\alpha : [n] \twoheadrightarrow [q]$ is a monotonic surjection then $\alpha^* : X_q \to X_n$ is an injection which has a weakly semialgebraic cosection. Thus α^* is a closed embedding and $\alpha^*(X_q)$ is a retract of X_n.

ii) If $\alpha : [n] \hookrightarrow [q]$ is a monotonic injection then $\alpha^* : X_q \to X_n$ is a surjection which has a weakly semialgebraic section. In particular α^* is strongly surjective.

All this is evident since α has a right inverse in the first case and a left inverse in the second case.

The first part of the proposition implies that $NX_n = X_n \setminus DX_n$ is an open subspace of X_n.

Remark 1.4. Keep n fixed. For every monotonic surjection $\alpha : [n] \twoheadrightarrow [q]$ let $X_{n,\alpha}$ denote the open subspace $\alpha^*(NX_q)$ of $\alpha^*(X_q)$. This is a locally closed subspace of X_n. It is well known [La, p. 7f] that every $x \in X_n$ has a <u>unique</u> description $x = \alpha^* y$ with $\alpha : [n] \twoheadrightarrow [q]$ a monotonic surjection and y nondegenerate. Thus X_n is the disjoint union of the subsets $X_{n,\alpha}$. Combining the isomorphisms $\alpha^* : NX_q \xrightarrow{\sim} X_{n,\alpha}$ we obtain a <u>semialgebraic bijection</u>

$$\bigsqcup_{q=0}^{n} \bigsqcup_{\alpha} NX_q \to X_n ,$$

with α running through the monotonic surjections from [n] to [q]. We may think of the family of all $X_{n,\alpha}$ as a stratification of the space X_n.

<u>Definition 4.</u> A <u>subspace</u> Y of X is a sequence $(Y_n | n \in \mathbb{N}_o)$ of subspaces Y_n of X_n such that $\alpha^*(Y_q) \subset Y_p$ for every monotonic map $\alpha : [p] \to [q]$, $p,q \in \mathbb{N}_o$. The subspace Y is called <u>closed</u> (open, locally closed) in X if every Y_n is closed (resp. open, resp. locally closed) in X_n.

This notion of subspace Y meets the usual expectations. Y is a simplicial space over R in the evident way, and the inclusion $i : Y \hookrightarrow X$ is a simplicial map. If $f : Z \to X$ is a simplicial map from a simplicial space Z to X, and if $f_n(Z_n) \subset Y_n$ for every $n \in \mathbb{N}_o$, then we have a unique factorization $f = i \circ g$ with g a simplicial map from Z to Y.

Let $n \in \mathbb{N}_o$ be given. We define, for every $k \in \mathbb{N}_o$, a subspace Y_k of X_k as follows. If $k \leq n$ then $Y_k = X_k$. If $k \geq n$ then Y_k is the union of the closed subspaces $\alpha^*(X_n)$ with α running through the finitely many monotonic surjections from [k] to [n].

<u>Proposition and Definition 1.5.</u> If $\beta : [p] \to [q]$ is monotonic then $\beta^*(Y_q) \subset Y_p$. Thus $(Y_k | k \in \mathbb{N}_o)$ is a closed subspace of X. We call this subspace the <u>n-skeleton</u> of X and write $Y = X^n$. If superscripts n will be needed for other reasons then we shall write more elaborately $Y = sk^n(X)$.

Proof. It suffices to study the cases that β is surjective or injective. The assertion is evident if $p \leq n$. Thus assume $p > n$. If β is surjective and $q \geq n$ then the assertion is again evident. If β is surjetive and $q < n$ then we have some factorization $\beta = \delta \circ \gamma$ of β with two monotonic surjections $\gamma : [p] \twoheadrightarrow [n]$, $\delta : [n] \twoheadrightarrow [q]$, and the assertion is again clear.

Assume now that β is injective. Let $x \in Y_q$ be given. Write $x = \alpha^*z$ with some $z \in X_n$ and a monotonic surjection $\alpha: [q] \twoheadrightarrow [n]$. We have a commuting square of monotonic maps

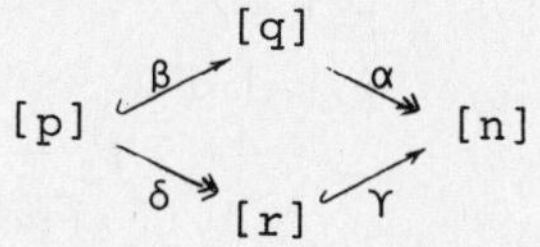

with γ injective and δ surjective. This implies $\beta^*x = \delta^*\gamma^*z$, and $\gamma^*z \in X_r = Y_r$. Since δ is surjective $\delta^*\gamma^*z \in Y_p$, as shown above.

N.B. If $k \leq n$ then $N(X^n)_k = NX_k$, but if $k > n$ then $N(X^n)_k$ is empty. If $n \leq m$ then X^n is a closed subspace of X^m.

It is easily seen that X is the inductive limit of the family $(X^n | n \geq 0)$ of its skeletons in the category sWSA(R), with the inclusions as transition maps.

§2. Realization of some simplicial spaces

Starting with the given simplicial space X we want to build a space $|X|$ over R by replacing each $x \in X_n$ by a true geometric n-simplex and gluing these simplices together.

Definition 1. a) We define a covariant functor ∇ from Ord to WSA(R) (i.e. a "cosimplicial space" over R) as follows. $\nabla([n])$ is the closed standard n-simplex in R^{n+1} (in the classical sense, with the vertices $e_o, e_1, \ldots, e_n$). The points of $\nabla([n])$ are the tuples $(t_o, \ldots, t_n) \in R^{n+1}$ with every $t_i \geq 0$ and $t_o + t_1 + \ldots + t_n = 1$. If $\alpha : [n] \to [m]$ is monotonic then $\nabla(\alpha)$ is the affine (= linear) map from $\nabla([n])$ to $\nabla([m])$ which sends e_i to $e_{\alpha(i)}$. Usually we write $\nabla(n)$ instead of $\nabla([n])$ and α_* instead of $\nabla(\alpha)$.[1)]

b) $\hat{X}$ denotes the direct sum $\bigsqcup(X_n \times \nabla(n) \mid n \in \mathbb{N}_o)$ of the spaces $X_n \times \nabla(n)$ over R.

c) We introduce on the set $\hat{X}$ the coarsest equivalence relation $\sim$ such that

$$(\alpha^* x, t) \sim (x, \alpha_* t)$$

for any $x \in X_m$, $t \in \nabla(n)$ and monotonic map $\alpha : [n] \to [m]$. Notice that this equivalence relation is also the coarsest one such that

$$(s_i x, t) \sim (x, (\sigma^i)_* t)$$

for $x \in X_n$, $0 \leq i \leq n, t \in \nabla(n+1)$, $n \geq 0$, and

$$(d_i x, t) \sim (x, (\delta^i)_* t)$$

for $x \in X_n$, $0 \leq i \leq n, t \in \nabla(n-1)$, $n \geq 1$.

d) $|X|$ denotes the <u>set</u> of equivalence classes of this equivalence relation, and $\eta_X : \hat{X} \twoheadrightarrow |X|$ denotes the natural projection from the set $\hat{X}$ to $|X|$. In later sections (starting from 3.6) we shall usually denote a value $\eta_X(x,t)$ more briefly by $|x,t|$ $(x \in X_n, t \in \nabla(n))$.

---

1) If different base fields are under consideration we write $\nabla(n)_R$ instead of $\nabla(n)$.

e) $\check{X}$ denotes the open subspace $\sqcup(NX_n \times \mathring{\nabla}(n) \mid n \in \mathbb{N}_0)$ of $\hat{X}$ and $\zeta_X : \check{X} \to |X|$ denotes the restriction of η_X to $\check{X}$. Here, as usual, $\mathring{\nabla}(n)$ means the interior of the geometric simplex $\nabla(n)$.

In the following the space $\check{X}$ will only play an auxiliary role in some proofs. We want to equip the set $|X|$ with the structure of a weakly semialgebraic space over R such that the map η_X is identifying (IV, §8). We shall succeed in doing this under an additional hypothesis on X and then shall call the space $|X|$ the realization of the simplicial space X.

We shall need two well known combinatorial facts.

Lemma 2.1 [La, p. 36]. The map ζ_X is bijective. In other words, every equivalence class in $\hat{X}$ contains a unique point (x,t) with $x \in NX_n$, $t \in \mathring{\nabla}(n)$, some $n \in \mathbb{N}_0$.

Lemma 2.2. If Z is a subspace of X then $Z_n \cap NX_n = NZ_n$ for every $n \in \mathbb{N}_0$.

I shall give the proof of the second lemma, since I did not find an appropriate reference. If an n-simplex $z \in Z_n$ is nondegenerate in X then certainly z is nondegenerate in Z. Thus $Z_n \cap NX_n \subset NZ_n$. Let now $z \in NZ_n$ be given. Write $z = \alpha^* x$ with $\alpha : [n] \twoheadrightarrow [p]$ a monotonic surjection and $x \in NX_p$. There exists a monotonic injection $\beta : [p] \hookrightarrow [n]$ with $\alpha \circ \beta = id_{[p]}$. We obtain $x = \beta^*\alpha^* x = \beta^* z$. Thus $x \in Z_p$. Since z is nondegenerate in Z we conclude that $p = n$, $\alpha = id_{[n]}$, hence $z \in NX_n$.

Let again Z be a subspace of X. Then $\hat{Z}$ is a subspace of $\hat{X}$. By Lemma 2.2 also $\check{Z}$ is a subspace of $\check{X}$. It is now clear from Lemma 2.1 that any two points in $\hat{Z}$ which are equivalent in $\hat{X}$ are already equivalent in $\hat{Z}$. Thus we may and shall regard $|Z|$ as a subset of $|X|$ {although only rarely $\hat{Z}$ will consist of full equivalence classes of $\hat{X}$}.

In particular we regard $|X^n|$ as a subset of $|X|$ for every $n \in \mathbb{N}_o$. Notice that $|X|$ is the union of these subsets.

For convenience we put $X^{-1} = \underline{\emptyset}$, the empty simplicial space, hence $|X^{-1}| = \emptyset$.

We shall need two more easy combinatorial lemmas.

<u>Lemma 2.3.</u> $\eta_X(X_n \times \mathring{\nabla}(n)) = \eta_X(X_n \times \nabla(n)) = |X^n|$ for every $n \in \mathbb{N}_o$.

<u>Proof.</u> Of course, η_X maps $X_n \times \nabla(n)$ into $|X^n|$. Let $\xi \in |X^n|$ be given. We write $\xi = \eta_X(x,t)$ with $x \in N(X^n)_k$, $t \in \mathring{\nabla}(k)$, both uniquely determined by ξ (Lemma 2.1). We have $k \leq n$. (Moreover $x \in NX_k$ by Lemma 2.2, but we shall not need this.) Choose some monotonic surjection $\alpha : [n] \twoheadrightarrow [k]$. There exists some $s \in \mathring{\nabla}(n)$ with $\alpha_*(s) = t$. We have

$$\xi = \eta_X(x,\alpha_* s) = \eta_X(\alpha^* x, s). \qquad \text{q.e.d.}$$

Let A_n denote the complement of $NX_n \times \mathring{\nabla}(n)$ in $X_n \times \nabla(n)$, i.e.

$$A_n := (DX_n \times \nabla(n)) \cup (X_n \times \partial\nabla(n)),$$

with $\partial\nabla(n)$ denoting the boundary of the simplex $\nabla(n)$, as usual. Notice that A_n is a closed subspace of $X_n \times \nabla(n)$.

<u>Lemma 2.4.</u> η_X maps A_n onto $|X^{n-1}|$.

<u>Proof.</u> We have $DX_n = (X^{n-1})_n$. Thus certainly $\eta_X(DX_n \times \nabla(n)) \subset |X^{n-1}|$. If $x \in X_n$ and $t \in \partial\nabla(n)$, i.e. $t = (\delta^i)_*(s)$ with some $i \in [n]$, $s \in \nabla(n-1)$, then $\eta_X(x,t) = \eta_X(d_i x, s)$ lies in $|X^{n-1}|$. Thus $\eta_X(A_n) \subset |X^{n-1}|$. Now choose some monotonic surjection $\alpha : [n] \twoheadrightarrow [n-1]$. Then $\alpha_*(\mathring{\nabla}(n)) = \mathring{\nabla}(n-1)$. Using Lemma 2.3 we obtain

$$|X^{n-1}| = \eta_X(X_{n-1} \times \mathring{\nabla}(n-1)) = \eta_X(\alpha^*(X_{n-1}) \times \mathring{\nabla}(n)) .$$

Thus certainly $\eta_X(A_n) = |X^{n-1}|$.

Definition 2. The simplicial space X is called partially proper (resp. proper) if all the face maps $d_i : X_n \to X_{n-1}$ are partially proper (resp. proper). This implies that, for every monotonic map $\alpha : [p] \to [q]$, the map $\alpha^* : X_q \to X_p$ is partially proper (proper).

Important example 2.5. Every discrete simplicial space X (i.e. $X = K_R$ for some simplicial set K, cf. Ex. 1.2.ix) is partially proper. But X is proper only if all the face maps $d_i : X_n \to X_{n-1}$ have finite fibres.

Theorem 2.6. Assume that the simplicial space X is partially proper. Then the set $|X|$ carries a unique structure of a weakly semialgebraic space over R such that $\eta_X : \hat{X} \to |X|$ is a strongly surjective partially proper map. In this structure the sets $|X^n|$ are closed subspaces of $|X|$ $(n \in \mathbb{N}_0)$, and the family of these subspaces is an admissible filtration (cf. V, §2, Def. 3) of $|X|$. η_X restricts to a strongly surjective partially proper map from $X_n \times \nabla(n)$ onto $|X^n|$ and to an isomorphism from $NX_n \times \mathring{\nabla}(n)$ onto $|X^n| \setminus |X^{n-1}|$. (N.B. These are open subspaces of $X_n \times \nabla(n)$ and $|X^n|$ respectively.) The restriction $\zeta_X : \check{X} \to |X|$ of η_X is bijective and semialgebraic.

The uniqueness of the space structure on $|X|$ is evident, since $|X|$ will be the strong quotient of $\hat{X}$ via η_X (cf. IV, §9), but we shall need some work to prove existence.

In the following we denote the map η_X briefly by η and the map η_{X^n} by η_n. We further denote the surjections obtained from η by restriction to $X_n \times \nabla(n)$ and A_n, according to the Lemmas 2.3 and 2.4 above, by

$$\psi_n : X_n \times \nabla(n) \twoheadrightarrow |X^n|,\ \varphi_n : A_n \twoheadrightarrow |X^{n-1}|.$$

Lemma 2.7. i) Every set $|X^n|$ carries a unique structure of a weakly semialgebraic space over R such that ψ_n is a strongly surjective and partially proper morphism. The space $|X^{n-1}|$ is a closed subspace of $|X^n|$, and (thus) the map φ_n is again partially proper. Also the map η_n is partially proper.

ii) The diagram

$$\begin{array}{ccc} A_n & \xrightarrow{\varphi_n} & |X^{n-1}| \\ \downarrow i & & \downarrow j \\ X_n \times \nabla(n) & \xrightarrow[\psi_n]{} & |X^n| \end{array} \qquad (*)$$

with inclusions i and j, is cocartesian in the category WSA(R), in short,

$$|X^n| = (X_n \times \nabla(n)) \cup_{\varphi_n} |X^{n-1}| .$$

Proof. a) It is evident from Lemma 2.1, applied to X^n and X^{n-1}, that η maps $NX_n \times \mathring{\nabla}(n) = (X_n \times \nabla(n)) \setminus A_n$ bijectively onto $|X^n| \setminus |X^{n-1}|$. This means that the diagram (*) is cocartesian on the set theoretic level. In particular $(n=0)$, ψ_o is a bijection from $X_o \times \nabla(0)$ to $|X^o|$.

b) We now prove the lemma by induction on n. Let $n=0$. We may identify $X_o = X_o \times \nabla(0)$. We transfer the space structure from X_o to $|X^o|$ by the bijection ψ_o, which thus becomes an isomorphism. It remains to prove that $\eta_o : (X^o)^\wedge \to X^o$ is partially proper, i.e. that $\eta_o | (X^o)_n \times \nabla(n)$ is partially proper for every $n \geq 1$.

We have $(X^o)_n = \alpha^* X_o$ with α the unique surjection from [n] to [0]. Moreover $\alpha^* : X_o \to (X^o)_n$ is an isomorphism. If $x \in X_o$, $t \in \nabla(n)$, then $\alpha_*(t)$ is the unique point e_o of $\nabla(0)$, hence $\eta(\alpha^* x, t) = \eta(x, e_o) = \psi_o(x)$. This means that the following square commutes

$$\begin{array}{ccc} X_o \times \nabla(n) & \xrightarrow{pr_1} & X_o \\ \downarrow {\scriptstyle \alpha^* \times id_{\nabla(n)}} & & \downarrow {\scriptstyle \psi_o} \\ (X^o)_n \times \nabla(n) & \xrightarrow[\eta_o|-]{} & |X^o| \end{array} \quad .$$

Here the vertical arrows are isomorphisms and the natural projection pr_1 is proper. We conclude that $\eta_o | (X^o)_n \times \nabla(n)$ is proper (!).

c) We study η_n for $n \geq 1$. By induction hypothesis we already have a space structure on the subset $|X^{n-1}|$ of $|X^n|$ such that $\eta_{n-1} : (X^{n-1})^\wedge \to |X^{n-1}|$ is partially proper and ψ_{n-1} is strongly surjective.

We first verify that $\varphi_n : A_n \to |X^{n-1}|$ is partially proper. Since $DX_n = (X^{n-1})_n$ it is already clear that $\varphi_n | DX_n \times \nabla(n)$ is partially proper. Thus it suffices to prove that $\varphi_n | X_n \times (\delta^i)_* \nabla(n-1)$ is partially proper for any given $i \in [n]$. We have the formula

$$\varphi_n(x, (\delta^i)_* t) = \psi_{n-1}(d_i(x), t)$$

for $x \in X_n$, $t \in \nabla(n-1)$. This means that the following square commutes:

$$\begin{array}{ccc} X_n \times \nabla(n-1) & \xrightarrow{d_i \times id} & X_{n-1} \times \nabla(n-1) \\ \downarrow {\scriptstyle id \times (\delta^i)_*} & & \downarrow {\scriptstyle \psi_{n-1}} \\ X_n \times (\delta^i)_* \nabla(n-1) & \xrightarrow[\varphi_n|-]{} & |X^{n-1}| \end{array} \quad .$$

The left vertical arrow is an isomorphism. The upper horizontal arrow is partially proper since by assumption d_i is partially proper. The right vertical arrow is partially proper by induction hypothesis. We conclude that the restriction of φ_n to $X_n \times (\delta^i)_* \nabla(n-1)$ is partially proper. Thus indeed φ_n is partially proper.

We look at the diagram (*) in the lemma. We equip $|X^n|$ with the structure of a weakly semialgebraic space which makes this diagram cocartesian in WSA(R). In more concrete terms, we glue $X_n \times \nabla(n)$ to $|X^{n-1}|$ along the closed subspace A_n by the partially proper map φ_n and take $|X^n|$ as the underlying set of this space via ψ_n. This is possible by IV, §8. We also know from IV, §8 that, for this space structure on $|X^n|$, the space $|X^{n-1}|$ is a closed subspace of $|X^n|$ and ψ_n is partially proper.

We now prove that $\eta_n|(X^n)_q \times \nabla(q)$ is partially proper for every $q \in \mathbb{N}_o$. This is already clear from above for $q \leq n$. Let $q > n$. $(X^n)_q$ is the union of the closed subspaces $\alpha^*(X_n)$ of X_q with α running through the monotonic surjections from $[q]$ to $[n]$. If α is such a surjection then, for $x \in X_n$, $t \in \nabla(q)$,

$$\eta_n(\alpha^*x,t) = \psi_n(x,\alpha_* t) .$$

From this formula we conclude, in much the same way as above, that the restriction of η_n to $\alpha^*(X_n) \times \nabla(q)$ is partially proper. Thus $\eta_n|(X^n)_q \times \nabla(q)$ is partially proper for every $q \in \mathbb{N}_o$, which means that the whole map η_n is partially proper.

ψ_{n-1}, and hence φ_n, is strongly surjective by induction hypothesis. ψ_n restricts to an isomorphism from $(X_n \times \nabla(n)) \smallsetminus A_n$ $(= NX_n \times \mathring{\nabla}(n))$ to $|X^n| \smallsetminus |X^{n-1}|$, cf. IV, §8. Thus ψ_n is again strongly surjective. This completes the proof of the lemma.

It is now easy to finish the proof of Theorem 2.6. Applying Theorem IV.7.1 to the set $|X|$ and the ordered family of spaces $(|X^n| \mid n \in \mathbb{N}_o)$ we obtain a structure of a weakly semialgebraic space over R on the set $|X|$ such that every $|X^n|$ is a closed subspace of $|X|$ and the family of these spaces is an admissible filtration of $|X|$. The family $((X^n)^\wedge \mid n \in \mathbb{N}_o)$ is clearly an admissible filtration of the space $\hat{X}$.

All the restrictions $\eta_n : (X^n)^\wedge \to |X^n|$ of η are partially proper. Thus η itself is partially proper. By the proof of the lemma we know that η restricts to an isomorphism from $(NX_n) \times \mathring{\nabla}(n)$ to $|X^n| \setminus |X^{n-1}|$. This implies that the restriction ζ_X of η is certainly semialgebraic (and bijective, as we know already). In particular, ζ_X is strongly surjective. A fortiori η is strongly surjective. Theorem 2.6 proved.

Remarks 2.8. i) We have seen in the course of the proof that the restriction $\eta_X | X_o \times \nabla(0)$ is an isomorphism from $X_o = X_o \times \nabla(0)$ to $|X^o|$.
ii) If the simplicial space X is proper then our arguments show that $\psi_n : X_n \times \nabla(n) \twoheadrightarrow |X^n|$ is proper for every $n \in \mathbb{N}_o$.
iii) If X is a simplicial weak polytope then $\hat{X}$ is a weak polytope. Since η_X is strongly surjective we conclude that $|X|$ is a weak polytope in this case (cf. IV.5.5.iii).
iv) If X is a semialgebraic simplicial space then all the spaces $|X^n|$ are semialgebraic since $X_n \times \nabla(n)$ is semialgebraic. If X is a simplicial polytope then every $|X^n|$ is a polytope but, usually, $|X|$ will not be a polytope. It will be a polytope iff there exists some n with NX_m empty for $m > n$.
v) In particular, if X is a discrete simplicial space (i.e. a simplicial set), then $|X|$ is a weak polytope.

Example 2.9. If X is a constant simplicial space $\underline{M}$ (cf. Ex. 1.2.iii) then η_X restricts to an isomorphism $M = X_o \xrightarrow{\sim} |X|$, as follows from our first remark above. We identify $|\underline{M}| = M$ by this map. Then the restriction of η_X to $X_n \times \nabla(n)$ turns out to be just the natural projection $M \times \nabla(n) \to M$.

Example 2.10. Given some $n \in \mathbb{N}_0$ let $\Delta(n)$ denote the obvious contravariant functor $\mathrm{Hom}(-,[n])$ from Ord to the category of sets. The elements of $\Delta(n)_q$ are the monotonic maps $\alpha : [q] \to [n]$. They may be written as sequences $\alpha = \langle a_0,\dots,a_q\rangle$ of integers with $0 \le a_0 \le a_1 \le \dots \le a_q \le n$. $\Delta(n)_n$ contains a distinguished element $i_n = \mathrm{id}_{[n]}$, and every element of $\Delta(n)_q$ has the form $\alpha^*(i_n)$ with a unique monotonic map $\alpha : [q] \to [n]$. The simplicial set $\Delta(n)$ gives us a discrete simplicial space $\Delta(n)_R$ (cf. Ex. 1.2.ix) which now - abusively - will again be denoted by $\Delta(n)$. The map $\eta = \eta_{\Delta(n)}$ from $\widehat{\Delta(n)}$ to $|\Delta(n)|$ restricts to a bijection from $\{i_n\} \times \nabla(n)$ to $|\Delta(n)|$, as is easily seen and well known [La, p. 37]. This bijection is a proper map and thus an isomorphism. We usually shall identify $\nabla(n)$ and $|\Delta(n)|$ by this isomorphism and the evident isomorphism $\nabla(n) \xrightarrow{\sim} \{i_n\} \times \nabla(n)$. Then we have the formula

$$\eta(\alpha^* i_n, t) = \alpha_*(t)$$

for $\alpha : [k] \to [n]$ monotonic and $t \in \nabla(k)$.

Every simplicial map $f : X \to Y$ between simplicial spaces gives us maps

$$f_n \times \mathrm{id}_{\nabla(n)} : X_n \times \nabla(n) \to Y_n \times \nabla(n) .$$

These maps combine into a map $\hat{f} : \hat{X} \to \hat{Y}$ which is compatible with the equivalence relations on $\hat{X}$ and $\hat{Y}$. Thus we obtain a set theoretic map $|f| : |X| \to |Y|$ such that the diagram

$$\begin{array}{ccc} \hat{X} & \xrightarrow{\hat{f}} & \hat{Y} \\ {\scriptstyle \eta_X}\downarrow & & \downarrow{\scriptstyle \eta_Y} \\ |X| & \xrightarrow[|f|]{} & |Y| \end{array} \qquad (2.11)$$

commutes. If both X and Y are partially proper then $|f|$ is a weakly semialgebraic map between the spaces $|X|$ and $|Y|$, since η_X is identifying. We then call $|f|$ the <u>realization</u> of the simplicial map f. Notice that $|f|$ maps $|X^n|$ into $|Y^n|$.

Example 2.12. Every monotonic map $\alpha : [p] \to [n]$ gives us a simplicial map $\Delta(\alpha) : \Delta(p) \to \Delta(n)$. It maps a q-simplex $\langle a_o,\dots,a_q\rangle$ of $\Delta(p)$ to the q-simplex $\langle\alpha(a_o),\dots,\alpha(a_q)\rangle$ of $\Delta(n)$. It is easily checked that, with the identifications $|\Delta(p)| = \nabla(p)$ and $|\Delta(n)| = \nabla(n)$ described in (2.10), the realization $|\Delta(\alpha)|$ is the same as our previous map $\alpha_* : \nabla(p) \to \nabla(n)$.

Given a real closed base field R we now have established a functor "realization" from the category of partially proper simplicial spaces over R (with all simplicial maps between them) to WSA(R).[1)] These functors behave well with respect to extension of the base field.

Proposition 2.13. Let $\tilde{R}$ be a real closed over field of R. If X is a partially proper simplicial space over R then $X(\tilde{R})$ (cf. Ex. 1.2.viii) is a partially proper simplicial space over $\tilde{R}$ and $|X(\tilde{R})| = |X|(\tilde{R})$. Also $X^n(R) = X(R)^n$ for every $n \in \mathbb{N}_o$, hence $|X^n|(\tilde{R}) = |X(\tilde{R})^n|$. If $f : X \to Y$ is a simplicial map into a second partially proper simplicial space Y over R then $|f_{\tilde{R}}| = |f|_{\tilde{R}}$.

All this is obvious.

The spaces $|X|$ are amenable to homology considerations. If h_* is a homology theory over R (VI, §4, Def. 1) and X is a simplicial space over R then, for every $q \in \mathbb{Z}$, we can form a chain complex

$$\dots \to h_q(X_2) \xrightarrow[\partial_2]{} h_q(X_1) \xrightarrow[\partial_1]{} h_q(X_o) \to 0$$

of abelian groups by defining

$$\partial_n := \sum_{i=0}^{n} (-1)^i\, h_q(d_i)$$

---

1) Perhaps it would be more reasonable to regard realization as a functor to the category of admissibly filtered weakly semi-algebraic spaces (cf. V, §2, Def. 3).

with d_i running through the face maps from X_n to X_{n-1}. We denote this chain complex by $h_q(X)$ and - as usual - its p-th homology group by $H_p(h_q(X))$.

Theorem 2.14. [Se] Assume that X is a simplicial weak polytope. There exists a homological spectral sequence, natural in X, which has the E^2-term

$$E^2_{p,q} = H_p(h_q(X))$$

and converges to $h_{p+q}(|X|)$.

Indeed, every admissible filtration $(M_n | n \in \mathbb{N}_o)$ of a space M over R gives us a homological spectral sequence with E^1-term

$$E^1_{p,q} = h_{p+q}(M_p, M_{p-1}) ,$$

as in the classical theory (e.g. [Sw, Chap. 15], [W, Chap. 13]). This sequence converges to $h_*(M)$ since $h_n(M, M_k)$ is the inductive limit of the groups $h_n(M_p, M_k)$ $(k \in \mathbb{N}_o, p \to \infty)$ by Theorem VI.6.6.

In the present case we choose the filtration $(|X^n| \; | n \in \mathbb{N}_o)$ of $|X|$. Since the $|X^n|$ are weak polytopes we can form the quotients $|X^n|/|X^{n-1}|$, and we have

$$E^1_{p,q} = h_{p+q}(|X^p|, |X^{p-1}|) = \tilde{h}_{p+q}(|X^p|/|X^{p-1}|).$$

Now recall from above (cf. 2.7) that $|X^p|$ can be obtained by gluing $X_p \times \nabla(p)$ to $|X^{p-1}|$ along the subspace $(X_p \times \partial\nabla(p)) \cup (DX_p \times \nabla(p))$. Thus we have a natural isomorphism

$$S^p \wedge (X_p/DK_p) \xrightarrow{\sim} |X^p|/|X^{p-1}|,$$

hence

$$E^1_{p,q} \cong \tilde{h}_p(X_p/DX_p) = h_p(X_p,DX_p).$$

The theorem now follows by a careful study of the differentials $d^1_{p,q}$, cf. [Se, p. 109f.].

If h^* is a cohomology theory over R then we obtain in the same way a cohomological spectral sequence with

$$E_2^{p,q} = H^p(h^q(X)) ,$$

but in general, as a consequence of VI, 6.11, this will converge to a quotient of $h^{p+q}(|X|)$, namely

$$\bar{h}^{p+q}(|X|) := h^{p+q}(|X|) / \varprojlim_p{}^{(1)} h^{p+q-1}(|X^p|)$$

(cf. [W, Chap. 13, §3], Segal seems to ignore this fact in [Se, §5]).

If h^* is ordinary cohomology $H^*(-,G)$ with coefficients in some abelian group G (i.e. $\tilde{h}^0(S^0) = G$, $\tilde{h}^q(S^0) = 0$ for $q \neq 0$, cf. VI, §3, Def. 2) then it turns out that the $\lim^{(1)}$-subgroup of $h^{p+q}(|X|)$ is zero (cf. the argument in [W, p. 631]). Thus for any G, we have a converging spectral sequence

$$H^p(H^q(X,G)) \underset{p}{\Longrightarrow} H^{p+q}(|X|,G) . \qquad (2.15)$$

<u>Example 2.16.</u> Assume that X is discrete, i.e. $X = K_R$ with K a simplicial set (cf. 1.2.ix). If h_* is ordinary homology $H_*(-,G)$ then $h_q(X) = 0$ for $q \neq 0$. Thus the homology spectral sequence collapses. It gives us an isomorphism from the classical "abstract" homology group $H_p(K,G)$ of the simplicial set K (cf. §7 below) to $H_p(|K_R|,G)$. The cohomology spectral sequence for $H^*(-,G)$ also collapses and gives us, for every p, an isomorphism from $H^p(K,G)$ to $H^p(|K_R|,G)$.

§3. Subspaces

In the whole section X is a partially proper simplicial space over R. If Z is a subspace of X (cf. §1, Def. 4), then we regard $|Z|$ as a subset of $|X|$, as explained in §2 (after Lemma 2.2).

More generally we will look at simplicial subsets Z of X, for technical reasons.

Definition 1. A simplicial subset Z of X is a sequence $(Z_n | n \in \mathbb{N}_0)$ of subsets Z_n of X_n such that $\alpha^*(Z_q) \subset Z_p$ for every monotonic map $\alpha : [p] \to [q]$.

Identifying simplicial sets with discrete simplicial spaces over R we may say more formally that the simplicial subsets of X are just the subspaces of the discretization X^δ of X (Ex. 1.2.x).

On the set theoretic level we have $|X^\delta| = |X|$. Thus every simplicial subset Z of X gives us a subset $|Z|$ of $|X|$. We define subsets $\hat{Z}$ and $\check{Z}$ of $\hat{X}$ and $\check{X}$ respectively as follows.

$$\hat{Z} := \sqcup(Z_n \times \nabla(n) | n \in \mathbb{N}_0) \quad ,$$
$$\check{Z} := \sqcup(NZ_n \times \mathring{\nabla}(n) | n \in \mathbb{N}_0) \quad .$$

We have $\check{Z} = \hat{Z} \cap \check{X}$ (cf. Lemma 2.2), and $\eta_X(\hat{Z}) = \zeta_X(\check{Z}) = |Z|$.

Definition 2. A simplicial subset Z of X is called closed (open, locally closed, semialgebraic, ...) in X if every Z_n is closed (open, ...) in X_n.

In particular, a weakly semialgebraic simplicial subset Z of X is nothing else than a subspace of X.

Proposition 3.1. Let Z be a subspace of X.

i) $|Z|$ is a weakly semialgebraic subset of $|X|$.

ii) $|Z^n| = |Z| \cap |X^n|$ for every $n \in \mathbb{N}_o$. {Recall that X^n means the n-skeleton of X, cf. 1.5.}

iii) If Z is closed in X then $|Z|$ is closed in $|X|$, and the set $|Z|$ with its subspace structure in $|X|$ is the realization of the partially proper simplicial space Z.

Proof. i) The subset $\check{Z}$ of $\hat{X}$ is weakly semialgebraic in $\check{X}$. Since the map $\zeta_X : \check{X} \to |X|$ between spaces is semialgebraic (Th. 2.6), we conclude that $|Z| = \zeta_X(\check{Z})$ is weakly semialgebraic in $|X|$ (cf. IV.5.1.ii).

ii) We have

$$|X^n| = \zeta_X \left(\bigsqcup_{k=0}^{n} NX_k \times \mathring{\nabla}(k) \right) ,$$

$$|Z^n| = \zeta_X \left(\bigsqcup_{k=0}^{n} NZ_k \times \mathring{\nabla}(k) \right) ,$$

and $NZ_k = Z_k \cap NX_k$ (Lemma 2.2). The assertion now follows by the bijectivity of ζ_X.

iii) We look at the commuting square of set theoretic maps

$$\begin{array}{ccc} \hat{Z} & \xhookrightarrow{\;\;i\;\;} & \hat{X} \\ {\scriptstyle \eta_Z}\downarrow & & \downarrow{\scriptstyle \eta_X} \\ |Z| & \xhookrightarrow[\;\;j\;\;]{} & |X| \end{array}$$

with i and j inclusion maps. We equip the set $|Z|$ with its subspace structure in $|X|$. Then all the maps in the square are morphisms between spaces. Now assume that Z is closed in X. Then $\hat{Z}$ is closed in $\hat{X}$. Thus $j \circ \eta_Z = \eta_X \circ i$ is partially proper. This implies that η_Z is partially proper. The set theoretic bijection $\zeta_Z : \check{Z} \to |Z|$ is a semialgebraic map since it can be obtained by restricting the semialgebraic map ζ_X to subspaces. Thus η_Z is strongly surjective. We conclude that our space $|Z|$ is the realization of Z.

Since $j \circ \eta_Z$ is partially proper and η_Z is strongly surjective the map j is partially proper. But a priori j is semialgebraic (since it is a subspace inclusion). Thus j is proper, hence a closed embedding. q.e.d.

Caution. If a subspace Z is open in X then it may happen that $\eta_X^{-1}(|Z|)$ is not open in $\hat{X}$, hence that $|Z|$ is not open in $|X|$.

Remark 3.2. If Z is merely a simplicial subset of X, then we can still speak of the n-skeleton Z^n of Z. (Regard Z as a discrete simplicial space!) Z^n is again a simplicial subset of X and the formula $|Z^n| = |Z| \cap |X^n|$ still holds. Formally this is contained in part ii) of the proposition, since Z is a subspace of X^δ.

Let Z and W be two simplicial subsets of X.

Definition 3. We call Z a simplicial subset of W, and write $Z \subset W$, if $Z_n \subset W_n$ for every $n \in \mathbb{N}_o$.

Proposition 3.3. If $|Z| \subset |W|$ then $Z \subset W$.

Proof. We have $\check{Z} = \zeta_X^{-1}(|Z|)$, $\check{W} = \zeta_X^{-1}(|W|)$, hence $\check{Z} \subset \check{W}$. This means that $NZ_n \subset NW_n$ for every n. We conclude by Remark 1.4 that $Z_n \subset W_n$ for every n. q.e.d.

The propositions 3.1 and 3.3 imply that the subspaces Z of X correspond uniquely to suitable subspaces A of $|X|$ by the relation $A = |Z|$. We take a closer look at this correspondence $Z \leftrightarrow |Z|$.

Definition 4. Let $(Z_\lambda | \lambda \in \Lambda)$ be a family of simplicial subsets of X. The union of this family is the simplicial subset

$$V := (\cup(Z_{\lambda n} | \lambda \in \Lambda) | n \in \mathbb{N}_o)$$

of X, and the <u>intersection</u> of this family is the simplicial subset

$$W := (\bigcap(Z_{\lambda n} \mid \lambda \in \Lambda) \mid n \in \mathbb{N}_o)$$

of X. We write $V = \bigcup(Z_\lambda \mid \lambda\in\Lambda)$ and $W = \bigcap(Z_\lambda \mid \lambda\in\Lambda)$.

We clearly have

$$\hat{V} = \bigcup(\hat{Z}_\lambda \mid \lambda\in\Lambda), \quad \hat{W} = \bigcap(\hat{Z}_\lambda \mid \lambda\in\Lambda) .$$

By use of Lemma 2.2 we obtain from this

$$\check{V} = \bigcup(\check{Z}_\lambda \mid \lambda\in\Lambda), \quad \check{W} = \bigcap(\check{Z}_\lambda \mid \lambda\in\Lambda) .$$

Then applying the bijection ζ_X to these equations we obtain

$$\begin{aligned} |\bigcup(Z_\lambda \mid \lambda\in\Lambda)| &= \bigcup(|Z_\lambda| \mid \lambda\in\Lambda) , \\ |\bigcap(Z_\lambda \mid \lambda\in\Lambda)| &= \bigcap(|Z_\lambda| \mid \lambda\in\Lambda) . \end{aligned} \tag{3.4}$$

Of course, the union and the intersection of a finite family $(Z_1,\ldots,Z_n)$ will also be denoted by $Z_1 \cup\ldots\cup Z_n$ and $Z_1 \cap \ldots \cap Z_n$ respectively. If the Z_i are subspaces of X then $Z_1 \cup \ldots \cup Z_n$ and $Z_1 \cap \ldots \cap Z_n$ are again subspaces of X. More generally we have the following fact.

<u>Proposition 3.5.</u> Assume that $(Z_\lambda \mid \lambda\in\Lambda)$ is a <u>partially finite</u> family of subspaces of X. By this we mean the family $(Z_{\lambda n} \mid \lambda\in\Lambda)$ is partially finite in X_n for every $n \in \mathbb{N}_o$ (cf. V, §1, Def. 5). Then the family $(|Z_\lambda| \mid \lambda\in\Lambda)$ is partially finite in $|X|$. Its union is the subspace $|\bigcup(Z_\lambda \mid \lambda\in\Lambda)|$ of $|X|$, and its intersection is the subspace $|\bigcap(Z_\lambda \mid \lambda\in\Lambda)|$ of $|X|$.

<u>Proof.</u> Here everything is obvious from the above except the first assertion. It suffices to verify for every $n \in \mathbb{N}_o$ that the family of intersections $(|X^n| \setminus |X^{n-1}|) \cap |Z_\lambda|$, with λ running through Λ, is partially finite in $|X^n| \setminus |X^{n-1}|$. Now η_X restricts to an isomorphism from

$NX_n \times \mathring{\nabla}(n)$ onto $|X^n| \smallsetminus |X^{n-1}|$ (cf. Th. 2.6). Thus we have to verify that $(NZ_{\lambda n} \times \mathring{\nabla}(n) \mid \lambda \in \Lambda)$ is partially finite in $NX_n \times \mathring{\nabla}(n)$. This is evident since $(NZ_{\lambda n} \mid \lambda \in \Lambda)$ is partially finite in NX_n. q.e.d.

Caution. If Z is a simplicial subset of X then, in general, $(X_n \smallsetminus Z_n \mid n \in \mathbb{N}_o)$ will not be a simplicial subset of X.

We turn attention to subspaces Z of X with special properties, in particular to semialgebraic subspaces.

We start to use the shorter notation $|x,t|$ instead of $\eta_X(x,t)\,(x \in X_n, t \in \nabla(n$
Also rather often we shall denote the map η_X more briefly by η.

Lemma 3.6. Let $t \in \mathring{\nabla}(n)$ be given, and let ψ_t denote the map from X_n to $|X|$ which sends x to $|x,t|$. If Z is a simplicial subset of X then $\psi_t^{-1}(|Z|) = Z_n$.

Proof. Let $x \in X_n$ be given with $|x,t| \in |Z|$. Write $x = \alpha^* y$ with $\alpha : [n] \twoheadrightarrow [m]$ a monotonic surjection and $y \in NX_m$, both uniquely determined by x. Our assumption $|x,t| \in |Z|$ means that $y \in NZ_m$. We conclude that $x = \alpha^* y \in Z_n$. Thus $\psi_t^{-1}(|Z|) \subset Z_n$. The reverse inclusion is trivial. q.e.d.

Exploiting merely the fact that ψ_t is a morphism between spaces, hence in particular continuous, we obtain from this lemma the following

Proposition 3.7. Let Z be a simplicial subset of X.

i) If $|Z|$ is a subspace of $|X|$ then Z is a subspace of X.

ii) If $|Z|$ is closed (open) in $|X|$ then Z is closed (resp. open) in X.

We head for an answer to the following problem: For which subspaces Z of X is the set $|Z|$ semialgebraic in $|X|$?

Lemma 3.8. For every $t \in \mathring{\nabla}(n)$ the map $\psi_t : X_n \to |X|$ (cf. 3.6) is a closed embedding.

Proof. We first verify that ψ_t is injective. Let x and y be points in X_n with $|x,t| = |y,t|$. We write $x = \alpha^* u$, $y = \beta^* v$ with monotonic surjections $\alpha : [n] \twoheadrightarrow [p]$, $\beta : [n] \twoheadrightarrow [q]$ and non degenerate simplices $u \in X_p$, $v \in X_q$. We have $|u, \alpha_* t| = |v, \beta_* t|$. Since $\alpha_* t$ and $\beta_* t$ are points of $\mathring{\nabla}(p)$ and $\mathring{\nabla}(q)$ we conclude (Lemma 2.1) that $p = q$, $u = v$, $\alpha_* t = \beta_* t$. The last equality implies $\alpha = \beta$. Thus $x = y$.

The map ψ_t is partially proper since η_X is partially proper and $X_n \times \{t\}$ is a closed subspace of $\hat{X}$. We now prove that ψ_t is semialgebraic. Then we will be done: ψ_t must be proper, hence a closed embedding.

Since the map $\zeta_X : \check{X} \to |X|$ is a semialgebraic bijection it is evident that every semialgebraic subset of $|X|$ is contained in the union of finitely many sets $\eta(L \times \mathring{\nabla}(k))$ with $L \in \gamma(NX_k)$, some k. Thus it suffices to prove that $\psi_t^{-1}\eta(L \times \mathring{\nabla}(k))$ is semialgebraic in X_n for such a set L. Let x be an element of this preimage. We have $|x,t| = |y,s|$ with some $y \in L$, $s \in \mathring{\nabla}(k)$. We write $x = \alpha^* u$ with $\alpha : [n] \twoheadrightarrow [m]$ a monotonic surjection and $u \in NX_m$. We have $|u, \alpha_* t| = |y,s|$ and conclude $m = k$, $u = y$, $\alpha_* t = s$. Thus $\psi_t^{-1}\eta(L \times \mathring{\nabla}(k))$ is the union of the finitely many semialgebraic sets $\alpha^* L$ with α running through the monotonic surjections from $[n]$ to $[k]$. (In particular, $\psi_t^{-1}\eta(L \times \mathring{\nabla}(k))$ is empty if $k > n$.)

q.e.d.

Lemma 3.9. If Z is a simplicial subset of X then $Z^n = Z \cap X^n$ for every n.

Proof. By 3.2 we have $|Z^n| = |Z| \cap |X^n| = |Z \cap X^n|$. We conclude by 3.3

that $Z^n = Z \cap X^n$. (Of course, one could prove the assertion in a more combinatorial way.)

<u>Proposition 3.10.</u> Let Z be a subspace of X. Then the set $|Z|$ is semialgebraic in $|X|$ iff Z is semialgebraic and $Z \subset X^n$ for some n.

<u>Proof.</u> Assume first that Z is semialgebraic and $Z \subset X^n$ for some n. Then $Z = Z^n$ by Lemma 3.9, and $|Z| = \eta(Z_n \times \nabla(n))$ by Lemma 2.3. Since Z_n is assumed to be semialgebraic this implies $|Z|$ to be semialgebraic.

Assume now that $|Z|$ is semialgebraic. Since $(|X^n| \mid n \in \mathbb{N}_o)$ is an admissible filtration of $|X|$ we have $|Z| \subset |X^n|$ for some n, hence $Z \subset X^n$ (cf. 3.3). Now look at the map $\psi_t : X_m \to |X|$ for some m and $t \in \mathring{\nabla}(m)$. This map is semialgebraic (even proper) by Lemma 3.8, and $\psi_t^{-1}(|Z|) = Z_m$ by Lemma 3.6. Thus Z_m is semialgebraic.

Given a nonnegative integer n we describe a procedure how to obtain a host of semialgebraic subspaces of X^n. Let A be a subset of X_n. For every $k \in \mathbb{N}_o$ we define a subset A_k^* of X_k by

$$A_k^* := \bigcup_\alpha \alpha^* A ,$$

with α running through the monotonic maps from $[k]$ to $[n]$. The sequence $(A_k^* \mid k \in \mathbb{N}_o)$ is a simplicial subset of X^n. We denote it by A^*. We have $NA_n^* \subset A \subset A_n^*$. Clearly A^* is contained in every simplicial subset Z of X with $A \subset Z_n$. We call A^* the <u>simplicial subset of</u> X <u>generated by</u> A.

Assume now that A is <u>semialgebraic</u> in X_n. Then A_k^* is semialgebraic in X_k for every k and thus A^* is a semialgebraic subspace of X^n. We do right to call A^* the <u>subspace of</u> X <u>generated by</u> A.

If $A \in \overline{\mathfrak{T}}(X_n)$ then A^* is a <u>closed</u> subspace of X since the maps α^* are assumed to be partially proper. If even $A \in \mathfrak{r}_c(X_n)$ then A^* is a simplicial polytope.

Lemma 3.11. If Z is any simplicial subset of X then $Z^n = (Z_n)^*$ for every n.

Proof. This is a purely combinatorial statement. We may assume that X is discrete and then without loss of generality that Z = X. If $k \geq n$ then $((X_n)^*)_k = (X^n)_k$ by 1.5. Let $k < n$. We choose a monotonic injection $\alpha : [k] \to [n]$. Then $\alpha^* : X_n \to X_k$ is surjective. Thus $((X_n)^*)_k = (X^n)_k$ also in this case. q.e.d.

All these easy observations together, starting with Proposition 3.7.ii, give us the following result.

Proposition 3.12. The simplicial subsets Z of X with $|Z|$ semialgebraic (resp. closed semialgebraic, resp. complete) in $|X|$, are precisely the subspaces A^* with $A \in \mathfrak{r}(X_n)$ (resp. $A \in \overline{\mathfrak{r}}(X_n)$, resp. $A \in \mathfrak{r}_c(X_n)$), some n.

Remark 3.13. Let A be a subset of X_m for some m and let $\alpha : [n] \twoheadrightarrow [m]$ be a monotonic surjection. Then

$$A^* = (\alpha^*(A))^*.$$

Proof. We choose a monotonic injection $\beta : [m] \hookrightarrow [n]$ with $\alpha \circ \beta = id_{[m]}$. Then $\beta^*\alpha^*(A) = A$. Thus, for any simplicial subset Z of X, we have $A \subset Z_m$ iff $\alpha^*A \subset Z_n$. This gives the result.

Remark 3.14. If A is a subset of X_n for some n then $|A^*| = \eta(A \times \nabla(n))$.

Proof. Clearly $|A^*|$ contains $\eta(A \times \nabla(n))$. Let $\xi \in |A^*|$ be given. Then

$\xi = |\alpha^*x,t|$ with some monotonic map $\alpha : [k] \to [n]$, some $x \in A$ and some $t \in \nabla(k)$. We have $\xi = |x,\alpha_* t| \in \eta(A \times \nabla(n))$. q.e.d.

Definition 5. If Z is a simplicial subset of X then $\overline{Z} := (\overline{Z}_n | n \in \mathbb{N}_o)$ is again a simplicial subset of X, as follows from the continuity of the transition maps $\alpha^* : X_q \to X_p$. {Of course, $\overline{Z}_n$ means the closure of Z_n in X_n.} We call $\overline{Z}$ the closure of the simplicial subset Z in X.

It is evident from the continuity of η_X that $|\overline{Z}|$ is contained in the closure $\overline{|Z|}$ of $|Z|$ in $|X|$.

Proposition 3.15. i) If Z is a semialgebraic subspace of X then $\overline{Z}$ is again a semialgebraic subspace of X and $|\overline{Z}| = \overline{|Z|}$.
ii) If A is a semialgebraic subset of X_n for some n then $(\overline{A})^* = \overline{A^*}$.

Proof. i) The sets $\overline{Z}_n$ are again semialgebraic. Thus $\overline{Z}$ is a closed semialgebraic subspace of X. The set $|\overline{Z}|$ is closed in $|X|$ by Proposition 3.1. We have $|Z| \subset |\overline{Z}|$, hence $\overline{|Z|} \subset |\overline{Z}|$. The reverse inclusion is trivial and has been observed above.
ii) For any subset A of X_n we have

$$\overline{A} \subset \overline{(A^*)_n} = (\overline{A^*})_n ,$$

hence $(\overline{A})^* \subset \overline{A^*}$. Assume now that $A \in \gamma(X_n)$. Then $\overline{A} \in \overline{\gamma}(X_n)$. As observed at the beginning of our study of subspaces A^*, this implies that $(\overline{A})^*$ is a closed subspace of X. We have $A^* \subset (\overline{A})^*$, hence $\overline{A^*} \subset (\overline{A})^*$. Thus indeed $\overline{A^*} = (\overline{A})^*$. q.e.d.

Having travelled that far it is of interest to look for admissible coverings by families of sets of type $|A^*|$.

Proposition 3.16. Let $(A_\lambda | \lambda \in \Lambda)$ be an admissible covering of X_n for some n by semialgebraic sets. Then $(|A_\lambda^*| \mid \lambda\in\Lambda)$ is an admissible covering of $|X^n|$.

Proof. As already observed above (Proof of 3.8), every semialgebraic subset of $|X^n|$ is contained in the union of finitely many sets $\eta(L \times \mathring{\nabla}(k))$ with $k \leq n$ and $L \in \gamma(NX_k)$. Thus it suffices to prove that a set $\eta(L \times \mathring{\nabla}(k))$ is contained in the union of finitely many sets $|A_\lambda^*|$. We choose a monotonic surjection $\alpha : |n| \twoheadrightarrow |k|$. Then $\alpha_*\mathring{\nabla}(n) = \mathring{\nabla}(k)$, hence

$$\eta(L \times \mathring{\nabla}(k)) = \eta(\alpha^*L \times \mathring{\nabla}(n)) .$$

The set α^*L is again semialgebraic. Thus there exists a finite subset Λ' of Λ with

$$\alpha^*L \subset \bigcup(A_\lambda \mid \lambda\in\Lambda') .$$

Now $\eta(A_\lambda \times \mathring{\nabla}(n))$ is contained in $|A_\lambda^*|$, hence

$$\eta(\alpha^*L \times \mathring{\nabla}(n)) \subset \bigcup(|A_\lambda^*| \mid \lambda \in \Lambda') ,$$

and we are done. q.e.d.

Let Y be a second partially proper simplicial space and $f : X \to Y$ a simplicial map.

Definition 6. i) If Z is a simplicial subset of X then

$$f(Z) := (f_n(Z_n) \mid n \in \mathbb{N}_0)$$

is a simplicial subset of Y, called the image of Z by f.
ii) If W is a simplicial subset of Y then

$$f^{-1}(W) := (f_n^{-1}(W_n) \mid n \in \mathbb{N}_0)$$

is a simplicial subset of X, called the preimage of W by f.

It is obvious from the definitions that

$$f(Z^n) = f(Z)^n \tag{3.17}$$

for every $n \in \mathbb{N}_o$, that, for any subset A of X_n,

$$f(A^*) = f_n(A)^*, \tag{3.18}$$

and that the subsets $|f(Z)|$ and $|f|(|Z|)$ of $|Y|$ are equal,

$$|f(Z)| = |f|(|Z|) \quad . \tag{3.19}$$

If Z is a semialgebraic subspace of X then f(Z) is a semialgebraic subspace of Y. If Z is any subspace of X and if f is semialgebraic (i.e. every f_n is semialgebraic, cf. §1), then f(Z) is a subspace of Y.

If W is a subspace of Y then always $f^{-1}(W)$ is a subspace of X. In some other respects preimages are more difficult to handle than images. For example, in general we shall only have $X^n \subset f^{-1}(Y^n)$ and $f_n^{-1}(B)^* \subset f^{-1}(B^*)$ for B a subset of Y_n.

<u>Proposition 3.20.</u> If W is a simplicial subset of Y then

$$|f^{-1}(W)| = |f|^{-1}(|W|) \quad .$$

<u>Proof.</u> It is evident that $|f^{-1}(W)|$ is contained in $|f|^{-1}(|W|)$. Let $\xi \in |f|^{-1}(|W|)$ be given. Write $\xi = |x,t|$ with $x \in NX_n$ and $t \in \mathring{\nabla}(n)$. Then $|f|(\xi) = |f_n(x),t|$. Write $f_n(x) = \alpha^* y$ with $\alpha : [n] \twoheadrightarrow [p]$ a monotonic surjection and $y \in NY_p$. Then $|f|(\xi) = |y,\alpha_*(t)|$ and $\alpha_*(t) \in \mathring{\nabla}(p)$. Our assumption that $|f|(\xi)$ lies in $|W|$ means that $y \in NW_p$. We conclude that $f_n(x) \in W_n$, hence $x \in f^{-1}(W)_n$, and $|x,t| \in |f^{-1}(W)|$. q.e.d.

<u>Proposition 3.21.</u> If the simplicial map f is <u>injective</u> then, for every $n \in \mathbb{N}_o$, f_n maps NX_n into NY_n, and $|f|$ is again injective. Moreover, in this case, for any simplicial subset W of Y,

$$f^{-1}(W)^n = f^{-1}(W^n) \ ,$$

and, for any subset B of Y_n,

$$f^{-1}(B^*) = (f_n^{-1}(B_n))^* \ .$$

These are purely combinatorial facts. To prove them we may replace X and Y by their discretizations. Then f is an isomorphism from X to the subspace f(X) of Y, and all assertions are clear from our subspace theory above.

<u>Proposition 3.22.</u> If f is injective and semialgebraic, then $|f|$ is again semialgebraic (and, of course, injective).

<u>Proof.</u> The map $\hat{f} : \hat{X} \to \hat{Y}$ restricts to a map $\check{f} : \check{X} \to \check{Y}$, since $f(NX_n) \subset NY_n$ for every n. This map $\check{f}$ is semialgebraic. We have a commuting square

$$\begin{array}{ccc} \check{X} & \xrightarrow{\check{f}} & \check{Y} \\ {\scriptstyle \zeta_X}\downarrow & & \downarrow{\scriptstyle \zeta_Y} \\ |X| & \xrightarrow[|f|]{} & |Y| \end{array} \quad .$$

with semialgebraic bijections ζ_X, ζ_Y. We conclude that $|f|$ is semialgebraic, as desired.

§4. Fibre products

We want to study the realizations of fibre products of partially proper simplicial spaces. We first deal with the special case of direct products.

Let X and Y be partially proper simplicial spaces (over R). Then the direct product $X \times Y$ (cf. Ex. 1.2.vi) is again a partially proper simplicial space. Let pr_1 and pr_2 denote the natural projection maps from $X \times Y$ to X and Y respectively. The maps $|pr_1|$ and $|pr_2|$ combine into a (weakly semialgebraic) map

$$h_{X,Y} = (|pr_1|,|pr_2|) : |X \times Y| \to |X| \times |Y| .$$

It is a well known - somewhat intriguing - combinatorial fact that $h_{X,Y}$ is bijective [La, p. 43ff.].

<u>Theorem 4.1.</u> $h_{X,Y}$ is an isomorphism.

The proof needs some preparations. We first consider the case that X is a constant simplicial space $\underline{M}$ (cf. 1.2.iii).

<u>Notations 4.2.</u> If M is a space over R and Y is a simplicial space over R then we write $M \times Y := \underline{M} \times Y$. Notice that $(M{\times}Y)_n = M \times Y_n$ and, for $x \in M$, $y \in Y_n$, $\alpha : [m] \to [n]$ monotonic,

$$\alpha^*(x,y) = (x,\alpha^* y) .$$

We identify $|X| = M$, as explained in 2.9. Then, for $x \in M$, $y \in Y_n$, $t \in \nabla(n)$,

$$|pr_1|(|(x,y),t|) = |x,t| = x ,$$
$$|pr_2|(|(x,y),t|) = |y,t| .$$

We have $(M{\times}Y)^\wedge = M \times \hat{Y}$. The map

$$\mathrm{id}_M \times \eta_Y : M \times \hat{Y} \to M \times |Y|$$

is strongly surjective and partially proper, and its fibres are the equivalence classes of $(M \times Y)^\wedge$. Thus we may and shall identify

$$|M \times Y| = M \times |Y|, \; \eta_{M\times Y} = \mathrm{id}_M \times \eta_Y .$$

For $x \in M$, $y \in Y_n$, $t \in \nabla(n)$ this means

$$|(x,y),t| = (x,|y,t|) . \tag{4.3}$$

After these identifications our map $h_{X,Y}$ is just the identity of $M \times |Y|$, as the formulas for $|pr_1|$, $|pr_2|$ above show. Thus $h_{X,Y}$ is certainly an isomorphism if X is a constant simplicial space.

In the following we shall keep these notations and identifications for $X = \underline{M}$ and any Y and, of course, also the analogous ones for X any simplicial space and Y a constant simplicial space $\underline{N}$.

We now look at the special case that $X = \Delta(p)$, $Y = \Delta(q)$ (cf. 2.10) with some $p,q \in \mathbb{N}_o$. We identify $|X| = \nabla(p)$, $|Y| = \nabla(q)$ (loc.cit.). The discrete simplicial space $X \times Y$ has only finitely many nondegenerate simplices. Indeed (loc.cit.), a simplex $x \in X_n$ can be written as a sequence $\langle a_o,a_1,\ldots,a_n\rangle$ with $0 \le a_o \le a_1 \le \ldots \le a_n \le p$. A simplex $(x,y) \in X_n \times Y_n$ can be written as a sequence $\langle (a_o,b_o),(a_1,b_1),\ldots,(a_n,b_n)\rangle$ with $0 \le a_o \le a_1 \le \ldots \le a_n \le p$ and $0 \le b_o \le b_1 \le \ldots \le b_n \le q$. This simplex is nondegenerate iff $a_i < a_{i+1}$ or $b_i < b_{i+1}$ for every $i \in [n]$. (Of course, all this is very well known.) Thus $|X \times Y|$ is a polytope (cf. 2.8.v). This forces the bijective map $h_{X,Y}$ to be an isomorphism.

<u>Lemma 4.4.</u> Let X,Y,Z,W be four partially proper simplicial spaces. Assume that $h_{X,Y}$, $h_{Z,W}$, $h_{X,Z}$, $h_{Y,W}$, and $h_{X\times Z,Y\times W}$ are isomorphisms. Then $h_{X\times Y,Z\times W}$ is an isomorphism.

This follows from the evident commuting diagram

$$\begin{array}{ccccc}
|X\times Z\times Y\times W| & \xrightarrow{\ |\sigma|\ } & |X\times Y\times Z\times W| & \xrightarrow{\ h_{X\times Y,Z\times W}\ } & |X\times Y|\times|Z\times W| \\
\Big\downarrow{\scriptstyle h_{X\times Z,Y\times W}} & & & & \Big\downarrow{\scriptstyle h_{X,Y}\times h_{Z,W}} \\
|X\times Z|\times|Y\times W| & \xrightarrow[\ h_{X,Z}\times h_{Y,W}\ ]{} & |X|\times|Z|\times|Y|\times|W| & \xrightarrow[\ \tau\ ]{} & |X|\times|Y|\times|Z|\times|W|
\end{array}$$

with switch isomorphisms σ and τ.

Lemma 4.5. If $X = M \times \Delta(p)$ and $Y = N \times \Delta(q)$ with spaces M,N and some $p,q \in \mathbb{N}_o$ then $h_{X,Y}$ is an isomorphism.

Proof. We know from above that $h_{\Delta(p),\Delta(q)}$, $h_{\underline{M},\underline{N}}$, $h_{\underline{M},\Delta(p)}$, $h_{\underline{N},\Delta(q)}$ and $h_{\underline{M\times N},\Delta(p)\times\Delta(q)}$ are isomorphisms. We conclude by the preceding lemma that $h_{X,Y}$ is an isomorphism. q.e.d.

In order to prove Theorem 4.1 in general we need two further results which will be of some use also later on.

Proposition 4.6. Assume that X and Y are partially proper simplicial spaces and $f : X \to Y$ is a simplicial map. If f is partially proper (i.e. every f_n is partially proper, cf. §1) then $|f| : |X| \to |Y|$ is partially proper. If f is strongly surjective then $|f|$ is strongly surjective.

Proof. Assume that f is partially proper, then also $\hat{f} : \hat{X} \to \hat{Y}$ is partially proper and $\eta_Y \circ \hat{f} = |f| \circ \eta_X$ (cf. 2.11). Since η_X is strongly surjective this implies that $|f|$ is partially proper. The same sort of argument works for "strongly surjective". q.e.d.

Definition 1. The deployment $\mathrm{De}\, X$ of a simplicial space X is the direct sum of the simplicial spaces $X_n \times \Delta(n) = \underline{X_n} \times \Delta(n)$ (cf. 4.2),

$$\mathrm{De}\,X := \coprod_{n\geq 0} X_n \times \Delta(n) .$$

Notice that $|\mathrm{De}\,X| = \hat{X}$.

Proposition 4.7. i) There exists a unique simplicial map

$$\chi = \chi_X : \mathrm{De}\,X \to X$$

with $\chi(x,i_n) = x$ for every $n \in \mathbb{N}_0$, $x \in X_n$. {As always, i_n denotes the distinguished element of $\Delta(n)_n$, cf. 2.10.} χ is strongly surjective.

ii) If X is partially proper then χ is partially proper.

iii) $|\chi| = \eta_X$. {This part of the proposition will not be needed in the proof of Theorem 4.1.}

Proof. i) We are forced to define $\chi_q : (\mathrm{De}\,X)_q \to X_q$ by

$$\chi_q(x,\alpha^*(i_n)) = \alpha^*(x)$$

with n running through $\mathbb{N}_0$, α running through the monotonic maps from [q] to [n] and $x \in X_n$ (cf. 4.2). This gives us indeed a simplicial map, as is easily checked. The restriction of χ_q to $X_q \times \{i_q\}$ is an isomorphism from the closed subspace $X_q \times \{i_q\}$ of $X_q \times \Delta(q)_q$ to X_q. Thus every χ_q is certainly strongly surjective, i.e. χ is strongly surjective.

ii) Assume that X is partially proper. The formula above shows that χ_q is partially proper. {Recall that $\Delta(n)_q$ is discrete, in fact finite!} Thus χ is partially proper.

iii) We have the identifications

$$|(x,\alpha^*i_n),t| = (x,|\alpha^*i_n,t|) = (x,\alpha_*t)$$

for x and α as above and $t \in \nabla(q)$ (cf. 2.10, 4.3). On the other hand

$$|\chi|(|(x,\alpha^*i_n),t|) = |\chi_q(x,\alpha^*i_n),t| = |\alpha^*x,t| = |x,\alpha_*t| .$$

Thus indeed $|\chi| = \eta_X$. q.e.d.

<u>Proof of Theorem 4.1.</u> Suppose that X is a direct sum $\bigsqcup(X_\alpha \mid \alpha \in I)$ of simplicial spaces X_α. Then

$$|X \times Y| = \bigsqcup_{\alpha \in I} |X_\alpha \times Y|,$$

$$|X| \times |Y| = \bigsqcup_{\alpha \in I} |X_\alpha| \times |Y|$$

and

$$h_{X,Y} = \bigsqcup_{\alpha \in I} h_{X_\alpha, Y} .$$

Thus if we know that $h_{X_\alpha, Y}$ is an isomorphism for every $\alpha \in I$ then we know that $h_{X,Y}$ is an isomorphism. An analogous remark holds if Y is a direct sum of simplicial spaces.

Let X and Y be arbitrary partially proper simplicial spaces. The deployments $\mathrm{De}\,X$ and $\mathrm{De}\,Y$ are direct sums of the simplicial spaces $X_n \times \Delta(n)$ and $Y_n \times \Delta(n)$ respectively. Thus we know from Lemma 4.5 that $h_{\mathrm{De}\,X, \mathrm{De}\,Y}$ is an isomorphism. Since the map $h_{X,Y}$ depends functorially on X and Y we have a commuting diagram

$$\begin{array}{ccc}
|\mathrm{De}\,X \times \mathrm{De}\,Y| & \xrightarrow{\;|\chi_X \times \chi_Y|\;} & |X \times Y| \\
{\scriptstyle h_{\mathrm{De}\,X,\mathrm{De}\,Y}} \downarrow & & \downarrow {\scriptstyle h_{X,Y}} \\
|\mathrm{De}\,X| \times |\mathrm{De}\,Y| & \xrightarrow{\;|\chi_X| \times |\chi_Y|\;} & |X| \times |Y|
\end{array} .$$

By Proposition 4.7 the simplicial maps χ_X and χ_Y are strongly surjective and partially proper, hence the same holds for $\chi_X \times \chi_Y$. By Proposition 4.6 the realizations $|\chi_X|$, $|\chi_Y|$, $|\chi_{X \times Y}|$ are again strongly surjective and partially proper. Thus in the diagram above the horizontal arrows are strongly surjective and partially proper. Since $h_{\mathrm{De}\,X, \mathrm{De}\,Y}$ is an isomorphism we conclude that $h_{X,Y}$ is strongly surjective and partially proper. Since $h_{X,Y}$ is also bijective it must be an isomorphism.

q.e.d.

The isomorphism $h_{X,Y}$ behaves well with respect to taking graphs of maps.

<u>Definition 2.</u> If $f : X \to Y$ is a simplicial map between simplicial spaces, then the sequence

$$\Gamma(f) := (\Gamma(f_n) \mid n \in \mathbb{N}_o)$$

of graphs of the maps $f_n : X_n \to Y_n$ is clearly a closed subspace of $X \times Y$. We call $\Gamma(f)$ the <u>graph</u> of f.

<u>Proposition 4.8.</u> i) The simplicial map

$$i := (id_X, f) : X \to X \times Y$$

yields an isomorphism from X to the closed subspace $\Gamma(f)$ of $X \times Y$.
ii) The projection $pr_1 : X \times Y \to X$ restricts to an isomorphism from $\Gamma(f)$ onto X.
iii) The bijection $h_{X,Y} : |X \times Y| \to |X| \times |Y|$ maps $|\Gamma(f)|$ onto the graph $\Gamma(|f|)$ of $|f|$. Thus, if X and Y are partially proper, $h_{X,Y}$ restricts to an isomorphism between the spaces $|\Gamma(f)|$ and $\Gamma(|f|)$.

<u>Proof.</u> We have $i(X) = \Gamma(f)$ and $pr_1 \circ i = id_X$. Thus i is a closed embedding and induces an isomorphism from X onto $\Gamma(f)$ with inverse map $pr_1 | \Gamma(f)$. The triangle

$$\begin{array}{ccc} |X| & \xrightarrow{\ |i|\ } & |X \times Y| \\ & \searrow^{(id_{|X|}, |f|)} & \downarrow h_{X,Y} \\ & & |X| \times |Y| \end{array}$$

commutes, since $h_{X,Y} = (|pr_1|, |pr_2|)$. This implies that the bijection $h_{X,Y}$ maps $|i|(|X|) = |\Gamma(f)|$ onto $\Gamma(|f|)$ (cf. 3.19). q.e.d.

<u>Example 4.9.</u> ($f = id_X$). Let Diag X denote the <u>diagonal</u> of X, i.e. the closed subspace of $X \times X$ with $(\mathrm{Diag}\, X)_n = \mathrm{Diag}(X_n)$, the diagonal of X_n

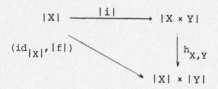

for every n. The bijection $h_{X,X}$ maps $|\mathrm{Diag}\, X|$ onto $\mathrm{Diag}\, |X|$. If X is partially proper then this is an isomorphism of the space $|\mathrm{Diag}\, X|$ to the space $\mathrm{Diag}\, |X|$.

We are ready to study cartesian squares in the category sWSA(R) and their behaviour under realization.

Let $f : X \to S$ and $g : Y \to S$ be simplicial maps between simplicial spaces, and let F denote the map $f \times g$ from $X \times Y$ to $S \times S$.

<u>Definition 3.</u> The <u>fibre product</u> of X and Y with respect to f and g is the closed subspace

$$X \times_S Y := F^{-1}(\mathrm{Diag}\, S)$$

of $X \times Y$.

In more concrete terms, $X \times_S Y$ is the subspace $(X_n \times_{S_n} Y_n \mid n \in \mathbb{N}_o)$ of $X \times Y$, where $X_n \times_{S_n} Y_n$ denotes the usual fibre product of the spaces X_n and Y_n with respect to $f_n : X_n \to S_n$ and $g_n : Y_n \to S_n$. Notice that, if X and Y are partially proper (resp. proper, semialgebraic, complete, partially complete) then $X \times_S Y$ has the same property.

We have a commuting square

$$\begin{array}{ccc} X \times_S Y & \xrightarrow{\;q\;} & Y \\ {\scriptstyle p}\downarrow & & \downarrow{\scriptstyle g} \\ X & \xrightarrow[\;f\;]{} & X \end{array} \qquad (4.10)$$

of simplicial spaces with p and q the restrictions to $X \times_S Y$ of the canonical projections $pr_1 : X \times Y \to X$, $pr_2 : X \times Y \to Y$. One checks in a straightforward way that this diagram is cartesian in sWSA(R).

Assume now that X,Y and S are partially proper.

Lemma 4.11. The isomorphism $h_{X,Y} : |X \times Y| \to |X| \times |Y|$ maps $|X \times_S Y|$ onto the fibre product $|X| \times_{|S|} |Y|$ of $|X|$ and $|Y|$ with respect to $|f|$ and $|g|$.

Proof. The isomorphism $h_{X,Y}$ behaves functorially with respect to X and Y. Thus we have a commuting square

$$\begin{array}{ccc} |X \times Y| & \xrightarrow{|F|} & |S \times S| \\ {\scriptstyle h_{X,Y}}\downarrow & & \downarrow{\scriptstyle h_{S,S}} \\ |X| \times |Y| & \xrightarrow[|f| \times |g|]{} & |S| \times |S| \end{array}$$

The preimage of $\text{Diag}\,|S|$ under $|f| \times |g|$ is $|X| \times_{|S|} |Y|$, while the preimage of $|\text{Diag}\,S|$ under $|F|$ is

$$|F^{-1}(\text{Diag}\,S)| = |X \times_S Y|,$$

(cf. 3.20). Finally, by 4.9, $h_{S,S}$ maps $|\text{Diag}\,S|$ onto $\text{Diag}\,|S|$. Thus $h_{X,Y}$ maps $|X \times_S Y|$ onto $|X| \times_{|S|} |X|$. q.e.d.

Theorem 4.12. Assume again that X,Y and S are partially proper. Then the diagram (cf. 4.10 above)

$$\begin{array}{ccc} |X \times_S Y| & \xrightarrow{|q|} & |Y| \\ {\scriptstyle |p|}\downarrow & & \downarrow{\scriptstyle |g|} \\ |X| & \xrightarrow[|f|]{} & |S| \end{array}$$

is cartesian in the category of spaces WSA(R).

Proof. We compare this diagram with the canonical cartesian square

$$\begin{array}{ccc} |X| \times_{|S|} |Y| & \xrightarrow{\pi_2} & |Y| \\ {\scriptstyle \pi_1}\downarrow & & \downarrow{\scriptstyle |g|} \\ |X| & \xrightarrow[|f|]{} & |S| \end{array}$$

{Of course, π_1, π_2 are the natural projections.} Since the diagram in the theorem commutes it gives us a unique map

$$h = h_{f,g} : |X \times_S Y| \to |X| \times_{|S|} |Y|,$$

with $\pi_1 \circ h = |p|$, $\pi_2 \circ h = |q|$. We have to verify that h is an isomorphism. It is easily checked that the diagram

$$\begin{array}{ccc} |X \times_S Y| & \xrightarrow{h} & |X| \times_{|S|} |Y| \\ {\scriptstyle i}\downarrow & & \downarrow{\scriptstyle j} \\ |X \times Y| & \xrightarrow[h_{X,Y}]{} & |X| \times |Y| \end{array} ,$$

with i and j inclusion mappings, commutes. {Recall that $h_{X,Y} = (|pr_1|, |pr_2|)$.} We learn from Lemma 4.11 above that h is indeed an isomorphism. q.e.d.

In the course of this proof we have seen

<u>Corollary 4.13.</u> The natural isomorphism $h_{f,g}$ from $|X \times_S Y|$ to $|X| \times_{|S|} |Y|$ is a restriction of the isomorphism $h_{X,Y}$.

We now are amply justified to identify $|X \times_S Y|$ with $|X| \times_{|S|} |Y|$ and shall also do so in later sections. Under this identification we shall have the equation

$$|(x,y),t| = (|x,t|,|y,t|) \tag{4.14}$$

for any $t \in \nabla(n)$, $x \in X_n$, $y \in Y_n$ with $f(x) = g(y)$. If $f : X \to X'$ and $g : Y \to Y'$ are simplicial maps between partially proper simplicial spaces over a common partially proper space S then $|f \times_S g| = |f| \times_{|S|} |g|$.

We present an application of Theorem 4.1.

Definition 4. A weakly semialgebraic simplicial group G over R is a group object G in the category sWSA(R). This means that G is a simplicial space with every G_n a weakly semialgebraic group (cf. IV, §11) and all maps $\alpha^* : G_n \to G_p$ group homomorphisms ($\alpha : [p] \to [n]$ monotonic).

In the following we use the shorter term "simplicial group" instead of "weakly semialgebraic simplicial group over R".

Let G be a simplicial group and let $m : G \times G \to G$ denote the (simplicial!) multiplication map. $\{m_n(x,y) = xy$ for $x \in G_n$, $y \in G_n$.$\}$ Let e_n denote the unit element of G_n. Assume that the simplicial space G is partially proper. We define a map $\mu : |G| \times |G| \to |G|$ by $\mu := |m| \circ h_{G,G}^{-1}$. {This is possible by Theorem 4.1.} It is easily checked that μ is an associative composition on $|G|$. The element $e := |e_o, 1|$ turns out to be a left and a right unit for μ. Also, for any $x \in G_n$, $t \in \nabla(n)$ we have

$$\mu(|x,t|, |x^{-1},t|) = |e_n,t| = |e_o,1|.$$

{Notice that $\alpha^* e_o = e_n$ for the map $\alpha : [n] \to [0]$.} Thus we have proved

Proposition 4.15. Let G be a simplicial group with multiplication map m. Assume that the simplicial space G is partially proper. We identify $|G \times G| = |G| \times |G|$ as above. Then the space $|G|$ together with the composition $|m|$ is a weakly semialgebraic group.

Examplex 4.16. Every group object Γ in the category of simplicial sets gives us a discrete simplicial group Γ_R and hence a weakly semialgebraic group $|\Gamma_R|$ over R. Such group objects abound in the literature on simplicial sets. For example, every simplicial set K gives us an abelian group object $\mathbb{Z}K$ in sSet with $(\mathbb{Z}K)_n = \mathbb{Z}[K_n]$, the free abelian group over

K_n. To give still another example, the famous Eilenberg-McLane simplicial sets $K(\pi,n)$ [EM, p. 86ff.] are abelian group objects in sSet. The realizations $|K(\pi,n)_R|$ for all n give us a spectrum for ordinary semialgebraic cohomology $H^*(-,\pi)$ over R.

Definition 5. A (left) simplicial G-space X is a simplicial space X together with a simplicial map $h : G \times X \to X$ such that $h_n : G_n \times X_n \to X_n$ is a (left) operation of G_n on X_n for every n.

By a similar straightforward discussion as above one obtains.

Proposition 4.17. Let X be a simplicial G-space. Assume that the simplicial spaces G and X are partially proper. We identify $|G \times X| = |G| \times |X|$. Then $|h|$ defines a (left) action of the weakly semialgebraic group $|G|$ on the space $|X|$.

§5. Quotients

We want to analyze the realization of the quotient of a simplicial space by a simplicial equivalence relation under favorable conditions. We first discuss a very special case: the gluing of a simplicial space to another one along a closed simplicial subspace. In this case we do not yet need Brumfiel's theorem IV.11.4.

Assume that A is a closed subspace of a simplicial space X and $f : A \to Y$ is a partially proper simplicial map from A to a second simplicial space. In this situation we define a simplicial space

$$Z := X \cup_f Y$$

as follows. Z_n is the space $X_n \cup_{f_n} Y_n$ obtained by gluing X_n to Y_n along A_n by the partially proper map f_n (cf. IV.§8). If $\alpha : [k] \to [l]$ is monotonic then we have commuting squares

$$\begin{array}{ccc} A_l & \xrightarrow{f_l} & Y_l \\ {\scriptstyle \alpha^*}\downarrow & & \downarrow{\scriptstyle \alpha^*} \\ A_k & \xrightarrow[f_k]{} & Y_k \end{array} , \qquad \begin{array}{ccc} A_l & \xrightarrow{i_l} & X_l \\ {\scriptstyle \alpha^*}\downarrow & & \downarrow{\scriptstyle \alpha^*} \\ A_k & \xrightarrow[i_k]{} & X_k \end{array} .$$

The transition maps $\alpha^* : X_l \longrightarrow X_k$ and $\alpha^* : Y_l \longrightarrow Y_k$ combine into the transition map $\alpha^* : Z_l \to Z_k$.

We call $Z = X \cup_f Y$ the <u>simplicial space obtained by gluing</u> X <u>to</u> Y <u>along</u> A <u>by</u> f. The simplicial space Y may and will be regarded as a closed subspace of Z in the obvious way. We have a commuting square

$$\begin{array}{ccc} A & \xrightarrow{f} & Y \\ {\scriptstyle i}\downarrow & & \downarrow{\scriptstyle j} \\ X & \xrightarrow[g]{} & Z \end{array} \qquad (5.1)$$

with i and j inclusion maps and g the obvious simplicial map from X to Z extending f. We know from IV, §8 that every component

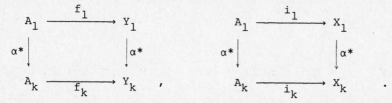

$g_n : X_n \to Z_n$ of g is partially proper and that $(g_n, j_n) : X_n \sqcup Y_n \longrightarrow Z_n$ is strongly surjective. This means, according to our terminology (cf. §1), that g is partially proper and $(g,j) : X \sqcup Y \to Z$ is strongly surjective (and partially proper). One checks in a straightforward way that the diagram (5.1) is cocartesian in the category sWSA(R) of simplicial spaces.

Theorem 5.2. If X and Y are partially proper then the simplicial space $Z := X \cup_f Y$ is again partially proper and the diagram (cf. 5.1)

$$\begin{array}{ccc} |A| & \xrightarrow{|f|} & |Y| \\ {\scriptstyle |i|}\downarrow & & \downarrow{\scriptstyle |j|} \\ |X| & \xrightarrow[|g|]{} & |Z| \end{array} \qquad (*)$$

is cocartesian in WSA(R). Recalling from Proposition 4.6 that $|f|$ is partially proper, we may and shall identify

$$|X \cup_f Y| = |X| \cup_{|f|} |Y|.$$

Proof. Let $\alpha : [k] \to [l]$ be monotonic. We have a commuting square

$$\begin{array}{ccc} X_l \sqcup Y_l & \xrightarrow{\alpha^* \sqcup \alpha^*} & X_k \sqcup Y_k \\ \downarrow{\scriptstyle p_l} & & \downarrow{\scriptstyle p_k} \\ Z_l & \xrightarrow[\alpha^*]{} & Z_k \end{array}$$

with strongly surjective partially proper maps p_l and p_k. By our assumption about X and Y the upper horizontal arrow $\alpha^* \sqcup \alpha^*$ is partially proper. This implies that the lower horizontal arrow α^* is partially proper (cf. the reasoning in the proof of 4.6). Thus Z is partially proper.

The map $(g,j) : X \sqcup Y \to Z$ is strongly surjective and partially proper. We conclude by Proposition 4.6 that

$$(|g|,|j|) : |X \sqcup Y| = |X| \sqcup |Y| \to |Z|$$

is again strongly surjective and partially proper, hence identifying. We are done if we know that the diagram (*) in the theorem is cocartesian on the set theoretic level. We know already that $(|g|,|j|)$ is surjective and $|j|$ is injective. Thus we only need to verify that $|g|$ maps $|X| \setminus |A|$ injectively into $|Z| \setminus |Y|$.

Let $\xi \in |X| \setminus |A|$ be given. Write $\xi = |x,t|$ with $x \in NX_n$, $t \in \overset{\circ}{\nabla}(n)$. Then

$$|g|(\xi) = |g_n(x),t|.$$

Write $g_n(x) = \alpha^*(z)$ with $\alpha : [n] \twoheadrightarrow [q]$ a monotonic surjection and $z \in NZ_q$. Then

$$|g|(\xi) = |z,\alpha_* t|,$$

and $\alpha_*(t) \in \overset{\circ}{\nabla}(q)$. Suppose that $|g|(\xi) \in |Y|$. By our subspace theory this means that $z \in NY_q$, and implies that

$$g_n(x) = \alpha^*(z) \in Y_n .$$

We conclude that $x \in A_n$. This contradicts our assumption that $\xi \notin A$. Thus $|g|$ maps indeed $|X| \setminus |A|$ into $|Z| \setminus |Y|$.

We return to the point $|g|(\xi)$. We have $z = g_q(x_1)$ with $x_1 \in X_q \setminus A_q$. We conclude from above that $g_n(x) = g_n(\alpha^* x_1)$, and then $x = \alpha^* x_1$. Since x is nondegenerate this implies $\alpha = id_{[n]}$, i.e. $g_n(x) \in NZ_n$.

Assume that ξ' is a second point in $|X| \setminus |A|$ with $|g|(\xi') = |g|(\xi)$. Write $\xi' = |x',t'|$ with $x' \in NX_m$, $t' \in \overset{\circ}{\nabla}(m)$. Then $g_n(x) \in NZ_n$, $g_m(x') \in NZ_m$, and $|g_n(x),t| = |g_m(x'),t'|$. We conclude by Lemma 2.1 that $m = n$, $t = t'$, $g_n(x) = g_n(x')$. Since g_n is injective on $X_n \setminus A_n$, we obtain $x = x'$, hence $\xi = \xi'$, as desired. q.e.d.

In the case that Y is the one-point simplicial space $\underline{\{*\}}$ the theorem means the following.

Example 5.3. Let X be a partially proper simplicial space and A a partially complete (closed) subspace of X. Then the simplicial space X/A {with $(X/A)_n = X_n/A_n$ for every n} is partially proper and the realization $|p| : |X| \to |X/A|$ of the natural simplicial projection map $p : X \to X/A$ induces an isomorphism from $|X|/|A|$ to $|X/A|$, in short,

$$|X/A| = |X|/|A| ,$$

We now extend the definitions and results of IV, §11 on equivalence relations to simplicial spaces. This can be done in a somewhat automatic way. Let X be a simplicial space.

Definition 1. An equivalence relation T on X is a simplicial subspace T of $X \times X$ such that T_n is an equivalence relation on the space X_n for every n. We call the equivalence relation T closed (partially proper, proper, ...) if the equivalence relation T_n on X_n is closed (partially proper, proper, ...) for every n.

Example 5.4. If $f : X \to Y$ is a simplicial map then the fibre product

$$E(f) := X \times_Y X$$

with respect to f in both the first and the second factor is a closed equivalence relation on X. We have $E(f)_n = E(f_n)$ for every n.

In the following T is an equivalence relation on X. We denote by p_1 and p_2 the two natural projections from T to X and by τ_T the switch automorphism of T (as in IV, §11). These are now simplicial maps. We denote by X/T the simplicial set defined as follows: $(X/T)_n$ is the set of equivalence classes X_n/T_n. If $\alpha : [p] \to [n]$ is monotonic, then $\alpha^* : X_n/T_n \to X_p/T_p$ is the set theoretic map induced by $\alpha^* : X_n \to X_p$. We denote by p_T the natural projection from the simplicial set X to the simplicial set X/T.

<u>Definition 2.</u> A simplicial map $f : X \to Y$ is called a <u>strong quotient</u> (a <u>partially proper quotient</u>, a <u>proper quotient</u>) <u>of</u> X <u>by</u> T, if $E(f) = T$ and every map $f_n : X_n \to Y_n$ is identifying (is strongly surjective and partially proper, is proper).

It is clear from IV, §8 that every partially proper quotient is a strong quotient. It is also evident (as in IV, §11) that, if there exists a strong quotient of X by T, then there is a unique structure of a simplicial space on the simplicial set X/T such that p_T is a strong quotient of X by T. In this case, of course, we mean by X/T this simplicial space, instead of the previous simplicial set.

Brumfiel's Theorem IV.11.4 extends immediately as follows.

<u>Theorem 5.5.</u> Assume that the equivalence relation T on X is closed and partially proper (resp. proper). Then the partially proper quotient (the proper quotient) of X by T exists.

We want to know how an equivalence relation behaves under realization. We assume that X is a partially proper simplicial space and that T is a closed equivalence relation on X. Then the simplicial space T is again partially proper, and $|T|$ is a closed subspace of $|X \times X| = |X| \times |X|$.

We verify that $|T|$ is an equivalence relation on $|X|$. Indeed, from $\text{Diag}\, X \subset T$ we conclude that $\text{Diag}\, |X| \subset |T|$, cf. 4.9. The realization of the switch automorphism of $X \times X$ is the switch automorphism of $|X| \times |X|$. We conclude that $|T|$ is mapped into itself by this automorphism, cf. 3.19. Finally, in the space $|X \times X \times X| = |X| \times |X| \times |X|$ we have

$$|(T \times X) \cap (X \times T)| = |T \times X| \cap |X \times T| = (|T| \times |X|) \cap (|X| \times |T|),$$

cf. 3.4, and then, again by 3.19,

$$|pr_{13}[(T\times X)\cap(X\times T)]| = |pr_{13}|[(|T|\times|X|)\cap(|X|\times|T|)],$$

with pr_{13} the natural projection from $X \times X \times X$ to the first and third factor. $|pr_{13}|$ is the natural projection from $|X| \times |X| \times |X|$ to the first and the third factor. It maps $(|T|\times|X|)\cap(|X|\times|T|)$ into $|T|$ since pr_{13} maps $(T\times X)\cap(X\times T)$ into T.

Proposition 5.6. If $f : X \to Y$ is a simplicial map from X to another partially proper space Y then $E(|f|) = |E(f)|$.

Proof. This follows from §4 (Th. 4.12 and Cor. 4.13), since $E(f) = X \times_Y X$ and $E(|f|) = |X| \times_{|Y|} |X|$, the fibre products using the maps f and $|f|$ respectively.

Example 5.7. Assume that X is partially proper. Then the deployment $\mathrm{De}\,X$ (cf. §4, Def. 1) is again a partially proper simplicial space and $\chi_X : \mathrm{De}\,X \to X$ is strongly surjective and partially proper (Prop. 4.7). Let $\mathrm{Rel}\,X$ denote the equivalence relation $E(\chi_X)$. We call $\mathrm{Rel}\,X$ the relation space of X. The map χ_X is a partially proper quotient of $\mathrm{De}\,X$ by $\mathrm{Rel}\,X$. The realization of χ_X is our previous map $\eta_X : \hat{X} \twoheadrightarrow |X|$ (cf. 4.7). It is again partially proper and strongly surjective by Proposition 4.6. We conclude that η_X is a partially proper quotient of $\hat{X}$ by $|\mathrm{Rel}\,X|$. Thus $|\mathrm{Rel}\,X|$ is just the equivalence relation on $\hat{X}$ used to define the realization $|X|$. {Of course, this also could be verified by a direct computation, and we know already from 2.6 that η_X is partially proper and strongly surjective.}

We are ready to state the main result of this section.

Theorem 5.8. Assume that T is a partially proper closed equivalence relation on the partially proper simplicial space X. Then $|T|$ is a partially proper closed equivalence relation on $|X|$. The simplicial space

X/T (cf. Th. 5.5) is again partially proper and the realization $|p_T|$ of the simplicial map $p_T : X \to X/T$ is a partially proper quotient of $|X|$ by $|T|$. In short, $|X/T| = |X|/|T|$.

<u>Proof.</u> If $\alpha : [q] \to [n]$ is monotonic then we have a commuting square ($p := p_T$)

$$\begin{array}{ccc} X_n & \xrightarrow{\alpha^*} & X_q \\ {\scriptstyle p_n}\downarrow & & \downarrow{\scriptstyle p_q} \\ X_n/T_n & \xrightarrow[\alpha^*]{} & X_q/T_q \end{array}$$

The vertical maps are partially proper and strongly surjective. The upper horizontal map α^* is partially proper. We conclude that the lower horizontal map α^* is partially proper. Thus the simplicial space X/T is partially proper. The other assertions in the theorem now follow from the Propositions 5.6 and 4.6.

We explicate the outcome of our theory in the case of group actions. Let G be a simplicial group and X a left simplicial G-space. We assume that we are in one of the following two cases. Case 1: G is complete, i.e. every G_n is a complete (hence semialgebraic) space, and X is partially proper. Case 2: The space X is discrete. In both these cases the image T(G) of the simplicial map $G \times X \to X \times X$, $(g,x) \mapsto (gx,x)$, is a closed simplicial subspace of $X \times X$, hence a closed equivalence relation of X. This equivalence relation is proper in the first case and partially proper in the second case. We denote the quotient X/T(G) more briefly by $G\backslash X$. Of course, $(G\backslash X)_n = G_n\backslash X_n$ for every n. Let $p : X \to G\backslash X$ denote the natural projection. It is proper in the first case and partially proper in the second case. We learn from Theorem 5.8 that $|p| : |X| \to |G\backslash X|$ is a proper quotient of $|X|$ by $|T(G)|$ in the first case and a partially proper quotient in the second one. The action of G on X gives us an action of $|G|$ on $|X|$ (cf. 4.17) with $T(|G|) = |T(G)|$

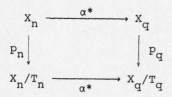

(cf. 3.19). Thus the partially proper quotient (in the first case even proper quotient) $|G|\backslash|X|$ exists, and $|G|\backslash|X| = |G\backslash X|$. Notice that usually $|G|$ will not be a semialgebraic group. Thus this result lies well beyond IV.11.8.

§6. Semialgebraic realization of simplicial sets

Let K be a simplicial set. We denote the realization of the associated discrete simplicial space K_R (cf. 1.2.ix and 2.5) by $|K|_R$ or more briefly by $|K|$. The notation $|K|_R$ instead of $|K_R|$ stresses the analogy with the realization of abstract simplicial complexes (cf. II, §3).

Similarly, if f is a simplicial map from K to a simplicial set L then we denote the realization of $f_R : K_R \to L_R$ by $|f|_R$ or by $|f|$. We call $|K|_R$ and $|f|_R$ the <u>semialgebraic realization over</u> R of K and f respectively. Notice that $|K|_R$ is a weak polytope (2.8.v).

If $\tilde{R}$ is a real closed field extension of R then clearly $|K|_{\tilde{R}} = (|K|_R)(\tilde{R})$ and $|f|_{\tilde{R}} = (|f|_R)_{\tilde{R}}$, cf. 2.13.

The topological realizations [Mi_1] of K and f will be denoted by $|K|_{top}$ and $|f|_{top}$ respectively. We have $|K|_{top} = (|K|_{\mathbb{R}})_{top}$ and $|f|_{top} = (|f|_{\mathbb{R}})_{top}$.

There exists an extensive literature on simplicial sets. We mention [Ca] for a pleasant introduction, the articles [Cu] and [Gu] for a concise survey, and the books [La] and [May] for thorough treatments of basic parts of the theory.

In principle we have every known theorem at our disposal which can be formulated entirely within the category of simplicial sets, but we have to check whether or not suitable results involving topological realizations remain true for our semialgebraic realizations, let alone results which involve other topological spaces.
We return to our simplicial set K. We start out to establish on K the structure of a normal (cf. V.1.3) CW-complex. For every $x \in K_n$ we define a <u>characteristic map</u> $\eta_x : \nabla(n) \to |K|$ by $\eta_x(t) := |x,t|$. We

denote the image of η_x by $|x|$ and the subset $\eta_x(\mathring{\nabla}(n))$ of $|x|$ by $|x|^o$. Notice that $|x|^o$ is a semialgebraic subset of $|K|$ and $|x|$ is a polytope.

If $\alpha : [p] \to [n]$ is a monotonic map then clearly the triangle

$$\begin{array}{ccc} \nabla(p) & \xrightarrow{\eta_{\alpha^*(x)}} & \\ \alpha_* \downarrow & & |K| \\ \nabla(n) & \xrightarrow{\eta_x} & \end{array} \qquad (6.1)$$

commutes.

From the diagram (6.1) we conclude that, if α is surjective, then $|\alpha^*(x)| = |x|$ and $|\alpha^*(x)|^o = |x|^o$, since in this case α_* maps $\nabla(p)$ onto $\nabla(n)$ and $\mathring{\nabla}(p)$ onto $\mathring{\nabla}(n)$.

If x is nondegenerate then η_x maps $\mathring{\nabla}(n)$ bijectively onto $|x|^o$ (cf. 2.1). Moreover, $\mathring{\nabla}(n)$ is the preimage of $|x|^o$ under η_x. Thus η_x is an isomorphism from $\mathring{\nabla}(n)$ to $|x|^o$. It is also clear from 2.1 that $\{|x|^o \mid x \in NK_n, n \in \mathbb{N}_o\}$ is a semialgebraic partition of $|K|$. Using again diagram (6.1), this time with α injective, we see that, for every $x \in NK_n$, the set $|x| \setminus |x|^o$ is a union of finitely many "cells" $|y|^o$, $y \in NK_p$, for some $p < n$.

As usual we denote the n-skeleton of the simplicial set K by K^n. Then $(K^n)_R$ is the n-skeleton of the simplicial space K_R, and $|K|$ is the inductive limit of the family of closed subspaces $(|K^n| \mid n \in \mathbb{N}_o)$. In §2 we have studied the relation between $|X^{n-1}|$ and $|X^n|$ for any partially proper simplicial space X. Let us spell out what this means in the special case $X = K_R$.

We shall identify any subset Z of K_n with the discrete space Z_R. The discrete space K_n is the direct sum of NK_n and DK_n. Thus we obtain the following cocartesian square from the diagram in Lemma 2.7

$$\begin{array}{ccc} NK_n \times \partial\nabla(n) & \xrightarrow{\varphi_n} & |K^{n-1}| \\ i \downarrow & & \downarrow j \\ NK_n \times \nabla(n) & \xrightarrow[\psi_n]{} & |K^n| \end{array}$$

for every $n \geq 0$. Here φ_n and ψ_n are restrictions of the natural maps φ_n and ψ_n there, and i and j are inclusion mappings. {In the case n = 0 the diagram gives us an isomorphism ψ_o from $K_o \times \nabla(0) \cong K_o$ to $|K^o|$.} Notice that $\psi_n(x,t) = \eta_x(t)$ for any $x \in NK_n$, $t \in \nabla(n)$. The following is now evident.

<u>Theorem 6.2.</u> The partition $\{|x|^o \mid x \in NK_n, n \in \mathbb{N}_o\}$ of $|K|$ is a patch decomposition of $|K|$ which gives $|K|$ the structure of a normal CW-complex. $|K^n|$ is the n-skeleton of this CW-complex and, for every $x \in NK_n$, the map $\eta_x : \nabla(n) \to |K|$ is a characteristic map for the cell $|x|^o$.

In the following we shall always regard $|K|$ as a CW-complex - not merely a weak polytope - in this way.

<u>Proposition 6.3.</u> Let B be a closed subcomplex of the CW-complex $|K|$. Then there exists a (unique) simplicial subset A of K such that $B = |A|$.

This is easily verified. We define A_n as the set of all n-simplices x of K such that $|x| \subset B$. If $\alpha : [p] \to [n]$ is monotonic, then it follows from the diagram (6.1) that $|\alpha^*x| \subset |x|$. This implies that the family $A := (A_n \mid n \geq 0)$ is a simplicial subset of K. Clearly $|A| = B$.

If f is a simplicial map from K to a simplicial set L then, for every $x \in NK_n$,

$$|f| \circ \eta_x = \eta_{f(x)} \tag{6.4}$$

hence $|f|(|x|) = |f(x)|$, and $|f|(|x|^o) = |f(x)|^o$. {We briefly write f(x) instead of $f_n(x)$.} Thus |f| maps every cell of |K| onto a cell of |L|. This implies, of course, that |f| is a cellular map.

As is well known - and somewhat at the roots of the theory of simplicial sets - there is a close relation between closed abstract simplicial complexes (cf. II, §3) and suitable simplicial sets which we recall now (cf. [Ca, §1]).

Convention. In this chapter a simplicial complex P means a closed (= classical) abstract simplicial complex. As in II, §3 we denote the set of vertices of P by E(P) and the set of simplices by S(P).

Definition 1. a) An ordered simplicial complex P is a simplicial complex P together with a total ordering of each simplex s of P such that, if t is a face of s (i.e. a non empty subset of s) the ordering of t is the restriction of the ordering of s to t. {This definition differs marginally from the one given in most textbooks.}
b) A simplicial map f from P to another ordered simplicial complex Q is a map $f : E(P) \to E(Q)$ which maps every simplex s of P onto a simplex f(s) of E(Q) in a monotonic (= weakly order preserving) way.

Example 6.5. We regard a totally ordered set L as an ordered simplicial complex in the following way: The vertices of the complex L are the elements of the set L. The simplices of the complex L are the finite non empty subsets of L equipped with the restrictions of the ordering of the set L. In particular we obtain an ordered simplicial complex

[n] for every $n \in \mathbb{N}_0$.

Definition 2. a) Given an ordered simplicial complex P we define a simplicial set $\tilde{P}$ as follows. The n-simplices of $\tilde{P}$ are the simplicial maps from [n] to P. If $\alpha : [p] \to [n]$ is monotonic and $x \in \tilde{P}_n$ then $\alpha^*(x)$ is the composite $x \circ \alpha$. {Notice that α is a simplicial map from the ordered simplicial complex [p] to [n].}
b) Given a simplicial map f from P to another ordered simplicial complex Q we define a simplicial map $\tilde{f} : \tilde{P} \to \tilde{Q}$ by $\tilde{f}_n(x) := f \circ x$. By categorial abstract nonsense we obtain in this way a bijection $f \mapsto \tilde{f}$ from the set of simplicial maps from P to Q to the set of simplicial maps from $\tilde{P}$ to $\tilde{Q}$.

We shall often denote an n-simplex $x \in \tilde{P}_n$ by $\langle v_0,\ldots,v_n\rangle$ with $v_i := x(i)$. Notice that then $\{v_0,\ldots,v_n\}$ is an m-simplex of P with $m \le n$ and $v_0 \le v_1 \le \ldots \le v_n$.

Example 6.6. $[n]^\sim = \Delta(n)$ (cf. 2.10).

Let us have a closer look at the simplicial set $\tilde{P}$ for a given ordered simplicial complex P! An n-simplex $x = \langle v_0,\ldots,v_n\rangle$ of $\tilde{P}$ is nondegenerate iff the v_i are all different. Then every face of x is again nondegenerate. Every n-simplex s of P gives us a nondegenerate n-simplex $\tilde{s}$ of $\tilde{P}$, namely $\tilde{s} := \langle v_0,\ldots,v_n\rangle$ with $v_0,\ldots,v_n$ the vertices (= elements) of s and $v_0 < v_1 < \ldots < v_n$. In this way we obtain a bijection $s \mapsto \tilde{s}$ from $S_n(P)$ to $N\tilde{P}_n$.

If Q is a (closed!) simplicial subcomplex of P, of course equipped with the "restriction" of the ordering of P, then $\tilde{Q}$ is a simplicial subset of $\tilde{P}$. Conversely if A is a simplicial subset of $\tilde{P}$ then there exists a (unique) simplicial subcomplex Q of P with $\tilde{Q} = A$. Identifying $S_n(P) = N\tilde{P}_n$ we have $S_n(Q) = NA_n$.

<u>Definition 3.</u> We call a simplicial set K <u>polyhedral</u> if K is isomorphic to $\tilde{P}$ for some ordered simplicial complex P.

In this case we can choose P and an isomorphism $\varphi : \tilde{P} \xrightarrow{\sim} K$ in a canonical way as follows. We put $E(P) := K_o$. For any $i \in [n]$ let $v_i : K_n \to K_o$ denote the i-th vertex map, i.e. the transition map induced by the map from [0] to [n] which sends 0 to i. The n-simplices of P are the sets $\{v_o(x), v_1(x), \ldots, v_n(x)\}$ with x running through NK_n, each such set being ordered by $v_o(x) \leq v_1(x) \leq \ldots \leq v_n(x)$. {Notice that, since K is polyhedral, every $x \in K_n$ is uniquely determined by the sequence $v_o(x), v_i(x), \ldots, v_n(x)$, and that x is nondegenerate iff the $v_i(x)$ are all different.} φ is the simplicial map from $\tilde{P}$ to K which sends an n-simplex $\langle \mu_o, \mu_1, \ldots, \mu_n \rangle$ of $\tilde{P}$ to the unique $x \in K_n$ with $v_o(x) = \mu_o$, $v_1(x) = \mu_1, \ldots, v_n(x) = \mu_n$.

Given an ordered simplicial complex P we now compare the realization $|P|$ over R (cf. II, §3, one forgets the ordering of P) with the realization $|\tilde{P}|$ of its associated simplicial set. Let s be an n-simplex of P, and let $e_o, e_1, \ldots, e_n$ be the vertices of s with $e_o < e_1 < \ldots < e_n$. Let $|s|$ denote the corresponding closed simplex of $|P|$. Recall from II, §3 that we have identified the vertices of P with the vertices of the geometric simplicial complex $|P|$. Then $|s|$ is the convex hull of $\{e_o, e_1, \ldots, e_n\}$. We define a semialgebraic map α_s from $|s|$ to $|\tilde{P}|$ by the formula

$$\alpha_s(t_o e_o + \ldots + t_n e_n) = |\tilde{s}, (t_o, \ldots, t_n)|$$

for any $(t_o, \ldots, t_n) \in \nabla(n)$. This is an isomorphism from $|s|$ onto $|\tilde{s}|$. If t is a face of s then α_t is the restriction of α_s to $|t|$. Thus the α_s fit together to a weakly semialgebraic map $\alpha : |P| \to |\tilde{P}|$. The map α is clearly bijective, partially proper and strongly surjective, hence an isomorphism of spaces. α maps the cells of the CW-complex $|P|$ onto the cells of $|\tilde{P}|$, hence is an isomorphism of CW-complexes.

In the following we shall always identify $|P| = |\tilde{P}|$ by this isomorphism α . If $f : P \to Q$ is a simplicial map to a second ordered simplicial complex Q then we have $|f| = |\tilde{f}|$.

Example 6.7. If we take $P = [n]$, hence $\tilde{P} = \Delta(n)$, then our present isomorphism from $\nabla(n) = |[n]|$ to $|\Delta(n)|$ is the same as the isomorphism described in 2.10.

We may identify the category $\mathcal{k}$ of ordered simplicial complexes with the full subcategory $\tilde{\mathcal{k}}$ of sSet with objects the polyhedral simplicial sets, and thus regard sSet as an enlargement of the category $\mathcal{k}$. This enlargement turns out to be very useful. On the one hand, the realization functor $\mathcal{k} \to WSA(R)$ extends to sSet in an agreeable way. On the other hand we can perform more constructions in sSet then in $\mathcal{k}$, notably quotients by arbitrary equivalence relations (cf. §5), and even more generally direct limits of arbitrary diagrams. These direct limits commute with the realization functor, cf. 7.2 below and, in the case of equivalence relations, §5. Thus we have gained a very useful combinatorial pattern beyond II, §3 (in the case of closed complexes) to create weak polytopes and to ensure the existence of direct limits of suitable diagrams of weak polytopes.

Notice also that every simplicial set K can be written canonically as a finite direct limit (= quotient) of polyhedral simplicial sets, namely as the coequalizer of the two projections $p_1, p_2 : \mathrm{Rel}\, K \rightrightarrows \mathrm{De}\, K$ from the relation set of K to the deployment of K (cf. 5.7). Here De K and Rel K are even direct sums of finite polyhedral simplicial sets. This is helpful to define various "subdivisions" of simplicial sets, starting from subdivisions of the standard simplices $\Delta(n)$. We shall use such subdivisions in an essential way but shall not be explicit about this, referring the reader to an extensive literature. {A thorough general treatment of subdivisions has been given by Fritsch [Frt_1].}

The realization of any simplicial set can be triangulated. More precisely we have the following important theorem.

Theorem 6.8. There exists an endomorphism D of the category sSet of simplicial sets, a natural transformation $\lambda : D \to id_{sSet}$ from D to the identity functor and, for every simplicial set K, a weakly semialgebraic map $t_K : |DK| \to |K|$ with the following properties.

T1) DK is polyhedral for every simplicial set K.

T2) t_K is an isomorphism from the space $|DK|$ to $|K|$, which maps every cell of $|DK|$ into a cell of $|K|$. {In other words, t_K is a simultaneous triangulation of $|K|$ and all its cells.}

T3) $|\lambda(K)|$ is homotopic to t_K by a homotopy F such that for each closed simplex $|x|$ of $|DK|$ the image $F(|x| \times [0,1])$ is contained in the smallest closed cell $|y|$ of $|K|$ which contains $t_K(|x|)$.

This theorem has been stated in the topological setting by Weingram following pioneering work of Barratt ([We], cf. also [LW]). Weingram's proof contained some gaps and errors. They have been bridged and corrected by Fritsch [Frt]. The proof of Weingram-Fritsch is completely of semialgebraic nature and thus also gives Theorem 6.8 over any real closed field R.

For later use we add the following obvious

Remark 6.9. Let A be a simplicial subset of a simplicial set K. In the situation of Theorem 6.8 the preimage $t_K^{-1}(|A|)$ is a closed subcomplex of $|DK|$, hence $t_K^{-1}(|A|) = |B|$ with some (polyhedral) simplicial subset

B of DK. We conclude from T3) that $|\lambda(K)|$ maps $|B|$ into $|A|$, hence $\lambda(K)$ maps B into A. We may read t_K as an isomorphism from the pair of spaces $(|DK|,|B|)$ to the pair of spaces $(|K|,|A|)$ and $\lambda(K)$ as a simplicial map from (DK,B) to (K,A). Moreover we may read the homotopy F in T3) as a homotopy from the map of pairs t_K to the map of pairs $|\lambda(K)|$.

§7. The space $|\mathrm{Sin}\, M|$ and singular homology

For any space M over R we define a simplicial set $\mathrm{Sin}\, M$ as follows. An n-simplex of $\mathrm{Sin}\, M$ is a (semialgebraic) map x from the geometric standard n-simplex $\nabla(n)$ to M. If $\alpha : [p] \to [n]$ is monotonic then $\alpha^*(x)$ is defined as the composite $x \circ \alpha_*$ of x and $\alpha_* : \nabla(p) \to \nabla(n)$. We call $\mathrm{Sin}\, M$ the <u>singular simplicial set of</u> M and the elements of $(\mathrm{Sin}\, M)_n$ the <u>singular</u> n-<u>simplices</u> of M. If x is a point of M then we denote the corresponding singular O-simplex, which maps $\nabla(0)$ to x, by $\tilde{x}$.

Every map from M to a second space N gives us a simplicial map $\mathrm{Sin}\, f : \mathrm{Sin}\, M \to \mathrm{Sin}\, N$ defined by $(\mathrm{Sin}\, f)_n(x) = f \circ x$ for x a singular n-simplex of M. Thus we have a functor Sin from the category WSA(R) to sSet.

There is a close relation between this functor Sin and the realization functor. For every simplicial set K we have a natural simplicial map $i_K : K \to \mathrm{Sin}|K|$ defined by $i_K(x) = \eta_x$ for any $x \in K_n$. {Recall from §6 that η_x is the characteristic map $t \mapsto |x,t|$ from $\nabla(n)$ to $|K|$.} On the other hand, we have, for every space M, a natural (weakly semialgebraic) map $j_M : |\mathrm{Sin}\, M| \to M$ defined by

$$j_M(|x,t|) = x(t), \qquad (x \in (\mathrm{Sin}\, M)_n, t \in \nabla(n)) .$$

In order to establish this map first define a map $\tilde{j}_M : (\mathrm{Sin}\, M)^\wedge \to M$ by $\tilde{j}_M(x,t) = x(t)$ for $(x,t) \in (\mathrm{Sin}\, M)_n \times \nabla(n)$ and then observe that $\tilde{j}_M(\alpha^* y,t) = y \circ \alpha_*(t) = \tilde{j}_M(y,\alpha_*(t))$ for y a singular p-simplex of M, $\alpha : [n] \to [p]$ monotonic, and $t \in \nabla(n)$.

One now verifies the following theorem precisely as in the topological setting [La, Chap. II, §6].

<u>Theorem 7.1.</u> For every simplicial set K we have

$$j_{|K|} \circ |i_K| = id_{|K|} , \tag{7.1a}$$

and for every space M we have

$$(\mathrm{Sin}\, j_M) \circ i_{\mathrm{Sin}\, M} = id_{\mathrm{Sin}\, M} . \tag{7.1b}$$

Thus the functor Sin : WSA(R) → sSet is right adjoint to the realization functor | | : sSet → WSA(R) via the adjunction maps j_M and i_K. More explicitly, given a space M and a simplicial set K, there is a one-to-one correspondence between the maps $f : |K| \to M$ and the simplicial maps $g : K \to \mathrm{Sin}\, M$, which can be characterized by either one of the two equations

$$f = j_M \circ |g|, \qquad g = (\mathrm{Sin}\, f) \circ i_K .$$

Here is a first application of Theorem 7.1 to the theory of realizations. Let $(K_\lambda | \lambda \in \Lambda)$ be any diagram of simplicial sets, i.e. a functor $\lambda \mapsto K_\lambda$ from a small category Λ to sSet. This gives us a diagram of spaces $(|K_\lambda| \; | \lambda \in \Lambda)$ by realization. In the category sSet there exists the direct limit (= colimit [Mt, II, §2]) $K := \varinjlim_\lambda K_\lambda$ of the first diagram. {Define $K_n := \varinjlim (K_\lambda)_n$ for every $n \in \mathbb{N}_0$.} For any $\lambda \in \Lambda$, let $\varphi_\lambda : K_\lambda \to K$ denote the canonical simplicial map from K_λ to K.

<u>Corollary 7.2.</u> |K| is the direct limit of the diagram of spaces $(|K_\lambda| \; | \lambda \in \Lambda)$ by the maps $|\varphi_\lambda| : |K_\lambda| \to |K|$.

This is an immediate consequence of the existence of a right adjoint of the realization functor. Indeed, for any space, in short hand notation,

$$\mathrm{Hom}(|K|,M) = \mathrm{Hom}(K,\mathrm{Sin}\, M) = \varprojlim_\lambda \mathrm{Hom}(K_\lambda,\mathrm{Sin}\, M) = \varprojlim_\lambda \mathrm{Hom}(|K_\lambda|,M).$$

Let us recall, for later use, the notion of simplicial homotopy.

Definition 1.

a) For any simplicial set K and $i \in \{0,1\}$ we denote by $\varepsilon_i(K)$, or ε_i for short, the simplicial map from K to $K \times \Delta(1)$ which sends an n-simplex x of K to the n-simplex $x \times \langle i,\ldots,i\rangle$ of $K \times \Delta(1)$. Notice that this is the composite of the evident simplicial isomorphism from K to $K \times \Delta(0)$ and the simplicial map $id_K \times \Delta(\delta^i)$ with $\delta^i : [0] \to [1]$ sending 0 to i. The realization $|\varepsilon_i|$ is the map $x \mapsto (x,i)$ from $|K|$ to $|K \times \Delta(1)| = |K| \times [0,1]$.

b) Let C be a simplicial subset of K and let $g_0, g_1 : K \rightrightarrows L$ be two simplicial maps from K to another simplicial set L with $g_0|C = g_1|C$. A simplicial homotopy from g_0 to g_1 relative C is a simplicial map $G : K \times \Delta(1) \to L$ such $G \circ \varepsilon_0 = g_0$, $G \circ \varepsilon_1 = g_1$ and $G|C \times \Delta(1)$ is a constant simplicial homotopy, i.e. the composite of the natural projection $pr_1 : C \times \Delta(1) \to C$ with $g_0|C$. Notice that then $|G| : |K| \times I \longrightarrow |L|$ is a homotopy relative $|C|$ from $|g_0|$ to $|g_1|$.

The one-to-one correspondence between simplicial maps and weakly semi-algebraic maps stated in Theorem 7.1 behaves well with respect to homotopy.

Proposition 7.3. Let K be a simplicial set, C a simplicial subset of K, and M a space. Let f_0 and f_1 be maps from $|K|$ to M which restrict to the same map from $|C|$ to M. Let F be a map from $|K| \times [0,1]$ to M. Let $g_0, g_1 : K \rightrightarrows \operatorname{Sin} M$ and $G : K \times \Delta(1) \to \operatorname{Sin} M$ be the left adjoints of f_0, f_1, F respectively, as explained in Theorem 7.1. Then $g_0|C = g_1|C$, and G is a homotopy from g_0 to g_1 relative C iff F is a homotopy from f_0 to f_1.

This is a straightforward consequence of the uniqueness statement in Theorem 7.1 (cf. [La, p. 47f]).

Definition and Remark 7.4. A (finite) system of simplicial sets is a tuple $(K,A_1,\dots,A_r)$, consisting of a simplicial set K and simplicial subsets $A_1,\dots,A_r$ of K. A simplicial map f from $(K,A_1,\dots,A_r)$ to another system $(L,B_1,\dots,B_r)$ means, of course, a simplicial map $f: K \to L$ with $f(A_i) \subset B_i$. Analogously to Definition 1 we have the notion of a simplicial homotopy between two simplicial maps from $(K,A_1,\dots,A_r)$ to $(L,B_1,\dots,B_r)$ relative to a simplicial subset C of K. Theorem 7.1 and Proposition 7.3 generalize immediately to systems of simplicial sets and systems of spaces.

We now come to the main result of this section.

<u>Theorem 7.5.</u> For any space M the map $j_M : |\mathrm{Sin}\, M| \to M$ is a homotopy equivalence.

In order to prove this it suffices, by "Whitehead's theorem" V.6.10, to verify for every $x \in M$ and every $n \in \mathbb{N}_0$ that the map

$$(j_M)_* : \pi_n(|\mathrm{Sin}\, M|, \tilde{x}) \to \pi_n(M,x)$$

is bijective. Here we have identified the vertex $\tilde{x}$ of $\mathrm{Sin}\, M$ (see above) with the point $|\tilde{x},1|$ of $|\mathrm{Sin}\, M|$. {Notice that every connected component of $|\mathrm{Sin}\, M|$ contains a point $\tilde{x}$.}

We shall essentially reproduce the arguments in the book of Lundell and Weingram [LW, p. 102ff.] (which in the topological setting only prove that j_M is a weak homotopy equivalence). For the convenience of the reader we shall give all details.

We identify the pointed n-sphere (S^n,∞), where ∞ denotes the north pole, with the realization $(|L|,\infty)$ of a suitable pointed polyhedral simplicial set (L,∞) arising from some triangulation of (S^n,∞). {Of course, ∞ has to be a vertex, i.e. a 0-simplex}.

In the following we shall omit the base points from our notation. All spaces and simplicial sets will be pointed, and all maps and homotopies will have to preserve the base points. We regard $\pi_n(M)$ as the set of homotopy classes of base point preserving maps from $S^n = |L|$ to M.

It is easily seen that $(j_M)_*$ is surjective. Indeed, let $f : S^n \to M$ be a given base point preserving map. By Theorem 7.1 there exists a pointed simplicial map $g : L \to \mathrm{Sin}\, M$ such that $f = j_M \circ |g|$. The map $(j_M)_*$ sends the homotopy class $[|g|]$ to $[f]$.

In order to prove the injectivity of $(j_M)_*$ we need a lemma which will be proved afterwards.

Lemma 7.6. Let K be a pointed simplicial set and $f : S^n \to |K|$ a pointed map. Then there exists a pointed polyhedral simplicial set T and an isomorphism $\varphi : |T| \xrightarrow{\sim} S^n$ (i.e. a triangulation of (S^n,∞)) such that $f \circ \varphi$ is homotopic (base point preserving) to the realization $|g|$ of a simplicial map $g : T \to K$.

Using this lemma the injectivity of $(j_M)_*$ can be seen as follows. We are given a base point preserving map $f : S^n \to |\mathrm{Sin}\, M|$ such that $j_M \circ f$ is null homotopic. We have to prove that f itself is null homotopic. We choose T, φ, g as in the lemma, with $K := \mathrm{Sin}\, M$. The map $j_M \circ |g|$ is again null homotopic. Let t denote the base point of T, $t \in T_o$. Let k denote the constant simplicial map from T to $\mathrm{Sin}\, M$, i.e. the unique simplicial map which sends T to the simplicial subset $\{\tilde{x}\}^*$ of $\mathrm{Sin}\, M$ generated by $\{\tilde{x}\}$. Then $j_M \circ |k|$ is the constant map from the pointed space $|T|$ to M. The map $j_M \circ |g|$ is homotopic relative $\{t\}$ to $j_M \circ |k|$. We conclude from Proposition 7.3 that g is simplicially homotopic to k relative $\{t\}^*$. This implies of course, that $|g|$ is homotopic to the constant map $|k|$ relative $\{t\}$, in other words, $|g|$ is null homotopic.

The map $|f|$ is homotopic to $|g|\circ\varphi^{-1}$, hence is null homotopic as well.

It remains to prove the lemma. We apply the triangulation theorem 6.8. We use the notations of that theorem. Let h denote the map $t_K^{-1}\circ f$ from S^n to $|DK|$. This map is base point preserving (cf. 6.9). $f = t_K \circ h$ is homotopic to $|\lambda(K)|\circ h$ respecting base points (cf. again 6.9). Now there exists a triangulation $\varphi : |T| \xrightarrow{\sim} S^n$ of the pointed sphere S^n such that $h\circ\varphi : |T| \to |DK|$ is homotopic respecting base points to the realization $|u|$ of a simplicial map $u : T \to DK$. This follows from our results on contiguity classes in Chapter III (cf. Theorem III.5.5 and Remark III.5.6) which are essentially an adaption of the classical theory of simplicial approximations [Spa, p. 126ff.] to the semialgebraic setting. {T is an iterated barycentric subdivision of L.} Theorem 7.5 is proved.

Let now $\mathfrak{M} := (M_o,\ldots,M_r)$ be a decreasing system of spaces. Every singular simplicial set $\operatorname{Sin} M_k$ is a simplicial subset of $\operatorname{Sin} M_{k-1}$ $(k=1,\ldots,r)$. Let $j_{\mathfrak{M}}$ denote the map from the decreasing system of CW-complexes $(|\operatorname{Sin} M_o|,\ldots,|\operatorname{Sin} M_r|)$ to $\mathfrak{M}$ whose components are the adjunction maps j_{M_k}. We immediately obtain the following generalization of Theorem 7.5 (cf. V.2.13 for the second statement).

<u>Theorem 7.7.</u> $j_{\mathfrak{M}}$ is a CW-approximation (cf. V, §7, Def. 4) of $\mathfrak{M}$. If the M_k are closed in M_o then $j_{\mathfrak{M}}$ is a homotopy equivalence between systems of spaces.

<u>Remark 7.8.</u> This CW-approximation is <u>natural</u> in $\mathfrak{M}$. Indeed, if $f : M \to N$ is a map between spaces, then it is easily checked that the square

$$\begin{array}{ccc} |\operatorname{Sin} M| & \xrightarrow{|\operatorname{Sin} f|} & |\operatorname{Sin} N| \\ \downarrow{\scriptstyle j_M} & & \downarrow{\scriptstyle j_N} \\ M & \xrightarrow[f]{} & N \end{array}$$

commutes. This implies the commutativity of the analogous square for a map between decreasing systems of spaces.

Corollary 7.9. Every closed decreasing system of spaces is homotopy equivalent to a decreasing system of closed geometric simplicial complexes.

This follows from our Theorem 7.7 and the triangulation Theorem 6.8 (cf. also Remark 6.9).

The case $r = 1$ of Theorem 7.7 gives us a description of ordinary homology (VI, §3, Def. 2) and cohomology by "singular chains" and "singular cochains" respectively, as we shall explain now.

We first recall some well known notions from simplicial algebra. Every simplicial set K gives us a simplicial abelian group $\mathbb{Z}[K]$ such that $\mathbb{Z}[K]_n$ is the free abelian group $\mathbb{Z}[K_n]$ generated by the set K_n for every n. One simply composes the functor $K : \text{Ord} \to \text{Set}$ with the functor "free abelian group" from Set to the category Ab of Abelian groups. Let $C.(K)$ denote the chain complex associated with the simplicial group $\mathbb{Z}[K]$. It is defined by $C_n(K) = \mathbb{Z}[K_n]$ for $n \geq 0$ and $C_n(K) = 0$ for $n < 0$, the boundary map from $C_n(K)$ to $C_{n-1}(K)$ being the alternating sum of the face maps from $\mathbb{Z}[K]_n$ to $\mathbb{Z}[K]_{n-1}$, if $n > 0$.

For G any abelian group we define the chain complex

$$C.(K,G) := C.(K) \otimes_{\mathbb{Z}} G$$

and the cochain complex

$$C^{\cdot}(K,G) := \text{Hom}_{\mathbb{Z}}(C.(K),G) .$$

The homology groups $H_n(C.(K,G))$ and the cohomology groups $H^n(C^{\cdot}(K,G))$ are, by definition, the homology groups $H_n(K,G)$ and cohomology groups

$H^n(K,G)$ of the simplicial set K with coefficients in G. We have mentioned these groups already in 2.16 and have observed there that they are naturally isomorphic to the ordinary homology and cohomology groups $H_n(|K|,G)$ and $H^n(|K|,G)$ of the space $|K|$, which we had defined in Chapter VI (cf. [LW, p. 192ff] for a perhaps more elementary proof than our proof in §2).

We define the reduced complexes $\tilde{C}.(K,G)$ and $\tilde{C}^{\cdot}(K,G)$ as the kernel and cokernel respectively of the homomorphisms $C.(K,G) \to C.(\Delta(0),G)$, $C^{\cdot}(\Delta(0),G) \to C^{\cdot}(K,G)$ induced by the simplicial map from K to $\Delta(0)$. Their homology and cohomology groups respectively are called the reduced homology groups $\tilde{H}_n(K,G)$ and reduced cohomology groups $\tilde{H}^n(K,G)$. These differ from the unreduced groups $H_n(K,G)$ and $H^n(K,G)$ only if $n = 0$, and, of course, they are naturally isomorphic to the reduced ordinary homology and cohomology groups $\tilde{H}_n(|K|,G)$, $\tilde{H}^n(|K|,G)$ defined in Chapter VI.

Finally, if L is a simplicial subset of K then we define the chain complex $C.(K,L;G)$ and the cochain complex $C^{\cdot}(K,L;G)$ as the cokernel and the kernel of the homomorphisms $C.(L,G) \to C.(K,G)$ and $C^{\cdot}(K,G) \to C^{\cdot}(L,G)$ induced by the inclusion $L \hookrightarrow K$. {The first homomorphism may again be regarded as an inclusion.} If L is not empty then we may identify

$$C.(K,L;G) = \tilde{C}.(K/L,G), \qquad C^{\cdot}(K,L;G) = \tilde{C}^{\cdot}(K/L,G) \ ,$$

while for $L = \underline{\emptyset}$, the empty simplicial set, we have $C.(K,\underline{\emptyset};G) = C.(K,G)$ and $C^{\cdot}(K,\underline{\emptyset};G) = C^{\cdot}(K,G)$. The n-th homology group of $C.(K,L;G)$ is denoted by $H_n(K,L;G)$, and the n-th cohomology group of $C^{\cdot}(K,L;G)$ is denoted by $H^n(K,L;G)$.

Definition 2. Let (M,A) be a pair of spaces over R. Then (Sin M, Sin A) is a pair of simplicial sets. We call the group $H_n(\mathrm{Sin}\, M, \mathrm{Sin}\, A; G)$ the n-th singular homology group of the pair (M,A), and we call $H^n(\mathrm{Sin}\,M, \mathrm{Sin}\,A; G)$ the n-th singular cohomology group of the pair (M,A). Further we call the elements of $C_n(\mathrm{Sin}\,M, \mathrm{Sin}\,A; G)$ the singular n-chains of (M,A) and the elements of $C^n(\mathrm{Sin}\,M, \mathrm{Sin}\,A; G)$ the singular n-cochains of (M,A) with coefficients in G.

This terminology is justified since it completely parallels the standard terminology in the topological setting. One simply uses semi-algebraic maps from $\nabla(n)$ to M instead of continuous maps. We could have defined singular homology and cohomology groups of any pair (M,A) of spaces already in Chapter IV, §3, but that would not have been of any use for us. Indeed, if the base field R is not archimedean, then - in contrast to the topological theory - it seems to be very difficult, if not impossible (cf. the introduction to this book), to prove in a direct elementary way that the singular homology groups, say, fit, into a homology theory in the technical sense (Chap. VI, §4), even if we restrict them to pairs of polytopes.

Now we are better off. First observe that the singular homology and cohomology groups are functors on the category of pairs of spaces, WSA(2,R). Indeed, a map $f : (M,A) \to (N,B)$ between pairs of spaces gives us a simplicial map Sin f from (Sin M, Sin A) to (Sin N, Sin B) and then homomorphisms

$$f_* : H_n(\mathrm{Sin}\,M, \mathrm{Sin}\,A; G) \to H_n(\mathrm{Sin}\,N, \mathrm{Sin}\,B; G) ,$$
$$f^* : H^n(\mathrm{Sin}\,N, \mathrm{Sin}\,B; G) \to H^n(\mathrm{Sin}\,M, \mathrm{Sin}\,A; G) .$$

Theorem 7.10. For every pair of spaces (M,A), every abelian group G and every $n \in \mathbb{N}_0$ there exist natural isomorphisms

$$H_n(\mathrm{Sin}\,M, \mathrm{Sin}\,A; G) \xrightarrow{\sim} H_n(M,A;G)$$

and

$$H^n(\mathrm{Sin}\,M, \mathrm{Sin}\,A; G) \xrightarrow{\sim} H^n(M,A;G) .$$

Proof. We prove this for homology. The arguments for cohomology will be analogous. We first consider the case that A is empty. Then Theorem 7.5 and 2.16 together give us isomorphisms

$$H_n(\mathrm{Sin}\,M, G) \xrightarrow{\sim} H_n(|\mathrm{Sin}\,M|, G) \xrightarrow[(j_M)_*]{} H_n(M,G) .$$

We now assume that A is notempty. Then $H_n(\mathrm{Sin}\,M, \mathrm{Sin}\,A; G)$ is naturally isomorphic to $\tilde{H}_n(\mathrm{Sin}\,M/\mathrm{Sin}\,A, G)$, as stated essentially above, and $H_n(|\mathrm{Sin}\,M|, |\mathrm{Sin}\,A|, G)$ is just the same as $\tilde{H}_n(|\mathrm{Sin}\,M|/|\mathrm{Sin}\,A|; G)$, cf. VI, §4. But $|\mathrm{Sin}\,M|/|\mathrm{Sin}\,A|$ may be identified with $|\mathrm{Sin}\,M/\mathrm{Sin}\,A|$, cf. 5.3. Thus we obtain from 2.16 a natural isomorphism from $H_n(\mathrm{Sin}\,M, \mathrm{Sin}\,A; G)$ to $H_n(|\mathrm{Sin}\,M|, |\mathrm{Sin}\,A|; G)$. {Notice that in 2.16, by applying the naturality of the isomorphism from $H_n(K,G)$ to $H_n(|K|,G)$ to the simplicial map from K to $\Delta(0)$, we obtain a natural isomorphism from $\tilde{H}_n(K,G)$ to $\tilde{H}_n(|K|,G)$.} On the other hand, Theorem 7.7 together with 7.8 gives us a natural isomorphism $(j_M)_*$ from $H_n(|\mathrm{Sin}\,M|, |\mathrm{Sin}\,A|; G)$ to $H_n(M,A;G)$. q.e.d.

Remark 7.11. It follows from Theorem 7.10 that the singular homology and cohomology groups can be read as functors on the homotopy category HWSA(2,R) instead of WSA(2,R).

Theorem 7.10, as it stands, leaves something to be desired. If L is a simplicial subset of a simplicial set K then the obvious short exact sequence

$$0 \to C.(L,G) \to C.(K,G) \to C.(K,L;G) \to 0$$

gives us a long exact sequence in homology with connecting homomorphisms

$$\partial_n(K,L) : H_n(K,L;G) \to H_{n-1}(L,G)$$

Similarly we have a canonical exact sequence in cohomology with connecting homomorphisms

$$\delta^n(K,L) : H^n(L,G) \to H^{n+1}(K,L;G).$$

The question arises whether in the case $(K,L) = (\operatorname{Sin} M, \operatorname{Sin} A)$ these connecting homomorphisms correspond, perhaps up to sign, to the homomorphism $\partial_n(M,A)$, $\delta^n(M,A)$ of ordinary homology and cohomology under the isomorphisms constructed above.

Starting from 2.16 it may be laborious to check whether this is true. We shall present in §8 a second proof of Theorem 7.10 (with perhaps other natural isomorphisms) where this problem disappears.

§8. Simplicial homotopy, and singular homology again

One obtains by the same easy argument as in topology (e.g. [La, p. 13f.])

Proposition 8.1. The singular simplicial set Sin M of any space M is a Kan set, i.e. fulfills Kan's extension condition.

We want to exploit this fact, first on the level of homotopy and then on the level of homology. We start with general results on simplicial homotopy some of which deserve independent interest.

Recall (cf. [La], [May], [Cu]) that if K and L are arbitrary simplicial sets then the relation "homotopic" on the set Map(K,L) of simplicial maps from K to L may not be transitive, but if L is a Kan set then it is transitive, hence an equivalence relation. In this case we denote the set of homotopy classes of simplicial maps from K to L by [K,L].

More generally we fix the following setting. $\mathcal{K} := (K_o,\dots,K_r)$ is a decreasing system of simplicial sets. By this we mean, of course, that the K_k are simplicial subsets of K_o with $K_o \supset K_1 \supset \dots \supset K_r$. Moreover $\mathcal{L} := (L_o,\dots,L_r)$ is a decreasing system of Kan sets. We use obvious notations concerning the functors Sin and $|\ |$. Thus $|\mathcal{K}|$ means the closed decreasing system of spaces $(|K_o|,\dots,|K_r|)$. $\mathrm{Sin}|\mathcal{K}|$ means, of course, the decreasing system of Kan sets $(\mathrm{Sin}|K_o|,\dots,\mathrm{Sin}|K_r|)$, and $i_{\mathcal{K}}$ means the simplicial map $(i_{K_o},\dots,i_{K_r})$ from $\mathcal{K}$ to $\mathrm{Sin}|\mathcal{K}|$ consisting of the adjunction maps i_{K_k} from K_k to $\mathrm{Sin}|K_k|$, $0 \le k \le r$. We always regard $\mathcal{K}$ as a subsystem of $\mathrm{Sin}|\mathcal{K}|$ via $i_{\mathcal{K}}$, which thus becomes the inclusion map from $\mathcal{K}$ to $\mathrm{Sin}|\mathcal{K}|$.

If $\mathcal{M} = (M_o,\dots,M_r)$ is a system of spaces then $j_{\mathcal{M}} : |\mathrm{Sin}\,\mathcal{M}| \to \mathcal{M}$ denotes the tuple $(j_{M_o},\dots,j_{M_r})$ consisting of the adjunction maps $j_{M_k} : |\mathrm{Sin}\, M_k| \longrightarrow M_k$. Recall from §7 that all these maps are homotopy equivalences.

We return to the systems $\mathfrak{K}$ and $\mathfrak{L}$ above. In the following C is a simplicial subset of K_o and $h : C \to L_o$ is a given simplicial map with $h(C \cap K_k) \subset L_k$ for $1 \leq k \leq r$. Since every L_k is Kan, the relation "homotopic relative C" on the set $\mathrm{Map}(\mathfrak{K},\mathfrak{L})$ of simplicial maps from $\mathfrak{K}$ to $\mathfrak{L}$ is an equivalence relation. Let $[\mathfrak{K},\mathfrak{L}]^h$ denote the set of homotopy classes relative C of those simplicial maps from $\mathfrak{K}$ to $\mathfrak{L}$ which coincide with h on C.

<u>Proposition 8.2.</u> Assume that $\mathfrak{L} = \mathrm{Sin}\,\mathcal{M}$ for some decreasing system of spaces $\mathcal{M} = (M_o,\dots,M_r)$. Then the map $[f] \mapsto [j_{\mathcal{M}} \circ |f|]$ from $[\mathfrak{K},\mathrm{Sin}\,\mathcal{M}]^h$ to $[|\mathfrak{K}|,\mathcal{M}]^{|h|}$ is a bijection, the inverse of this map being given by $[g] \mapsto [(\mathrm{Sin}\,g) \circ i_{\mathfrak{K}}]$.

<u>Proof.</u> This follows from the adjunction identities 7.1 and Proposition 7.3.

<u>Theorem 8.3</u> (Simplicial approximation theorem).
Let f be a map from $|\mathfrak{K}|$ to $|\mathfrak{L}|$ which extends $|h|$. Then there exists a simplicial map g from $\mathfrak{K}$ to $\mathfrak{L}$ extending h such that f is homotopic to $|g|$ relative $|C|$.

<u>Proof.</u> We invoke the topological analogue of the theorem in the case $r = 0$. This can be found in standard texts on simplicial homotopy theory, e.g. [La, p. 48]. We obtain the claim for $r = 0$ from the topological result in the customary way by use of the main theorems on homotopy sets (Th. V.5.2; first consider the case $R = \mathbb{R}$, then $R = R_o$, then R arbitrary.)

Let now $r = 1$ and $f = (f_o, f_1)$ with maps $f_i : K_i \to L_i$. By the case $r = 0$ there exists a simplicial map $g_1 : K_1 \to L_1$ extending $h|C \cap K_1$ together with a homotopy $H : |K_1| \times I \to |L_1|$ relative $|C \cap K_1|$ from f_1 to $|g_1|$. By use of the homotopy extension theorem V.2.9 we obtain a homotopy $\tilde{H} : |K_o| \times I \to |L_o|$ relative $|C \cup K_1|$ which starts with f_o and extends H. Let $\tilde{f}_o := \tilde{H}(-,1)$. This map from $|K_o|$ to $|L_o|$ extends $|h|$ and $|g_1|$. Again by the case $r = 0$ there exists a simplicial map $g_o : K_o \to L_o$ extending h and g_1 together with a homotopy $G : |K_o| \times I \to |L_o|$ relative $|C \cup K_1|$ from $\tilde{f}_o$ to $|g_o|$. We have found a simplicial map $g := (g_o, g_1)$ from $\mathcal{K}$ to $\mathcal{L}$. We may read the composed homotopy $\tilde{H} * G$ as a homotopy relative $|C|$ from f to $|g|$. Thus we have proved the claim for $r = 1$. In general one proves the claim by the same argument and induction on r. q.e.d.

Corollary 8.4. The evident map $[f] \to [|f|]$ from $[\mathcal{K}, \mathcal{L}]^h$ to $[|\mathcal{K}|, |\mathcal{L}|]^{|h|}$ is a bijection.

Proof. Theorem 8.3 tells us that this map is surjective. Injectivity now follows in the usual way by applying the surjectivity result to $\mathcal{K}$ and $(L_o \times \Delta(1), \dots, L_r \times \Delta(1))$. q.e.d.

Remark. As I learned from letters of Ronnie Brown and Rainer Vogt it is possible to deduce a result analogous to 8.4 for much more general diagrams of simplicial sets than our decreasing systems by applying more advanced techniques using simplicial function sets. For the purposes I have in mind up to now Corollary 8.4 will be sufficient.

Corollary 8.5. Assume that K_k (and, as before, L_k) is a Kan set for every $k \in \{0, \dots, n\}$. Let f be a simplicial map from $\mathcal{K}$ to $\mathcal{L}$ all whose components $f_k : K_k \to L_k$ are homotopy equivalences. Then f itself is a homotopy equivalence.

Proof. This follows from Corollary 8.4 since we know that the analogous fact for closed decreasing systems of spaces is true. {V, 2.13; here we work over an arbitrarily chosen real closed field R, say $R = R_o$. We could use equally well topological realizations.}

Corollary 8.6. Assume again that the K_k are Kan sets. Assume further that K_o is a simplicial subset of L_o, that $K_k = K_o \cap L_k$, for $1 \le k \le r$, and that the components of the inclusion map $i : \mathcal{k} \hookrightarrow \mathcal{L}$ are homotopy equivalences. Then $\mathcal{k}$ is a strong deformation retract of $\mathcal{L}$.

Proof. The components of $|i|$ are again homotopy equivalences, and $|K_k| = |K_o| \cap |L_k|$. Thus we know from Proposition V.2.16 that $|\mathcal{k}|$ is a strong deformation retract of $|\mathcal{L}|$. In other terms, we have a retraction map r from $|\mathcal{L}|$ to $|\mathcal{k}|$ such that $|i| \circ r \simeq |id_{\mathcal{L}}|$ rel. $|K_o|$. By our Theorem 8.3 there exists a simplicial map $\rho : \mathcal{L} \to \mathcal{k}$ extending $id_{\mathcal{k}}$ (i.e. a retraction from $\mathcal{L}$ to $\mathcal{k}$) such that $|\rho| \simeq r$ rel. $|K_o|$. Then

$$|i \circ \rho| \simeq |i| \circ r \simeq |id_{\mathcal{L}}| \text{ rel. } |K_o| .$$

We conclude by Corollary 8.4 that $i \circ \rho \simeq id_{\mathcal{L}}$ rel. K_o, and we are done.
q.e.d.

We now are well prepared to prove two rather satisfying results about systems of singular simplicial sets, Proposition 8.8 and Theorem 8.10 below.

Lemma 8.7. Let B be a simplicial set and A a simplicial subset of B. Then

$$B \cap \mathrm{Sin}|A| = A .$$

{Recall that we consider B as a simplicial subset of Sin|B| via i_B.}

Proof. Let $x \in B_n$ be an n-simplex of B. We have identified x with the singular n-simplex η_x of Sin|B|. Now assume that $x \in (\mathrm{Sin}|A|)_n$. This means that η_x maps $\nabla(n)$ into $|A|$. Write $x = \alpha^*(u)$ with $\alpha : [n] \twoheadrightarrow [p]$ a monotonic epimorphism and $u \in NB_p$. Then $\eta_x(\mathring{\nabla}(n)) = \eta_u(\mathring{\nabla}(p)) \subset |A|$. This implies that $u \in A_p$ (cf. 1.4 and 2.2). We conclude that $x \in A_n$, as desired.

Proposition 8.8. $\mathcal{L}$ is a strong deformation retract of $\mathrm{Sin}|\mathcal{L}|$. {Recall that every L_k is assumed to be Kan.}

Proof. We have a commuting triangle

$$\begin{array}{ccc} [\mathcal{K},\mathcal{L}] & \xrightarrow{\ \mu\ } & [|\mathcal{K}|,|\mathcal{L}|] \\ & \underset{\alpha}{\searrow} \quad \underset{\beta}{\nearrow} & \\ & [\mathcal{K},\mathrm{Sin}|\mathcal{L}|] & \end{array}$$

Here α is induced by the inclusion $i_{\mathcal{L}}$ from $\mathcal{L}$ to $\mathrm{Sin}|\mathcal{L}|$, β is the adjunction isomorphism in Proposition 8.2 (with $\mathcal{M} = |\mathcal{L}|$, $C = \emptyset$), and μ the bijection in 8.4 (with $C = \emptyset$). We conclude that α is bijective for any $\mathcal{K}$. This means that $i_{\mathcal{L}}$ is a homotopy equivalence, and gives us the claim by 8.6 and 8.7. q.e.d.

Let $\mathcal{M} = (M_o,\ldots,M_r)$ be a decreasing system of spaces over R. If S is a real closed field extension of R then $\mathrm{Sin}\,\mathcal{M}$ is a subsystem of $\mathrm{Sin}\,\mathcal{M}(S)$. If $R = \mathbb{R}$ then $\mathrm{Sin}\,\mathcal{M}$ is a subsystem of $\mathrm{Sin}\,\mathcal{M}_{top}$. All these systems consist of Kan sets. {Of course, $\mathcal{M}(S)$ means $(M_o(S),\ldots,M_r(S))$, and $\mathcal{M}_{top}$ means $((M_o)_{top},\ldots,(M_r)_{top})$.}

Lemma 8.9. For every $k \in \{1,\ldots,r\}$ we have $\mathrm{Sin}\,M_o \cap \mathrm{Sin}\,M_k(S) = \mathrm{Sin}\,M_k$. If $R = \mathbb{R}$ then we also have $\mathrm{Sin}\,M_o \cap \mathrm{Sin}(M_k)_{top} = \mathrm{Sin}\,M_k$.

Proof. Let $\varphi : \nabla(n) \to M_o$ be an n-simplex of $\mathrm{Sin}\,M_o$. Assume that φ_S maps $\nabla(n)_S$ into $M_k(S)$. Then, of course, φ maps $\nabla(n)$ into M_k. This proves the first claim. The proof of the second one is even more trivial.

Theorem 8.10. i) $\mathrm{Sin}\,\mathcal{M}$ is a strong deformation retract of $\mathrm{Sin}\,\mathcal{M}(S)$.
ii) If $R = \mathbb{R}$ then $\mathrm{Sin}\,\mathcal{M}$ is a strong deformation retract of $\mathrm{Sin}\,\mathcal{M}_{top}$.

Proof. By the preceding lemma and 8.6 it suffices to prove that, for any space M, the inclusion $\alpha : \operatorname{Sin} M \hookrightarrow \operatorname{Sin} M(S)$ is a homotopy equivalence and, in case $R = \mathbb{R}$, also the inclusion $\beta : \operatorname{Sin} M \hookrightarrow \operatorname{Sin} M_{top}$ is a homotopy equivalence.

Let us look at the first inclusion $\alpha : \operatorname{Sin} M \to \operatorname{Sin} M(S)$. It is easily checked that the following triangle commutes.

$$
\begin{array}{ccc}
|\operatorname{Sin} M|_S & \xrightarrow{\;|\alpha|_S\;} & |\operatorname{Sin} M(S)|_S \\
 & {\scriptstyle (j_M)_S} \searrow \quad \swarrow {\scriptstyle j_{M(S)}} & \\
 & M(S) &
\end{array}
\quad .
$$

We know (from Theorem 7.5) that j_M and $j_{M(S)}$ are homotopy equivalences. We conclude that $|\alpha|_S$ is a homotopy equivalence and then, by Corollary 8.4, that α is a homotopy equivalence. {N.B. We constantly exploit the fact 8.1.}

Let us now look at the second inclusion β. It is again easily checked that the following triangle commutes.

$$
\begin{array}{ccc}
|\operatorname{Sin} M|_{top} & \xrightarrow{\;|\beta|_{top}\;} & |\operatorname{Sin} M_{top}|_{top} \\
 & {\scriptstyle (j_M)_{top}} \searrow \quad \swarrow {\scriptstyle j_{M_{top}}} & \\
 & M_{top} &
\end{array}
\quad .
$$

It is known from topology that $j_{M_{top}}$ is a weak homotopy equivalence ([Mi_1]; this can be proved as our Theorem 7.5 [LW]). We conclude that $|\beta|_{top}$ is a weak homotopy equivalence and then that $|\beta|_{\mathbb{R}}$ is a weak homotopy equivalence, hence a homotopy equivalence (V.6.10). By Corollary 8.4 this implies that β is a homotopy equivalence. q.e.d.

<u>Remark 8.11.</u> Alternatively we can conclude directly that, for every space M over $\mathbb{R}$ the continuous map $j_{M_{top}}$ is a (topological) homotopy equivalence instead of just a weak homotopy equivalence. Indeed, this

follows from the topological Whitehead theorem since, by 7.5, M is homotopy equivalent to a CW-complex.

Theorem 8.10 gives us a new approach to singular homology and cohomology. We fix some abelian group G.

Definition 1. We call two simplicial maps f and g from $\mathcal{k}$ to $\mathcal{L}$ pseudo-homotopic, if, for every k in {0,...,r}, the component $f_k : K_k \to L_k$ is homotopic to g_k. Similarly we call two maps $f,g : \mathcal{M} \rightrightarrows \mathcal{N}$ between decreasing systems of spaces pseudohomotopic, if the components of f are homotopic to the corresponding components of g.

Lemma 8.12. Let r = 1. Assume that $f,g : \mathcal{k} \rightrightarrows \mathcal{L}$ are pseudohomotopic simplicial maps. Then the induced homomorphisms

$$f_*, g_* : H_*(K_o,K_1;G) \to H_*(L_o,L_1;G)$$

in simplicial homology are equal.

Proof. It is well known that the induced maps $C.(f_o)$ and $C.(g_o)$ from $C.(K_o)$ to $C.(L_o)$ are chain homotopic and also the induced maps $C.(f_1)$ and $C.(g_1)$ from $C.(K_1)$ to $C.(L_1)$ are chain homotopic. The claims follow by use of the five-lemma. q.e.d.

Lemma 8.13. Any two pseudohomotopic maps f and g from a pair of spaces (M,A) to a pair of spaces (N,B) induce, for every n, the same homomorphism

$$f_* = g_* : H_n(\mathrm{Sin}\,M, \mathrm{Sin}\,A; G) \to H_n(\mathrm{Sin}\,N, \mathrm{Sin}\,B; G) .$$

Proof. From the commutativity of the square

$$\begin{array}{ccc} (|\operatorname{Sin} M|, |\operatorname{Sin} A|) & \xrightarrow{|\operatorname{Sin} f|} & (|\operatorname{Sin} N|, |\operatorname{Sin} B|) \\ {\scriptstyle j_{(M,A)}} \downarrow & & \downarrow {\scriptstyle j_{(N,B)}} \\ (M,A) & \xrightarrow[f]{} & (N,B) \end{array}$$

and of the analogous square for g we see, by use of Theorem 7.7, that the maps $|\operatorname{Sin} f|$ and $|\operatorname{Sin} g|$ from $(|\operatorname{Sin} M|, |\operatorname{Sin} A|)$ to $(|\operatorname{Sin} N|, |\operatorname{Sin} B|)$ are pseudohomotopic. Then we conclude from Corollary 8.4 (with $r = 0$ and $C = \emptyset$) that the simplicial maps $\operatorname{Sin} f$ and $\operatorname{Sin} g$ from $(\operatorname{Sin} M, \operatorname{Sin} A)$ to $(\operatorname{Sin} N, \operatorname{Sin} B)$ are pseudohomotopic. This gives the desired result.

q.e.d.

Definition 2. Clearly the functors $(M,A) \mapsto H_n(\operatorname{Sin} M, \operatorname{Sin} A; G)$ from HWSA(2,R) to Ab, together with the connecting homomorphisms $\partial_n(M,A)$ described at the end of §7 constitute a prehomology theory on the space category WSA(R), as defined in VI, §5. We call this theory *singular homology over* R *with coefficients in* G.

Theorem 8.14. Singular homology over R with coefficients in G is an ordinary homology theory (cf. Chapter VI, starting from VI, §3, Def. 2) with coefficient group G.

N.B. Recall that, up to isomorphism, there exists only one ordinary homology theory on WSA(R) with coefficient group G. Thus Theorem 8.4 gives us an interpretation by singular chains of the ordinary homology theory constructed in Chapter VI.

Proof. The analogous result in algebraic topology is very well known to be true. If $R = \mathbb{R}$ then we obtain from Theorem 8.10.ii canonical isomorphisms

$$H_n(\operatorname{Sin} M, \operatorname{Sin} A; G) \xrightarrow{\sim} H_n(\operatorname{Sin} M_{top}, \operatorname{Sin} A_{top}; G)$$

which are compatible with the connecting homomorphisms. Thus the theorem holds for $R = \mathbb{R}$. We now obtain from Theorem 8.10.i (with $R = R_o$, $S = \mathbb{R}$) in the same way that the theorem holds for $R = R_o$.

Let finally R be an arbitrary real closed field. We denote the singular homology theory over R_o with coefficients in G by h_*. We extend this homology theory to a homology theory h_*^R over R as we have learned in Chapter VI. Since h_* is an ordinary homology theory we know that h_*^R again is an ordinary homology theory with the same coefficient group G. If (M,A) is any pair of spaces over R then now - in contrast to Chapter VI - we have the canonical CW-approximation $j_{(M,A)} : (|K|_R, |L|_R) \to (M,A)$ with $K := \operatorname{Sin} M$ and $L := \operatorname{Sin} A$ at our disposal. Thus we may write

$$h_n^R(M,A) = h_n(|K|_{R_o}, |L|_{R_o}) = H_n(\operatorname{Sin} |K|_{R_o}, \operatorname{Sin} |L|_{R_o}; G) .$$

By Proposition 8.8, applied to the ground field R_o, this group is canonically isomorphic to $H_n(K,L;G) = H_n(\operatorname{Sin} M, \operatorname{Sin} A; G)$. For different values of n all these canonical isomorphisms come from the same homotopy equivalence of pairs of simplicial sets. Thus they are compatible with the connecting homomorphisms. We have found an isomorphism from h_*^R to singular homology over R with coefficients in G and thus know that the latter theory is again an ordinary homology theory with coefficient group G. q.e.d.

In just the same way one verifies that singular cohomology is an ordinary cohomology theory.

Final Remarks. In order to prove Theorem 8.14 it would have been sufficient to work in the category $\mathcal{P}(2,R)$ of pairs of weak polytopes instead of WSA(2,R), once we know that $(M,A) \mapsto H_*(\operatorname{Sin} M, \operatorname{Sin} A; G)$ is a prehomology theory on the whole of WSA(R). Notice also that it was - or would have been - sufficient to use Theorem 8.10 in the case $r = 0$. The notion of pseudohomotopy enables us to avoid a serious use of homotopy theory for pairs of spaces or Kan sets. But our results about systems up to 8.10 deserve interest on their own for $r > 0$.

§9. A group of automorphisms of [0,1]

In this section, apart from the appendix the last one of the present volume, we deviate from the main lines of thought in this chapter. We want to construct a sufficiently large weakly semialgebraic group of automorphisms of the unit interval [0,1] over R, which acts in a natural way on the realization $|X|$ of every partially proper simplicial space X over R. This will give us a new occasion to apply the principles for constructing weakly semialgebraic spaces gained in Chapter IV.

Definition 1. A monotonic PL-automorphism of [0,1] is a bijective monotonic map $g : [0,1] \to [0,1]$ such that there exists a sequence $t_{-1} = 0 \le t_o \le \dots \le t_n = 1$ in R with g linear on each closed subinterval $[t_{i-1},t_i]$ of [0,1] $\{0 \le i \le n$; put $[t_{i-1},t_i] = \{t_i\}$ if $t_{i-1} = t_i\}$.

In this situation, the points $s_i := g(t_i)$ form again a sequence $s_{-1} = 0 \le s_o \le \dots \le s_n = 1$, with $s_{i-1} < s_i$ iff $t_{i-1} < t_i$, and, for $t \in [t_{i-1},t_i]$,

$$g(t) = \begin{cases} s_{i-1} + (t-t_{i-1})(t_i-t_{i-1})^{-1}(s_i-s_{i-1}) & \text{if } t_{i-1} < t_i \,, \\ s_i & \text{if } t_{i-1} = t_i \,. \end{cases} \qquad (9.1)$$

Remark. At present it looks stupid that we allow $t_{i-1} = t_i$. We could avoid this by throwing out some t_i. The reason why we do not do this will become apparent soon.

Every such map g is an automorphism of the semialgebraic space [0,1]. The inverse g^{-1} is again a monotonic PL-automorphism of [0,1], and all these maps together form a subgroup of the group of all automorphisms of the semialgebraic space [0,1]. We denote this subgroup by $PL\ Aut^+([0,1])$, in the present section also more briefly by G. {The sign $^+$ reflects that our maps preserve orientation.} Our goal is to equip the abstract group G with the structure of a weak polytope such that it becomes a weakly semialgebraic group. {We shall succeed only if the field R is sequential.}

Definition 2. Let $g \in PL\ Aut^+([0,1]) = G$. We call an element $c > 1$ of R a two sided Lipschitz constant for g if, for any two points $t < u$ in [0,1], we have

$$c^{-1}(u-t) \leq g(u) - g(t) \leq c(u-t) \, .$$

Here the first inequality means that c is a Lipschitz constant for the map g^{-1}.

Given some $c > 1$, we denote the set of all maps $g \in G$ for which c is a two sided Lipschitz constant by M(c). The group G is the union of all these subsets M(c).

We now fix a constant $c > 1$ for some time and start out to equip the set $M := M(c)$ with the structure of a weak polytope. For any non negative integer n let M_n denote the set of all $g \in M$ which have a description (9.1) with this number n. This gives us a filtration

$$M_o = \{id\} \subset M_1 \subset M_2 \subset \dots$$

of the set M.

In the following we describe a point t of the standard n-simplex $\nabla(n)$ by its so called "sum coordinates",

$$t = (t_{-1}, t_o, t_1, \dots, t_n)$$

with $t_{-1} = 0 \leq t_o \leq t_1 \leq \dots \leq t_n = 1$. Here t_i is the sum of the first i+1 barycentric coordinates of t. {If $t = \sum_{i=0}^{n} u_i e_i$, then $t_i = \sum_{j \leq i} u_j$.} Let $\hat{M}_n$ denote the set of all $(s,t) \in \nabla(n) \times \nabla(n)$ such that $(0 \leq i \leq n)$

$$c^{-1}(t_i - t_{i-1}) \leq s_i - s_{i-1} \leq c(t_i - t_{i-1}) \ .$$

This is a closed semialgebraic subset of $\nabla(n) \times \nabla(n)$, hence a polytope.

We have an obvious map $\eta_n : \hat{M}_n \twoheadrightarrow M_n$ from the set $\hat{M}_n$ onto the set M_n, which sends a point (s,t) of $\hat{M}_n$ to the map g described by (9.1).

<u>Proposition 9.2.</u> On the set $M = M(c)$ there exists a unique structure of a space (= weakly semialgebraic space over R) such that $(M_n \mid n \in \mathbb{N}_o)$ is an exhaustion of M, and, for every $n \in \mathbb{N}_o$, the map $\eta_n : \hat{M}_n \longrightarrow M_n$ is proper (hence η_n is identifying and M is a weak polytope of countable type).

N.B. Of course, the index set $\mathbb{N}_o$ is equipped with its natural total ordering.

<u>Proof.</u> The uniqueness of such a space structure on M is evident: M_n is the proper quotient of $\hat{M}_n$ via η_n and M is the inductive limit of the family of spaces $(M_n \mid n \in \mathbb{N})$. In order to prove existence we shall equip every M_n with the structure of a polytope such that η_n is a semialgebraic (hence proper, hence identifying) map and, if $n > 0$, M_{n-1} is a closed subspace of M_n. If this has been done, Theorem IV.1.6 will give us the desired result.

We equip M_o with its unique structure as a one-point space. Assume now that $n \geq 1$. For every $i \in \{0,\dots,n-1\}$ let $D_i\hat{M}_n$ denote the set of all $(s,t) \in \hat{M}_n$ with $(t_{i+1}-t_i)(s_i-s_{i-1}) = (t_i-t_{i-1})(s_{i+1}-s_i)$. It is a closed semialgebraic subset of $\hat{M}_n$. We have a semialgebraic map from $D_i\hat{M}_n$ onto $\hat{M}_{n-1}$ which sends a point (s,t) of $D_i\hat{M}_n$ to the point (s',t') in which the i-th sum coordinates of s and t are omitted. We denote this map by $\pi_{n,i}$. Notice that, for every $(s,t) \in D_i\hat{M}_n$,

$$\eta_{n-1}\circ\pi_{n,i}(s,t) = \eta_n(s,t) \quad . \qquad (*)$$

We introduce the closed subspace

$$D\hat{M}_n := D_o\hat{M}_n \cup \ldots \cup D_{n-1}\hat{M}_n$$

of $\hat{M}_n$. It is mapped by η_n onto the subset M_{n-1} of M_n. Let $\varphi_n : D\hat{M}_n \twoheadrightarrow M_{n-1}$ denote the set theoretic map obtained by restriction of η_n. Now M_{n-1} is already a semialgebraic space and $\eta_{n-1} : \hat{M}_{n-1} \twoheadrightarrow M_{n-1}$ is semialgebraic. By the equation (*) above, $\varphi_n | D_i\hat{M}_n = \eta_{n-1}\circ\pi_{n,i}$ is semialgebraic for every $i \in \{0,\ldots,n-1\}$. Thus φ_n is semialgebraic, hence proper.

We now glue the polytope $\hat{M}_n$ to the polytope M_{n-1} along $D\hat{M}_n$ by the map φ_n (cf. IV, §8) and obtain a polytope $\hat{M}_n \cup_{\varphi_n} M_{n-1}$. The map η_n and the inclusion $M_{n-1} \hookrightarrow M_n$ combine into a set theoretic map

$$\hat{M}_n \cup_{\varphi_n} M_{n-1} \to M_n \ .$$

(Recall equation (*) above.) It turns out that this map is a bijection (!). We transfer the space structure of $\hat{M}_n \cup_{\varphi_n} M_{n-1}$ to M_n by this bijection. Now M_n is a polytope which contains M_{n-1}, with its given space structure, as a closed subspace. Also $\eta_n : \hat{M}_n \to M_n$ is semialgebraic, as desired. q.e.d.

We now allow the Lipschitz constant c to vary. We denote the spaces $M_n, \hat{M}_n$ more precisely by $M_n(c)$, $\hat{M}_n(c)$, and the map η_n by $\eta_{n,c}$. We shall use the space

$$\hat{M}(c) := \bigsqcup (\hat{M}_n(c) \mid n \geq 0)$$

and the map $\eta_c : \hat{M}(c) \to M(c)$ defined by $\eta_c | \hat{M}_n(c) = \eta_{n,c}$. The map η_c is strongly surjective and partially proper, hence identifying.

If $1 < c < d$ then $\hat{M}_n(c)$ is a closed subspace of $\hat{M}_n(d)$ for every n, hence

$\hat{M}(c)$ is a closed subspace of $\hat{M}(d)$. The preimage of the subset M(c) of M(d) under the map η_d is $\hat{M}(c)$, and the map η_c is a restriction of η_d. We conclude that M(c), with its given space structure, is a closed subspace of M(d).

Assume now that the field R is sequential. We choose a null sequence $(\varepsilon_n | n \in \mathbb{N})$ in $]0,1[$ which is strictly monotonically decreasing and put $c_n := \varepsilon_n^{-1}$. Then G is the union of the family of subsets $(M(c_n) | n \in \mathbb{N})$, and $M(c_{n-1})$ is a closed subspace of $M(c_n)$. We conclude by Theorem IV.7.1 that there exists on G a unique structure of a space over R such that every $M(c_n)$, with its given space structure, is a closed subspace of G and $(M(c_n) | n \in \mathbb{N})$ is an admissible filtration of G. We equip G with this space structure. It is a weak polytope of countable type. For any $c > 1$ there exists some $n \in \mathbb{N}$ with $c < c_n$. The space M(c) is a closed subspace of $M(c_n)$ and hence of G. Thus G is the inductive limit of the family $(M(c) | c > 1)$ in the category Space(R) of function ringed spaces over R, and $(M(c) | c > 1)$ is an admissible covering of G. Also $(M_n(c) | n \in \mathbb{N}_0, c > 1)$ is an admissible covering of G.

Now G is also an abstract group. This group acts on each standard simplex $\nabla(n)$ by semialgebraic automorphisms, the action being given by the formula

$$g(t_{-1}, t_0, \ldots, t_n) := (g(t_{-1}), g(t_0), \ldots, g(t_n))$$
$$(0 = t_{-1} \leq t_0 \leq \ldots \leq t_n = 1) .$$

<u>Theorem 9.3.</u> i) The involution $\iota : G \to G$, $g \mapsto g^{-1}$, is weakly semialgebraic, hence an automorphism of the space G.
ii) The multiplication map $\mu : G \times G \to G$, $(g,h) \mapsto gh$, is weakly semialgebraic.
iii) The map $\alpha : G \times \nabla(p) \to \nabla(p)$, $(g,t) \mapsto gt$, is weakly semialgebraic

for every $p \in \mathbb{N}_0$.

In short, G is a weakly semialgebraic group which acts weakly semialgebraically on each $\nabla(p)$.

<u>Proof.</u> a) We first study the involution ι on G. For every $n \in \mathbb{N}_0$ and $c > 1$ we have a commuting square

$$\begin{array}{ccc} \hat{M}_n(c) & \xrightarrow{\sigma_{n,c}} & \hat{M}_n(c) \\ {\scriptstyle \eta_{n,c}}\downarrow & & \downarrow{\scriptstyle \eta_{n,c}} \\ M_n(c) & \xrightarrow[\iota_{n,c}]{} & M_n(c) \end{array}$$

with $\iota_{n,c}$ a restriction of ι and $\sigma_{n,c}$ the switch automorphism $(s,t) \mapsto (t,s)$ of $\hat{M}_n(c)$. Since $\eta_{n,c}$ is identifying we conclude that $\iota_{n,c}$ is semialgebraic. It follows that the map $\iota : G \to G$ is weakly semialgebraic. {It is even semialgebraic since it is an automorphism of the space G.}

b) We now prove iii) for $p = 1$. It suffices to verify that the restriction

$$\alpha_{n,c} : M_n(c) \times [0,1] \to [0,1], \quad (g,t) \mapsto g(t),$$

is semialgebraic for every $n \in \mathbb{N}_0$, $c > 1$. Since $\eta_{n,c} : \hat{M}_n(c) \twoheadrightarrow M_n(c)$ is identifying we only need to verify that

$$\alpha_{n,c} \circ (\eta_{n,c} \times \mathrm{id}_{[0,1]}) : \hat{M}_n(c) \times [0,1] \to [0,1]$$

is semialgebraic.

Fixing some $n \in \mathbb{N}_0$ and some $c > 1$ we denote this map by f and the space $\hat{M}_n(c) \times [0,1]$ by L for short. The points of L we denote as triples (s,t,u) with $s \in \nabla(n)$, $t \in \nabla(n)$, $u \in [0,1]$ and, of course,

$$c^{-1}(t_i - t_{i-1}) \le s_i - s_{i-1} \le c(t_i - t_{i-1})$$

for every $i \in \{0,\dots,n\}$. For any $j \in \{0,\dots,n\}$ let L_j denote the set of all $(s,t,u) \in L$ with $t_{j-1} < t_j$ and $t_{j-1} \le u \le t_j$. It is semialgebraic in L, and its closure $\overline{L}_j$ in L consists of all points $(s,t,u) \in L$ with $t_{j-1} \le t_j$ and $t_{j-1} \le u \le t_j$. Since L is the union of the sets $\overline{L}_j$ it suffices to prove that $f|\overline{L}_j$ is semialgebraic for every $j \in \{0,\dots,n\}$.

Henceforth we fix an index j in $\{0,\dots,n\}$. For $(s,t,u) \in L_j$ we have

$$f(s,t,u) = s_{j-1} + (u-t_{j-1})(t_j-t_{j-1})^{-1}(s_j-s_{j-1})$$

while for $(s,t,u) \in \overline{L}_j \setminus L_j$ we have $f(s,t,u) = s_j$. Clearly $f|\overline{L}_j$ has a semialgebraic graph and f is continuous on the semialgebraic (open) dense subset L_j of $\overline{L}_j$. The map $f|\overline{L}_j$ is also continuous at any point $x \in \overline{L}_j \setminus L_j$ since the ratio $(t_j-t_{j-1})^{-1}(s_j-s_{j-1})$ is a bounded function on L_j. Thus $f|\overline{L}_j$ is indeed a semialgebraic function.

c) The map $\beta : G\times[0,1]^{p+2} \to [0,1]^{p+2}$, defined by

$$\beta(g,(t_{-1},t_0,\dots,t_p)) = (gt_{-1},gt_0,\dots,gt) \ ,$$

is again weakly semialgebraic. This map sends $G \times \nabla(p)$ to $\nabla(p)$. Thus its restriction $\alpha : G \times \nabla(p) \to \nabla(p)$ is weakly semialgebraic.

d) Now i) and iii) are proved. It remains to prove ii). We shall need a "shuffle map"

$$\sigma_{m,n} : \nabla(m) \times \nabla(n) \to \nabla(m+n)$$

for every $m \in \mathbb{N}_0$, $n \in \mathbb{N}_0$, defined as follows. If $t = (0,t_0,\dots,t_{m-1},1)$ and $u = (0,u_0,\dots,u_{n-1},1)$ are points in $\nabla(m)$ and $\nabla(n)$ then $\sigma_{m,n}(t,u)$ is the sequence $(0,v_0,\dots,v_{m+n-1},1)$ with $(v_0,\dots,v_{m+n-1})$ obtained by re-ordering the sequence $(t_0,\dots,t_{m-1},u_0,\dots,u_{n-1})$ according to the size of the coordinates. It is easily verified that $\sigma_{m,n}$ is semialgebraic. We denote the point $\sigma_{m,n}(t,u)$ more briefly by $t * u$.

Let constants $c > 1$ and $d > 1$ in R and numbers m,n in $\mathbb{N}_0$ be given. Then the multiplication map $\mu : G \times G \to G$ restricts to a map

$$\mu_{m,n,c,d} : M_m(c) \times M_n(d) \to M_{m+n}(cd) .$$

We want to verify that this map is semialgebraic. Then we shall know that μ is weakly semialgebraic.

We shall invent a semialgebraic map

$$\zeta = \zeta_{m,n,c,d} : \hat{M}_m(c) \times \hat{M}_n(d) \to \hat{M}_{m+n}(cd)$$

such that the diagram

$$(*) \qquad \begin{array}{ccc} \hat{M}_m(c) \times \hat{M}_n(d) & \xrightarrow{\zeta} & \hat{M}_{m+n}(cd) \\ \big\downarrow{\scriptstyle \eta_{m,c} \times \eta_{n,d}} & & \big\downarrow{\scriptstyle \eta_{m+n,cd}} \\ M_m(c) \times M_n(d) & \xrightarrow[\mu_{m,n,c,d}]{} & M_{m+n}(cd) \end{array}$$

commutes. Then it will be evident that $\mu_{m,n,c,d}$ is semialgebraic, since the vertical arrows in the diagram are identifying semialgebraic maps.

Let points $(u,v) \in \hat{M}_m(c)$ and $(s,t) \in \hat{M}_n(d)$ be given. We define new pairs $(x,v*s) \in \hat{M}_{m+n}(c)$, $(v*s,y) \in \hat{M}_{m+n}(d)$ as follows:

$$x := [\eta_{m,c}(u,v)]\ (v*s) ,$$
$$y := [\eta_{n,d}(s,t)]^{-1}(v*s) .$$

Then

$$\eta_{m+n,c}(x,v*s) = \eta_{m,c}(u,v) ,$$
$$\eta_{m+n,d}(v*s,y) = \eta_{n,d}(s,t) .$$

We define the desired map ζ by $\zeta((u,v),(s,t)) := (x,y)$. Then the diagram (*) commutes. It follows from the previous steps a) and c) of the proof that ζ is indeed semialgebraic. q.e.d.

The orbits of G on $\nabla(n)$ are easily seen to be the open faces of $\nabla(n)$. Indeed, if s and t are two points in the same open face of $\nabla(n)$ {for example, in $\mathring{\nabla}(n)$}, then the element $g \in G$ given by the formula (9.1) takes the point t to s. Also the following is easily checked.

Lemma 9.4. For every monotonic map $\alpha : [p] \to [n]$ the semialgebraic map $\alpha_* : \nabla(p) \longrightarrow \nabla(n)$ is G-equivariant.

We now are ready to prove a beautiful result.

Theorem 9.5. As before assume that R is sequential. There exists a unique weakly semialgebraic action of $G = PL\, Aut^+([0,1])$ on the realization $|X|$ of every partially proper simplicial space X such that the following three properties hold.

A1) If $X = \underline{M}$ is constant then the action of G on $|X| = M$ is trivial.

A2) If $X = \Delta(n)$ then the action of G on $\nabla(n)$ is as just described.

A3) If $f : X \to Y$ is a simplicial map then $|f| : |X| \to |Y|$ is G-equivariant.

This action of G on X is given by the formula $(x \in X, t \in \nabla(n))$

$$g|x,t| = |x,gt|. \qquad (*)$$

If X is discrete then the orbits of G on $|X|$ are the open cells of the CW-complex $|X|$.

Proof. If, for any two partially proper simplicial spaces X and Y, we have already defined the action of G on $|X|$ and $|Y|$ then we are forced to define the action of G on $|X \times Y| = |X| \times |Y|$ by the formula $g(x,y) = (gx,gy)$, since the projections from $|X| \times |Y|$ to $|X|$ and $|Y|$ must be G-equivariant. Thus, for any product $\underline{M} \times \Delta(n)$ of a constant space $\underline{M}$ and a standard simplicial set $\Delta(n)$ we have to define the G-action on

$|\underline{M} \times \Delta(n)| = M \times \nabla(n)$ by the formula $g(x,y) = (x,gy)$.

Let now X be any partially proper simplicial space. We have a canonical simplicial map χ_X from the deployment De X to X, and the realization of this map is the partially proper strongly surjective map $\eta_X : \hat{X} \to |X|$ (cf. Prop. 4.7). We define the G-action on the direct sum $\hat{X}$ of the spaces $X_n \times \nabla(n)$ by the formula $g(x,t) = (x,gt)$, as we are forced to. This is indeed a weakly semialgebraic action. It follows from Lemma 9.4 above that this action is compatible with the equivalence relation $E(\eta_X)$ on $\hat{X}$. Since η_X is identifying we obtain a weakly semialgebraic action of G on |X| such that formula (*) in the theorem holds, and we are forced to use this formula, since η_X has to become G-equivariant. We now have established a weakly semialgebraic G-action on the realization |X| of any partially proper simplicial space X. It is clear from the formula (*) that these actions fulfill A3. It can be verified in a straightforward way that they also fulfill A1 and A2. The last assertion in the theorem is evident from the formula (*) and our description in §6 of the open cells of |X| for X discrete. q.e.d.

Remark 9.6. If R is not sequential then we still have an action of the abstract group $G = PL\,Aut^+[0,1]$ on |X| by semialgebraic automorphisms. If X is discrete then this action is transitive on the cells of |X|, a fact which is sometimes useful. For example, let M be a space over R and assume that there exists some isomorphism $\alpha : |K| \xrightarrow{\sim} M$ with K a simplicial set. {This means that M can be triangulated, cf. 6.8}. Let P be some local intrinsic property formulated for points of M. Then the set N of all points of M, for which P holds, is a weakly semialgebraic subset of M. Indeed, since our property P is intrinsic, the set N must be stable under every automorphism of M, hence $\alpha^{-1}(N)$ must be a union of open cells of |K|.

Epilogue. It looks somewhat artificial that, in order to establish a space structure on PL $Aut^+([0,1])$, we had to assume that our base field R is sequential, although in practice this trouble might be surmounted by passing to a sequential real closed field extension of R, a trick we have used in Chapters V and VI at various occasions.

Anyway, our construction of the weakly semialgebraic group PL $Aut^+([0,1])$ seems to reveal a deficiency of our definition of weakly semialgebraic spaces in Chapter IV, §1 for some purposes. The trouble comes from the exhaustion axiom E3. On the one hand, this axiom seems to be crucial for our patch constructions in Chapter V and thus seems to be largely responsible for some of our best results, as for example the strong version of Whitehead's theorem in Chapter V, §6. On the other hand, why not sometimes admit more general inductive limits of semialgebraic spaces than we did in this book?

It seems that PL $Aut^+([0,1])$ is an honest space in some general sense which is still reasonable. Niels Schwartz recently started an investigation of "abstract" weakly semialgebraic spaces based on his theory of real closed spaces (cf. [Sch1] and [LSA, App. A]). He gained evidence that a theory of weakly semialgebraic spaces without axiom E3 is still feasible [oral communication].

Appendix C (to Chapter IV): When is $\mathring{\mathfrak{T}}(M)$ a basis of open sets of M_{top}?

We shall discuss this question mostly by examples. In the following R_+ denotes the set of positive elements of R.

Example C.1. Let M be a countable or uncountable comb, cf. IV.4.8 and IV.4.9. Then M is not a locally semialgebraic space. Nevertheless is $\mathring{\mathfrak{T}}(M)$ a basis of open sets of M_{top}.

Example C.2. Let $R = \mathbb{R}$ and let M be the subset $\mathbb{R}_+ \times \mathbb{R}_+ \times \mathbb{R}_+ \cup \mathbb{R}_+ \times \mathbb{R}_+ \times \{0\} \cup \{(0,0,0)\}$ of $\mathbb{R}^3$. For any finite subset J of $\mathbb{R}_+$ let M_J be the semialgebraic subspace $J \times \mathbb{R}_+ \times \mathbb{R}_+ \cup \mathbb{R}_+ \times \mathbb{R}_+ \times \{0\} \cup \{(0,0,0)\}$ of $\mathbb{R}^3$. Using IV, 1.6 we equip M with the unique structure of a weakly semialgebraic space such that every M_J, in its given structure, is a closed semialgebraic subspace of M and $(M_J \mid J \subset \mathbb{R}_+, J \text{ finite})$ is an exhaustion of M. $U = \{(x,y,z) \in M \mid z < e^{-\frac{1}{y}}\} \cup \{(0,0,0)\}$ is an open subset of M. There exists no $V \in \mathring{\mathfrak{T}}(M)$ with $(0,0,0) \in V \subset U$. Hence $\mathring{\mathfrak{T}}(M)$ is not a basis of the strong topology of M. The reason for this is that M is not polytopic, as the following proposition shows.

Proposition C.3. Let M be a polytopic space over $\mathbb{R}$. Then $\mathring{\mathfrak{T}}(M)$ is a basis of the strong topology of M.

Proof. Let U be an open subset of M and a an element of U. We have to show that there exists a set $V \in \mathring{\mathfrak{T}}(M)$ with $a \in V \subset U$. Let $(M_\alpha \mid \alpha \in I)$ be an exhaustion of M. Let E be the set of all pairs (J, V_J), where J is a subset of I and V_J is a subset of $M_J = \cup(M_\alpha \mid \alpha \in J)$ such that $a \in V_J$ and, for every $\alpha \in J$, $V_J \cap M_\alpha$ is an open semialgebraic subset of M_α whose closure is complete and contained in $U \cap M_\alpha$. For elements (J, V_J) and (K, V_K) of E we put $(J, V_J) \le (K, V_K)$ if $J \subset K$ and $V_J = V_K \cap M_J$. E is not empty. By Zorn's Lemma there exists a maximal element (L, V_L) of E.

Notice that, if γ and δ are indices in I with $\gamma < \delta$ and $\delta \in L$, then $\gamma \in L$.

Assume that $L \neq I$. Let β be an element of $I \setminus L$. By I, 7.5 we may assume that M_β is a closed subspace of some $\mathbb{R}^n$. Let d be the euclidean distance function on $\mathbb{R}^n$. The closure of $V_L \cap M_\beta$ is complete and contained in $U \cap M_\beta$. Therefore $t := d(V_L \cap M_\beta, M_\beta \setminus U) > 0$. {Here we use that $R = \mathbb{R}$!} Put $V_o := \{x \in M_\beta \mid d(x, V_L \cap M_\beta) < \frac{1}{2}t\}$. V_o is an open semialgebraic subset of M_β containing $V_L \cap M_\beta$ and the closure of V_o is complete and contained in $U \cap M_\beta$. Let V_1 be an open semialgebraic subset of M_β with $V_1 \cap (M_L \cap M_\beta) = V_L \cap M_\beta$. Then $(L \cup \{\beta\}, V_L \cup (V_o \cap V_1))$ is an element of E and $(L, V_L) < (L \cup \{\beta\}, V_L \cup (V_o \cap V_1))$. This contradiction shows that $L = I$. Hence V_L is an open weakly semialgebraic subset of M with $a \in V_L \subset U$.

The space $\mathbb{R}^\infty$ (cf. IV.6.2) is polytopic. Hence $\mathring{\mathcal{T}}(\mathbb{R}^\infty)$ generates the strong topology of $\mathbb{R}^\infty$. $\mathbb{R}$ is the only sequential real closed field with this property.

<u>Example C.4.</u> Let R be a sequential real closed field different from $\mathbb{R}$. Then $\mathring{\mathfrak{J}}(R^\infty)$ is not a basis of the strong topology of R^∞.

<u>Proof.</u> We choose a sequence $(a_n \mid n \in \mathbb{N})$ in R_+ with $\lim_{n\to\infty} a_n = 0$. We also choose a sequence $(b_n \mid n \in \mathbb{N})$ in $]0,1[$ consisting of pairwise different elements such that the set $\{b_n \mid n \in \mathbb{N}\}$ is discrete and closed in $[0,1]$. This is possible since $R \neq \mathbb{R}$. Indeed, if R is archimedean, $R \subset \mathbb{R}$, we may choose $(b_n \mid n \in \mathbb{N})$ as a sequence converging in $\mathbb{R}$ to a point $\mathfrak{J} \in \mathbb{R} \setminus R$ with $0 < \mathfrak{J} < 1$. If R is not archimedean, take $b_n = n\varepsilon$ with $\varepsilon > 0$ and ε smaller than every positive element of $\mathbb{Q}$.

We now consider the set F consisting of all tuples $(x_i) \in R^\infty$ such that there exist natural numbers n,m with $x_1 = a_n b_m$, $x_{n+1} = a_m$ and all other $x_i = 0$. For every $k \in \mathbb{N}$ the set $F \cap (R^k \times 0)$ consists of the tuples $(x_i) \in R^\infty$ with $x_1 = a_n b_m$ for some $m \in \mathbb{N}$ and some $n < k$, $x_{n+1} = a_m$, and all other $x_i = 0$. Since the sets $\{a_n b_m \mid m \in \mathbb{N}\}$ are closed and discrete in R, we conclude that $F \cap (R^k \times 0)$ is closed in $R^k \times 0$ for every k, hence that F is closed in R^∞.

Let $U := R^\infty \setminus F$. This open set contains the origin $a = (0,0,\ldots)$. Suppose that there exists a set $V \in \mathring{\mathfrak{J}}(R^\infty)$ with $a \in V \subset U$. We choose an element $\delta > 0$ in R such that the set A consisting of all tuples (x_i) with $x_1 \in [0,\delta]$, $x_i = 0$ for $i > 1$, is contained in V. Then we choose a natural number k with $a_k < \delta$. Let Q denote the closed semialgebraic subspace $R^{k+1} \times \{0\}$ of R^∞, and let d denote the euclidean distance function on Q. The set $V \cap Q$ is open and semialgebraic in Q and contains the complete semialgebraic space A. Hence

$$0 < d(A, Q \setminus V) \leq d(A, Q \setminus U) .$$

But $Q \setminus U = Q \cap F$ contains the points $(a_k b_m, 0, \ldots, 0, a_m, 0, 0, \ldots)$ with coordinate a_m at the place k+1. Thus $d(A, Q \setminus U) = 0$, a contradiction. This proves our claim.

References

References which occur in this book and in the first volume [LSA] have been denoted by the same sigle here and there.

[LSA] H. Delfs, M. Knebusch, "Locally semialgebraic spaces". Lecture Notes Math. 1173, Springer 1985.

[SFC] M. Knebusch, "Semialgebraic fibrations and covering maps", in preparation.

[Ad] J.F. Adams, "Stable homotopy and generalized homology". The University of Chicago Press 1974 (Chicago Lectures in Math.).

[Ad_1] J.F. Adams, A variant of E.H. Brown's representability theorem. Topology 10, 185-198 (1971).

[AM] M. Artin, B. Mazur, "Etale homotopy". Lecture Notes Math. 100, Springer 1969.

[Ba] M.G. Barratt, Track groups I, II. Proc. London Math. Soc. 5, 71-106, 285-329 (1955).

[BCR] J. Bochnak, M. Coste, M.F. Coste-Roy, "Géométrie algébrique réelle". Ergebn. Math. Grenzgeb. (3) 12, Springer 1987.

[BD] T. Bröcker, T. tom Dieck, "Kobordismentheorie". Lecture Notes Math. 178, Springer 1970.

[Bn] E.H. Brown, Cohomology theories. Ann. Math. 75, 467-484 (1962); with correction, Ann. Math. 78, 201 (1963).

[B] G.W. Brumfiel, "Partially ordered rings and semialgebraic geometry". London Math. Soc. Lecture Notes 37, Cambridge University Press 1979.

[B_2] G.W. Brumfiel, Quotient spaces for semialgebraic equivalence relations. Math. Z. 195, 69-78 (1987).

[Ca] P. Cartier, Structure simpliciales. Sém. Bourbaki Exp. 199, 1-12 (1960).

[Ch] C. Chevalley, "Theory of Lie groups I". Princeton University Press 1946.

[Cu] E.B. Curtis, Simplicial homotopy theory. Adv. Math. 6, 107-209 (1971).

[D] H. Delfs, Kohomologie affiner semialgebraischer Räume. Dissertation Regensburg 1981.

[D_1] H. Delfs, The homotopy axiom in semialgebraic cohomology. J. reine angew. Math. 355, 108-128 (1985) [D_2] p. 358

[DK_2] H. Delfs, M. Knebusch. Semialgebraic topology over a real closed field II: Basic theory of semialgebraic spaces. Math. Z. 178, 175-213 (1981).

[DK_3] H. Delfs, M. Knebusch, On the homology of algebraic varieties over real closed fields. J. reine angew. Math. 335, 122-163 (1981).

[DK_4] H. Delfs, M. Knebusch, Zur Theorie der semialgebraischen Wege und Intervalle über einem reell abgeschlossenen Körper. In: "Géométrie algébrique et formes quadratiques", Ed. J.L. Colliot-Thélène et al., Lecture Notes Math. 959, 299-323, Springer 1982.

[DK_5] H. Delfs, M. Knebusch, Separation, retractions and homotopy extension in semialgebraic spaces. Pacific J. Math. 114, 47-71 (1984).

[DK_6] H. Delfs, M. Knebusch, An introduction to locally semialgebraic spaces. Rocky Mountain J. Math. 14, 945-963 (1984).

[De] P. Deligne, Théorie de Hodge, III. Publ. Math. IHES 44, 5-78 (1974).

[tD] T. tom Dieck, Klassifikation numerierbarer Bündel. Archiv Math. 17, 395-399 (1966).

[DKP] T. tom Dieck, K.H. Kamps, D. Puppe, "Homotopietheorie". Lecture Notes Math. 157, Springer 1970.

[Do] A. Dold, "Halbexakte Homotopiefunktoren", Lecture Notes Math. 12, Springer 1966.

[Du] D.W. Dubois, Real algebraic curves. Technical Report No. 227, Univ. New Mexico, Albuquerque 1971.

[EM] S. Eilenberg, S. MacLane, On the groups $H(\Pi,n)$, I. Annals Math. 58, 55-106 (1953).

[ES] S. Eilenberg, N. Steenrod, "Foundations of algebraic topology". Princeton Univ. Press 1952.

[Frd] E.M. Friedlander, "Etale homotopy of simplicial schemes". Annals Math. Studies 104, Princeton Univ. Press 1982.

[Frt] R. Fritsch, Some remarks on S. Weingram: On the triangulation of semisimple complex. Illinois J. Math. 14, 529-535 (1970).

[Frt_1] R. Fritsch, Zur Unterteilung semisimplizialer Mengen. I: Math. Z. 108, 329-367 (1969); II: Math. Z. 109, 131-152 (1969).

[Gr] A. Grothendieck, Téchnique de descente et théorèmes d'existence en géométrie algébrique, I. Sém. Bourbaki Exp. 190 (1959, W.A. Benjamin 1966.

[Gu] V.K.A.M. Gugenheim, Semisimplicial homotopy theory. In: "Studies in modern topology" (ed. P.J. Hilton), 99-133, Studies in Math. vo. 5, Math. Ass. Am. 1968.

[Kan] D.M. Kan, On c.s.s. complexes. Amer. J. Math. 79, 449-476 (1957).

[KW] M. Knebusch, M.J. Wright, Bewertungen mit reeller Henselisierung J. reine angew. Math. 286/287, 314-321 (1976).

[L] T.Y. Lam, "The theory of ordered fields". Ring theory and algebra III. (ed. B. McDonald), Lecture Notes Pure Appl. Math. 55, 1-152, Dekker 1980.

[La] K. Lamotke, "Semisimpliziale algebraische Topologie". Grundlehren math. Wiss. 147, Springer 1968.

[LW] A.T. Lundell, S. Weingram, "The topology of CW complexes". Van Nostrand Reinhold Co., New York 1969.

[May] J.P. May, "Simplicial objects in algebraic topology". Van Nostrand Math. Studies 11, D. Van Nostrand Co., Princeton 1967.

[Mi] J. Milnor, On axiomatic homology theory. Pacific J. Math. 12, 337-341 (1962).

[Mi_1] J. Milnor, The geometric realization of a semi-simplicial complex. Annals Math. 65, 357-362 (1957).

[Mt] B. Mitchell, "Theory of categories". Academic Press 1965.

[P] A. Prestel, "Lectures on formally real fields". Lecture Notes Math. 1093, Springer 1984.

[Pu] D. Puppe, Homotopiemengen und ihre induzierten Abbildungen I. Math. Z. 69, 299-344 (1958).

[Q] B. von Querenburg, "Mengentheoretische Topologie". Springer 1973.

[R] R. Robson, Embedding semialgebraic spaces. Math. Z. 183, 365-370 (1983).

[Schd] C. Scheiderer, Quotients of semi-algebraic spaces.Math.Z.,to appear.

[Sch_1] N. Schwartz, "The basic theory of real closed spaces". Regensburger Math. Schriften Nr. 15, 1987.

[Se] G. Segal, Classifying spaces and spectral sequences. Publ. Math. IHES 34, 105-112 (1986).

[Spa] E.H. Spanier, "Algebraic topology", McGraw-Hill 1966.

[Sw] R.M. Switzer, "Algebraic topology - homotopy and homology". Grundlehren math. Wiss. 212, Springer 1975.

[V] A. Verona, "Stratified mappings - Structure and Triangulability". Lecture Notes Math. 1102, Springer 1984.

[We] S. Weingram, On the triangulation of the realization of a semi-simplicial complex. Illinois J. Math. 12, 403-414 (1968).

[W] G.W. Whitehead, "Elements of homotopy theory". Graduate Texts Math. 61, Springer 1978.

[W_2] G. Whitehead, Generalized homology theories. Trans. Amer. Math. Soc. 102, 227-283 (1962).

[D_2] H. Delfs, Sheaf theory and Borel-Moore homology on locally semialgebraic spaces. Habilitationsschrift Regensburg 1984.

<u>Symbols</u>

Glossary

LOCALLY SEMIALGEBRAIC SPACES H. Delfs, M. Knebusch (Lecture Notes in Mathematics Vol. 1173)

TABLE OF CONTENTS

LECTURE NOTES IN MATHEMATICS
Edited by A. Dold and B. Eckmann

Some general remarks on the publication of monographs and seminars

In what follows all references to monographs, are applicable also to multiauthorship volumes such as seminar notes.

§1. Lecture Notes aim to report new developments - quickly, informally, and at a high level. Monograph manuscripts should be reasonably self-contained and rounded off. Thus they may, and often will, present not only results of the author but also related work by other people. Furthermore, the manuscripts should provide sufficient motivation, examples and applications. This clearly distinguishes Lecture Notes manuscripts from journal articles which normally are very concise. Articles intended for a journal but too long to be accepted by most journals, usually do not have this "lecture notes" character. For similar reasons it is unusual for Ph.D. theses to be accepted for the Lecture Notes series.

Experience has shown that English language manuscripts achieve a much wider distribution.

§2. Manuscripts or plans for Lecture Notes volumes should be submitted either to one of the series editors or to Springer-Verlag, Heidelberg. These proposals are then refereed. A final decision concerning publication can only be made on the basis of the complete manuscripts, but a preliminary decision can usually be based on partial information: a fairly detailed outline describing the planned contents of each chapter, and an indication of the estimated length, a bibliography, and one or two sample chapters - or a first draft of the manuscript. The editors will try to make the preliminary decision as definite as they can on the basis of the available information.

§3. Lecture Notes are printed by photo-offset from typed copy delivered in camera-ready form by the authors. Springer-Verlag provides technical instructions for the preparation of manuscripts, and will also, on request, supply special staionery on which the prescribed typing area is outlined. Careful preparation of the manuscripts will help keep production time short and ensure satisfactory appearance of the finished book. Running titles are not required; if however they are considered necessary, they should be uniform in appearance. We generally advise authors not to start having their final manuscripts specially tpyed beforehand. For professionally typed manuscripts, prepared on the special stationery according to our instructions, Springer-Verlag will, if necessary, contribute towards the typing costs at a fixed rate.

The actual production of a Lecture Notes volume takes 6-8 weeks.

.../...

§4. Final manuscripts should contain at least 100 pages of mathematical text and should include

- a table of contents
- an informative introduction, perhaps with some historical remarks. It should be accessible to a reader not particularly familiar with the topic treated.
- a subject index; this is almost always genuinely helpful for the reader.

§5. Authors receive a total of 50 free copies of their volume, but no royalties. They are entitled to purchase further copies of their book for their personal use at a discount of 33.3 %, other Springer mathematics books at a discount of 20 % directly from Springer-Verlag.

Commitment to publish is made by letter of intent rather than by signing a formal contract. Springer-Verlag secures the copyright for each volume.

Vol. 1259: F. Cano Torres, Desingularization Strategies for Three-Dimensional Vector Fields. IX, 189 pages. 1987.

Vol. 1260: N.H. Pavel, Nonlinear Evolution Operators and Semigroups. VI, 285 pages. 1987.

Vol. 1261: H. Abels, Finite Presentability of S-Arithmetic Groups. Compact Presentability of Solvable Groups. VI, 178 pages. 1987.

Vol. 1262: E. Hlawka (Hrsg.), Zahlentheoretische Analysis II. Seminar, 1984–86. V, 158 Seiten. 1987.

Vol. 1263: V.L. Hansen (Ed.), Differential Geometry. Proceedings, 1985. XI, 288 pages. 1987.

Vol. 1264: Wu Wen-tsün, Rational Homotopy Type. VIII, 219 pages. 1987.

Vol. 1265: W. Van Assche, Asymptotics for Orthogonal Polynomials. VI, 201 pages. 1987.

Vol. 1266: F. Ghione, C. Peskine, E. Sernesi (Eds.), Space Curves. Proceedings, 1985. VI, 272 pages. 1987.

Vol. 1267: J. Lindenstrauss, V.D. Milman (Eds.), Geometrical Aspects of Functional Analysis. Seminar. VII, 212 pages. 1987.

Vol. 1268: S.G. Krantz (Ed.), Complex Analysis. Seminar, 1986. VII, 195 pages. 1987.

Vol. 1269: M. Shiota, Nash Manifolds. VI, 223 pages. 1987.

Vol. 1270: C. Carasso, P.-A. Raviart, D. Serre (Eds.), Nonlinear Hyperbolic Problems. Proceedings, 1986. XV, 341 pages. 1987.

Vol. 1271: A.M. Cohen, W.H. Hesselink, W.L.J. van der Kallen, J.R. Strooker (Eds.), Algebraic Groups Utrecht 1986. Proceedings. XII, 284 pages. 1987.

Vol. 1272: M.S. Livšic, L.L. Waksman, Commuting Nonselfadjoint Operators in Hilbert Space. III, 115 pages. 1987.

Vol. 1273: G.-M. Greuel, G. Trautmann (Eds.), Singularities, Representation of Algebras, and Vector Bundles. Proceedings, 1985. XIV, 383 pages. 1987.

Vol. 1274: N. C. Phillips, Equivariant K-Theory and Freeness of Group Actions on C*-Algebras. VIII, 371 pages. 1987.

Vol. 1275: C.A. Berenstein (Ed.), Complex Analysis I. Proceedings, 1985–86. XV, 331 pages. 1987.

Vol. 1276: C.A. Berenstein (Ed.), Complex Analysis II. Proceedings, 1985–86. IX, 320 pages. 1987.

Vol. 1277: C.A. Berenstein (Ed.), Complex Analysis III. Proceedings, 1985–86. X, 350 pages. 1987.

Vol. 1278: S.S. Koh (Ed.), Invariant Theory. Proceedings, 1985. V, 102 pages. 1987.

Vol. 1279: D. Ieşan, Saint-Venant's Problem. VIII, 162 Seiten. 1987.

Vol. 1280: E. Neher, Jordan Triple Systems by the Grid Approach. XII, 193 pages. 1987.

Vol. 1281: O.H. Kegel, F. Menegazzo, G. Zacher (Eds.), Group Theory. Proceedings, 1986. VII, 179 pages. 1987.

Vol. 1282: D.E. Handelman, Positive Polynomials, Convex Integral Polytopes, and a Random Walk Problem. XI, 136 pages. 1987.

Vol. 1283: S. Mardešić, J. Segal (Eds.), Geometric Topology and Shape Theory. Proceedings, 1986. V, 261 pages. 1987.

Vol. 1284: B.H. Matzat, Konstruktive Galoistheorie. X, 286 pages. 1987.

Vol. 1285: I.W. Knowles, Y. Saitō (Eds.), Differential Equations and Mathematical Physics. Proceedings, 1986. XVI, 499 pages. 1987.

Vol. 1286: H.R. Miller, D.C. Ravenel (Eds.), Algebraic Topology. Proceedings, 1986. VII, 341 pages. 1987.

Vol. 1287: E.B. Saff (Ed.), Approximation Theory, Tampa. Proceedings, 1985–1986. V, 228 pages. 1987.

Vol. 1288: Yu. L. Rodin, Generalized Analytic Functions on Riemann Surfaces. V, 128 pages, 1987.

Vol. 1289: Yu. I. Manin (Ed.), K-Theory, Arithmetic and Geometry. Seminar, 1984–1986. V, 399 pages. 1987.

Vol. 1290: G. Wüstholz (Ed.), Diophantine Approximation and scendence Theory. Seminar, 1985. V, 243 pages. 1987.

Vol. 1291: C. Mœglin, M.-F. Vignéras, J.-L. Waldspurger, Corre dances de Howe sur un Corps p-adique. VII, 163 pages. 1987

Vol. 1292: J.T. Baldwin (Ed.), Classification Theory. Procee 1985. VI, 500 pages. 1987.

Vol. 1293: W. Ebeling, The Monodromy Groups of Isolated Sin ties of Complete Intersections. XIV, 153 pages. 1987.

Vol. 1294: M. Queffélec, Substitution Dynamical Systems – Sp Analysis. XIII, 240 pages. 1987.

Vol. 1295: P. Lelong, P. Dolbeault, H. Skoda (Réd.), Sém d'Analyse P. Lelong – P. Dolbeault – H. Skoda. Seminar, 1985/ VII, 283 pages. 1987.

Vol. 1296: M.-P. Malliavin (Ed.), Séminaire d'Algèbre Paul Dub Marie-Paule Malliavin. Proceedings, 1986. IV, 324 pages. 1987.

Vol. 1297: Zhu Y.-l., Guo B.-y. (Eds.), Numerical Methods for Differential Equations. Proceedings. XI, 244 pages. 1987.

Vol. 1298: J. Aguadé, R. Kane (Eds.), Algebraic Topology, Barc 1986. Proceedings. X, 255 pages. 1987.

Vol. 1299: S. Watanabe, Yu.V. Prokhorov (Eds.), Probability T and Mathematical Statistics. Proceedings, 1986. VIII, 589 p 1988.

Vol. 1300: G.B. Seligman, Constructions of Lie Algebras and Modules. VI, 190 pages. 1988.

Vol. 1301: N. Schappacher, Periods of Hecke Characters. XV pages. 1988.

Vol. 1302: M. Cwikel, J. Peetre, Y. Sagher, H. Wallin (Eds.), Fun Spaces and Applications. Proceedings, 1986. VI, 445 pages. 19

Vol. 1303: L. Accardi, W. von Waldenfels (Eds.), Quantum Prob and Applications III. Proceedings, 1987. VI, 373 pages. 1988.

Vol. 1304: F.Q. Gouvêa, Arithmetic of p-adic Modular Forms. VIII pages. 1988.

Vol. 1305: D.S. Lubinsky, E.B. Saff, Strong Asymptotics for Extr Polynomials Associated with Weights on ℝ. VII, 153 pages. 1988

Vol. 1306: S.S. Chern (Ed.), Partial Differential Equations. Proc ings, 1986. VI, 294 pages. 1988.

Vol. 1307: T. Murai, A Real Variable Method for the Cauchy Transf and Analytic Capacity. VIII, 133 pages. 1988.

Vol. 1308: P. Imkeller, Two-Parameter Martingales and Their Qua tic Variation. IV, 177 pages. 1988.

Vol. 1309: B. Fiedler, Global Bifurcation of Periodic Solutions Symmetry. VIII, 144 pages. 1988.

Vol. 1310: O.A. Laudal, G. Pfister, Local Moduli and Singularities 117 pages. 1988.

Vol. 1311: A. Holme, R. Speiser (Eds.), Algebraic Geometry, dance 1986. Proceedings. VI, 320 pages. 1988.

Vol. 1312: N.A. Shirokov, Analytic Functions Smooth up to Boundary. III, 213 pages. 1988.

Vol. 1313: F. Colonius, Optimal Periodic Control. VI, 177 pa 1988.

Vol. 1314: A. Futaki, Kähler-Einstein Metrics and Integral Invariants 140 pages. 1988.

Vol. 1315: R.A. McCoy, I. Ntantu, Topological Properties of Space Continuous Functions. IV, 124 pages. 1988.

Vol. 1316: H. Korezlioglu, A.S. Ustunel (Eds.), Stochastic Analysis Related Topics. Proceedings, 1986. V, 371 pages. 1988.

Vol. 1317: J. Lindenstrauss, V.D. Milman (Eds.), Geometric Aspec Functional Analysis. Seminar, 1986–87. VII, 289 pages. 1988.

Vol. 1318: Y. Felix (Ed.), Algebraic Topology – Rational Homot Proceedings, 1986. VIII, 245 pages. 1988

Vol. 1319: M. Vuorinen, Conformal Geometry and Quasireg Mappings. XIX, 209 pages. 1988.

Vol. 1201: Curvature and Topology of Riemannian Manifolds. Proceedings, 1985. Edited by K. Shiohama, T. Sakai and T. Sunada. VII, 336 pages. 1986.

Vol. 1202: A. Dür, Möbius Functions, Incidence Algebras and Power Series Representations. XI, 134 pages. 1986.

Vol. 1203: Stochastic Processes and Their Applications. Proceedings, 1985. Edited by K. Itô and T. Hida. VI, 222 pages. 1986.

Vol. 1204: Séminaire de Probabilités XX, 1984/85. Proceedings. Edité par J. Azéma et M. Yor. V, 639 pages. 1986.

Vol. 1205: B.Z. Moroz, Analytic Arithmetic in Algebraic Number Fields. VII, 177 pages. 1986.

Vol. 1206: Probability and Analysis, Varenna (Como) 1985. Seminar. Edited by G. Letta and M. Pratelli. VIII, 280 pages. 1986.

Vol. 1207: P.H. Bérard, Spectral Geometry: Direct and Inverse Problems. With an Appendix by G. Besson. XIII, 272 pages. 1986.

Vol. 1208: S. Kaijser, J.W. Pelletier, Interpolation Functors and Duality. IV, 167 pages. 1986.

Vol. 1209: Differential Geometry, Peñíscola 1985. Proceedings. Edited by A.M. Naveira, A. Ferrández and F. Mascaró. VIII, 306 pages. 1986.

Vol. 1210: Probability Measures on Groups VIII. Proceedings, 1985. Edited by H. Heyer. X, 386 pages. 1986.

Vol. 1211: M.B. Sevryuk, Reversible Systems. V, 319 pages. 1986.

Vol. 1212: Stochastic Spatial Processes. Proceedings, 1984. Edited by P. Tautu. VIII, 311 pages. 1986.

Vol. 1213: L.G. Lewis, Jr., J.P. May, M. Steinberger, Equivariant Stable Homotopy Theory. IX, 538 pages. 1986.

Vol. 1214: Global Analysis – Studies and Applications II. Edited by Yu.G. Borisovich and Yu.E. Gliklikh. V, 275 pages. 1986.

Vol. 1215: Lectures in Probability and Statistics. Edited by G. del Pino and R. Rebolledo. V, 491 pages. 1986.

Vol. 1216: J. Kogan, Bifurcation of Extremals in Optimal Control. VIII, 106 pages. 1986.

Vol. 1217: Transformation Groups. Proceedings, 1985. Edited by S. Jackowski and K. Pawalowski. X, 396 pages. 1986.

Vol. 1218: Schrödinger Operators, Aarhus 1985. Seminar. Edited by E. Balslev. V, 222 pages. 1986.

Vol. 1219: R. Weissauer, Stabile Modulformen und Eisensteinreihen. III, 147 Seiten. 1986.

Vol. 1220: Séminaire d'Algèbre Paul Dubreil et Marie-Paule Malliavin. Proceedings, 1985. Edité par M.-P. Malliavin. IV, 200 pages. 1986.

Vol. 1221: Probability and Banach Spaces. Proceedings, 1985. Edited by J. Bastero and M. San Miguel. XI, 222 pages. 1986.

Vol. 1222: A. Katok, J.-M. Strelcyn, with the collaboration of F. Ledrappier and F. Przytycki, Invariant Manifolds, Entropy and Billiards; Smooth Maps with Singularities. VIII, 283 pages. 1986.

Vol. 1223: Differential Equations in Banach Spaces. Proceedings, 1985. Edited by A. Favini and E. Obrecht. VIII, 299 pages. 1986.

Vol. 1224: Nonlinear Diffusion Problems, Montecatini Terme 1985. Seminar. Edited by A. Fasano and M. Primicerio. VIII, 188 pages. 1986.

Vol. 1225: Inverse Problems, Montecatini Terme 1986. Seminar. Edited by G. Talenti. VIII, 204 pages. 1986.

Vol. 1226: A. Buium, Differential Function Fields and Moduli of Algebraic Varieties. IX, 146 pages. 1986.

Vol. 1227: H. Helson, The Spectral Theorem. VI, 104 pages. 1986.

Vol. 1228: Multigrid Methods II. Proceedings, 1985. Edited by W. Hackbusch and U. Trottenberg. VI, 336 pages. 1986.

Vol. 1229: O. Bratteli, Derivations, Dissipations and Group Actions on C*-algebras. IV, 277 pages. 1986.

Vol. 1230: Numerical Analysis. Proceedings, 1984. Edited by J.-P. Hennart. X, 234 pages. 1986.

Vol. 1231: E.-U. Gekeler, Drinfeld Modular Curves. XIV, 107 pages. 1986.

Vol. 1232: P.C. Schuur, Asymptotic Analysis of Soliton Problems. VIII, 180 pages. 1986.

Vol. 1233: Stability Problems for Stochastic Models. Proceedings, 1985. Edited by V.V. Kalashnikov, B. Penkov and V.M. Zolotarev. VI, 223 pages. 1986.

Vol. 1234: Combinatoire énumérative. Proceedings, 1985. Edité par G. Labelle et P. Leroux. XIV, 387 pages. 1986.

Vol. 1235: Séminaire de Théorie du Potentiel, Paris, No. 8. Directeurs: M. Brelot, G. Choquet et J. Deny. Rédacteurs: F. Hirsch et G. Mokobodzki. III, 209 pages. 1987.

Vol. 1236: Stochastic Partial Differential Equations and Applications. Proceedings, 1985. Edited by G. Da Prato and L. Tubaro. V, 257 pages. 1987.

Vol. 1237: Rational Approximation and its Applications in Mathematics and Physics. Proceedings, 1985. Edited by J. Gilewicz, M. Pindor and W. Siemaszko. XII, 350 pages. 1987.

Vol. 1238: M. Holz, K.-P. Podewski and K. Steffens, Injective Choice Functions. VI, 183 pages. 1987.

Vol. 1239: P. Vojta, Diophantine Approximations and Value Distribution Theory. X, 132 pages. 1987.

Vol. 1240: Number Theory, New York 1984–85. Seminar. Edited by D.V. Chudnovsky, G.V. Chudnovsky, H. Cohn and M.B. Nathanson. V, 324 pages. 1987.

Vol. 1241: L. Gårding, Singularities in Linear Wave Propagation. III, 125 pages. 1987.

Vol. 1242: Functional Analysis II, with Contributions by J. Hoffmann-Jørgensen et al. Edited by S. Kurepa, H. Kraljević and D. Butković. VII, 432 pages. 1987.

Vol. 1243: Non Commutative Harmonic Analysis and Lie Groups. Proceedings, 1985. Edited by J. Carmona, P. Delorme and M. Vergne. V, 309 pages. 1987.

Vol. 1244: W. Müller, Manifolds with Cusps of Rank One. XI, 158 pages. 1987.

Vol. 1245: S. Rallis, L-Functions and the Oscillator Representation. XVI, 239 pages. 1987.

Vol. 1246: Hodge Theory. Proceedings, 1985. Edited by E. Cattani, F. Guillén, A. Kaplan and F. Puerta. VII, 175 pages. 1987.

Vol. 1247: Séminaire de Probabilités XXI. Proceedings. Edité par J. Azéma, P.A. Meyer et M. Yor. IV, 579 pages. 1987.

Vol. 1248: Nonlinear Semigroups, Partial Differential Equations and Attractors. Proceedings, 1985. Edited by T.L. Gill and W.W. Zachary. IX, 185 pages. 1987.

Vol. 1249: I. van den Berg, Nonstandard Asymptotic Analysis. IX, 187 pages. 1987.

Vol. 1250: Stochastic Processes – Mathematics and Physics II. Proceedings 1985. Edited by S. Albeverio, Ph. Blanchard and L. Streit. VI, 359 pages. 1987.

Vol. 1251: Differential Geometric Methods in Mathematical Physics. Proceedings, 1985. Edited by P.L. García and A. Pérez-Rendón. VII, 300 pages. 1987.

Vol. 1252: T. Kaise, Représentations de Weil et GL_2 Algèbres de division et GL_n. VII, 203 pages. 1987.

Vol. 1253: J. Fischer, An Approach to the Selberg Trace Formula via the Selberg Zeta-Function. III, 184 pages. 1987.

Vol. 1254: S. Gelbart, I. Piatetski-Shapiro, S. Rallis. Explicit Constructions of Automorphic L-Functions. VI, 152 pages. 1987.

Vol. 1255: Differential Geometry and Differential Equations. Proceedings, 1985. Edited by C. Gu, M. Berger and R.L. Bryant. XII, 243 pages. 1987.

Vol. 1256: Pseudo-Differential Operators. Proceedings, 1986. Edited by H.O. Cordes, B. Gramsch and H. Widom. X, 479 pages. 1987.

Vol. 1257: X. Wang, On the C*-Algebras of Foliations in the Plane. V, 165 pages. 1987.

Vol. 1258: J. Weidmann, Spectral Theory of Ordinary Differential Operators. VI, 303 pages. 1987.